RISK

RISK

A Practical Guide for Deciding
What's Really Safe
and What's Really Dangerous
in the World Around You

DAVID ROPEIK
AND
GEORGE GRAY

Houghton Mifflin Company
Boston • New York
2002

For information about permission to reproduce selections from
this book, write to Permissions, Houghton Mifflin Company,
215 Park Avenue South, New York, New York 10003.

Visit our Web site: www.houghtonmifflinbooks.com.

Library of Congress Cataloging-in-Publication Data
Ropeik, David.
Risk : a practical guide for deciding what's really safe and what's really
dangerous in the world around you / David Ropeik and George Gray.
p.   cm.
Includes index.
ISBN 0-618-14372-6
1. Risk assessment. 2. Environmental risk assessment.
3. Health risk assessment. I. Gray, George (George M.) II. Title.
T174.5 .R66 2002
613.6—dc21 2002075934

Book design by Joyce C. Weston

Printed in the United States of America

QUM 10 9 8 7 6 5 4 3 2 1

"The carpal tunnel," "The anatomy of sciatica," and "Particulates in
the respiratory system" illustrations © 2002 Fairman Studios, LLC.

To Michael Fisher for the vision from which this book was born; to Toby and Ann and everyone in our families for their ideas, encouragement, and patience; and to Laura for her inspirational courage and example at keeping risk in perspective.

～

CONTENTS

PART III. MEDICINE

RISK

INTRODUCTION

"I've developed a new philosophy . . . I only dread
one day at a time."

— Charlie Brown

WE LIVE in a dangerous world. Yet it is also a world far safer in many
ways than it has ever been. Life expectancy is up. Infant mortality
is down. Diseases that only recently were mass killers have been all
but eradicated. Advances in public health, medicine, environmental regulation, food safety, and worker protection have dramatically reduced
many of the major risks we faced just a few decades ago.

Yet new risks have arisen. Hazardous waste. Nuclear power. Genetically modified foods. Mad cow disease. Ozone depletion. Artificial
sweeteners. For all the unquestionable benefits of the modern technological world and its scientific power, the march of progress that has
given us longer, healthier lives has subjected us to new perils.

We often react to this conflict, of progress on the one hand and risk
on the other, with fear. Most of us are more afraid than we have ever
been. And not just from any single risk that happens to be grabbing the
headlines at a given point in time, whether it's terrorism or West Nile
virus. We are afraid, cumulatively, of all the new bogeymen to which
our modern existence has exposed us. Many polls find that people feel
the world today is more dangerous for humans than it has ever been.

It is true that the industrial and information ages have spawned a
whole new range of risks, and raised awareness of those that were lurking all the time. But research suggests that our fears may not match the
facts. We may be too afraid of lesser risks and not concerned enough
about bigger ones. Polls show a wide gap between what the public and
the "experts" think is actually dangerous and what is considered relatively safe. Who's right? There are no simple answers.

But information can help us begin to sort things out. Some basic
facts about the risks we face, or *think* we face, can help us make more

sense of just what we need to worry about. The intent of this book is to provide that information. We want to empower you to make better judgments about how to protect yourself and your family and friends. Our goal is to help you put the risks you face into perspective.

~

Risk issues are often emotional. They are contentious. Disagreement is often deep and fierce. This is not surprising, given that how we perceive and respond to risk is, at its core, about nothing less than survival. The perception of and response to danger is a powerful and fundamental driver of human behavior, thought, and emotion.

In writing this book, we tried to stay as neutral about these controversial issues as we could. We think that information devoid of advocacy is a tough commodity to come by these days, and will be more useful to you. We do not tell you what you should think. Nor do we make judgments about whether a risk is big or small for you as an individual. We offer numbers for society as a whole, but there is no overarching single conclusion about any risk that can be drawn for each reader. Each of you has unique circumstances that make any given risk higher or lower for you than it might be for the next person. Ultimately, how you perceive a given risk is a decision for you to make in the context of your own life. We simply hope that you are more able to make more informed choices after reading the information we present. As Arthur Conan Doyle wrote in *The Hound of the Baskervilles,* "That which is clearly known hath less terror than that which is but hinted at and guessed."

We have gathered and analyzed the basic information available on major risk issues and synthesized from all that research a fair presentation that you can use to make up your mind about the risks we examine. Of course we have made judgments along the way, about which risks to include or omit, about what information to offer and what information to leave out. But we have done so in an honest effort to get to the basic core truths about each risk as we see it, in as fair a way as possible. You may well disagree with some of the judgments we've made. That's a risk we run in taking on a subject fraught with so much emotion.

We encourage you to use this book in two ways. Reading it all the way through will let you see how each risk compares with the others and will help you put them all in perspective. There are a lot of statistics in this book. They are provided to give you an idea of how big or small each individual risk might be. But they will also let you compare similar statistics for various risks from chapter to chapter. Together, these numbers should help you gain a larger view of many of the risks you face.

But we also encourage you to use this guide as you would an encyclopedia, as a reference work you will turn to over time, whenever there's something about a particular risk you want to know. Each chapter, for example, begins with a useful explanation of the specific hazard: What *is* radon? How *do* air bags or nuclear power plants or cell phones work? What are the most common forms of sexually transmitted or food-borne diseases?

We hope this book remains valuable to you for some time. Yes, the numbers of victims for various risks may change from year to year, and we will certainly learn more about some risks than we know now. But the nature of the consequences of alcohol consumption or radon exposure will stay the same. Years from now the use of caffeine, the prevalence of heart disease, the mechanics of the way radiation or lead or pesticides affects us, will all be pretty much the same.

We also hope you find this book useful no matter where you live. While the numbers and exposure patterns we cite are focused on the United States, the details of most of the risks we explain are the same in Europe or Asia or South America. The effects of mercury, the science of genetic modification of food, the persistence of some chemicals in the environment, the way X rays work, are the same whether you live in Canada or France or Japan. We recognize that the relative scale of risks varies from place to place. The public health risk from cigarette smoke, for example, is higher in Europe, where more people smoke, than in the United States. Firearms risks are higher for U.S. residents than citizens of any other country. At the time of this writing, mad cow disease is a higher risk in some nations than others. So the data we use for exposure levels and numbers of victims, based on statistics for the United States, may well vary for citizens of different countries. But the general explanations of many of the risks we explore are applicable for anyone, anywhere.

WHAT IS RISK?

Of all the wonders that I yet have heard,
It seems to me most strange that men should fear;
Seeing that death, a necessary end,
Will come when it will come.

— William Shakespeare, *Julius Caesar*

An anonymous writer once observed, "To risk living is to risk dying." Risk is, indeed, inescapable. But just what is risk? How do you define it? To a stockbroker it means the prospect of losing, or making, money. Same thing for a person at the racetrack or at a blackjack table. For a skier or a bungee jumper or a skydiver, on the other hand, risk has more to do with physical than fiscal health. To the person taking a pill with known side effects, it's about choice. To the person eating food with potentially harmful ingredients that aren't listed on the label, it's about no choice.

At it's simplest, risk is the idea that something might happen, usually something bad. But within that simple notion are some important components that you need to understand in order to have a better basis on which to make your personal risk judgments.

You may be hoping that this book answers the common question we all have about most risks: "What are the chances that . . . ?" If you are like most people, you think that risk means *probability,* the likelihood that something will happen, as in "Your risk of dying from X is one in a million." But there is more to risk than just calculating the statistical chances of a certain outcome.

There is also the issue of *consequences,* as in "The likelihood of a nuclear plant meltdown may be low, but it's a risk because it's disastrous if it does happen." A full definition of risk must take into account not just the probability of an outcome, but its severity. Generally, risk involves an outcome that is negative. You might say, "The odds of winning the lottery are . . ." but you wouldn't say that winning the lottery is a risk. And the more severe the outcome, the higher we judge the risk to be.

A complete definition of risk must also include the presence of a *hazard,* as in "That compound is a risk. It causes cancer in lab animals." If something to which we're exposed isn't hazardous, it isn't a risk. We're all exposed to a lot of cotton in the clothes we wear. So what.

Which brings up the fourth major component of risk, *exposure,* as in "Flooding isn't a risk. I live on a hilltop." If a substance is harmful to test subjects, but we're never exposed to it, it doesn't pose a risk. The risk of being eaten by a shark doesn't exist in Kansas. A hazard can't do you any harm if you are out of harm's way.

So a more complete way of thinking about risk might read: *Risk is the probability that exposure to a hazard will lead to a negative consequence.*

It's helpful to keep all these elements in mind when thinking about risk. Take out any one of those components, and the definition is incomplete. Each one involves characteristics that help you understand risks more completely and keep them in clearer perspective.

As an illustration, let's consider that dreaded common risk: ketchup. If we are exposed to ketchup, that *exposure* alone doesn't make it a risk. As far as we know, ketchup isn't a hazard, except for the chance of spilling some on your clothes.

But let's say somebody discovers that ketchup is hazardous. It still isn't much of a risk if the consequence of exposure to this hazard is, say, an increased taste for pickles on your hamburger. The nature and severity of the consequence has a lot to do with judging whether a risk is big or small.

But let's say that you're allergic to pickles, so anything that entices you to eat them could indeed be dangerous to your health. Ketchup still isn't much of a risk if the *probability* of its leading to increased pickle consumption is one in a million. You may have exposure to a hazard, but the level of risk still depends on the likelihood, the chance, that a negative consequence might occur.

In other words, we can make better judgments about how to think about risks if we keep in mind the ideas of *hazard, exposure, consequence,* and *probability.* These characteristics help to define and explain the risks in this book. Accordingly, most chapters are laid out as follows:

The Hazard: Just what is the agent we're talking about? (What *is* asbestos?) How does this hazard come to be in the world around us? (How does mercury get into our fish?) What is the biological or physical mechanism by which the hazard does its supposed harm? (How does radiation affect us?)

The Range of Exposures: How are we exposed to this risk? Where? When? How do exposures vary over time, by location, or by population subgroup?

The Range of Consequences: How much harm does the hazard do? In what ways? To how many people? To what kinds of people? Who is most at risk? Is the harm short-term or long-term, fatal or not? What is the probability of harm? How many people are injured or killed by the risk?

Reducing Your Risk: In this section we offer some general suggestions about what you can do to minimize the risk we're discussing.

For More Information: Each chapter ends with a list of resources to provide you with more information.

Perhaps the biggest risk we take as authors is offering our perspective and judgment of whether the risk is big or small, with visual guides at the beginning of each chapter. This estimate is our best effort to synthesize what we've learned on your behalf and to give you our opinion. You will find two "risk meters" in each chapter. One will offer our assessment of the general *likelihood of exposure to hazardous levels,* taking into account the factors of exposure and hazard from our definition. The other meter will indicate our assessment of the risk's *consequences—including severity and number of people affected.*

Here are a few examples of what you will see in each chapter. At the beginning of Chapter 1, "Accidents," the first risk meter will look like this:

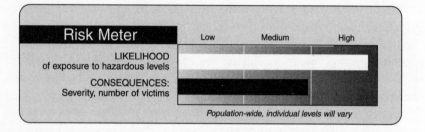

The upper bar indicates that the likelihood of exposure to accidents in a way that will probably cause harm is high. The lower bar indicates that the consequences of the risk of accidents—the severity of the outcome and how many people suffer these consequences—are also high. But not quite as high as the first meter, because the majority of accidents are not fatal, so the severity of the consequences brings the rating down a bit.

Here's what the meter will look like for Chapter 35, "Radon."

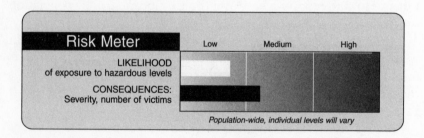

The likelihood of exposure to levels that will probably cause harm is, in general, pretty low. Lots of people are exposed, but the levels in most cases are fortunately not usually enough to cause harm. So there is exposure, but not to levels that present a hazard in a lot of cases. But since the consequences of radon exposure at levels high enough to do harm are potentially severe, and several thousand Americans a year suffer those consequences, the lower bar for radon takes account of both those factors.

These meters require several cautions. First, they refer to the population as a whole. Your risk is almost certainly different from that of the general population because of your age, gender, genetics, income, education, location, and other factors that make you unique. These risk meters offer only a general reference to where we think the risk falls on the high-low scale. Second, these are estimates. They are not scientific. They are the result of our analysis of the information we've collected and are not statements of fact and truth. And since there is a lot of uncertainty about many of the risks in this book, the meters are ballpark estimates that offer only a general range of where we think the risk falls. That's why we don't give our ratings specific numbers between 0 and 10. (In Appendix 2 we discuss our thinking behind each of the ratings. The appendix does offer our ratings numerically, though some of them are given as a range rather than as one specific number.)

Further, these risk gauges don't take into account the benefits that come from the hazard being discussed. Air bags can be harmful, for example, but clearly they save many more lives than they take. Some people suffer serious side effects from vaccines, but vaccination's benefits far outweigh the risks. We leave that risk-benefit accounting out of our judgment. Since our definition or risk presumes that most people think of risk as resulting in a negative consequence, that's what we rate.

Not every risk meter offers a ranking. Instead, a question mark acknowledges that some risks are too new or poorly studied to rate. As an example, for endocrine disrupters, the meter looks like this:

Risk Meter	Low	Medium	High
LIKELIHOOD of exposure to hazardous levels	?	?	?
CONSEQUENCES: Severity, number of victims	?	?	?

Population-wide, individual levels will vary

In a few cases the risk meter will be blank. That's because there is no specific hazard, so the first bar is irrelevant. A few of the risks we discuss, like cancer and heart disease, are really outcomes. That is, you're not exposed to cancer or heart disease. You just end up with these ill-

nesses as a result of other processes. But because they kill so many people, and are so often the result of exposure to many of the hazards we
discuss, we think they deserve explanation in a book about risk.

WHERE DO "THE FACTS" ABOUT RISK COME FROM?

"There is something fascinating about science. One gets such
wholesale returns of conjecture out of such a trifling invest-
ment of fact."

— Mark Twain

As we try to judge what's risky and what's not, we look to science for
answers. But even with all the facts that science can provide, much uncertainty remains, for a number of reasons. First, the sciences by which
risk is investigated—toxicology, epidemiology, and statistical analysis—
are inherently imprecise. Second, there are a lot of risk questions science
simply hasn't asked yet. New risks like using a cell phone while driving
or eating genetically modified food haven't been studied nearly enough
for us to have all the answers. And third, even for risks that have been
studied, the facts as we know them are constantly changing as scientific answers to one set of questions reveal more questions.

A lot of information in this book comes from the findings of the
three major risk sciences. It's important for anyone trying to make informed judgments about risk to understand what these sciences can,
and cannot, tell us.

Toxicology

Most simply described, toxicology is the study of poisons. But because
of that very definition, you can understand why toxicologists usually
can't test the agent they're investigating on human subjects. So animals are used as surrogates. But toxicologists admit that they can't say
for sure what a compound will do in humans based on evidence of what
it does in animals. As one toxicologist says, "With stuff that might kill
people, animal testing, as imprecise as it is, is the best we can do. But
despite what you might think of your boss or some people you don't
like, humans aren't rats." Toxicologists don't know which lab animal
species serve as the best indicators of what would happen in people, nor
do they know which species are better indicators for which kinds of
hazards. So extrapolating from lab animals to humans is imprecise. As
one example, cyclamate, an artificial sweetener, causes one type of liver

tumor in only one species of rat, and then only in males, and doesn't cause it in any other test animals. Yet test data from the experiments on those rats caused the food additive to be banned for human consumption.

Another imprecision from toxicology arises because testing of lab animals often involves subjecting the animals to massive doses of an agent. In testing for carcinogenicity, animals routinely get doses, each day, far greater than you would be exposed to in your entire lifetime. Toxicologists call this dose the MTD, for "maximum tolerated dose." They use this technique when testing for cancer in order to maximize the chance that they'll find any effect that might occur and that might not show up from a milder dose.

Using these MTDs, toxicologists presume that if the substance they're testing causes an effect at a high dose, it might cause the same effect at a lower dose. This approach seems like a rational way to deal with potentially dangerous chemicals and other agents; if high doses cause harm, assume that low doses might too. But sometimes the size of the dose is what's really causing the harm. Think of aspirin, for example. One or two aspirin are fine. Too many will kill you. The standard toxicological approach of subjecting lab animals to high doses of a test compound can reveal subtle effects, but it can also produce misleading results.

(Toxicological tests for noncancer health problems, like developmental hazard or cognitive impairment, don't rely on the MTD approach. For these outcomes, scientists assume that higher levels cause worse effects and lower levels cause weaker effects, and below a certain level the hazard might not cause an effect at all. So they subject the test animals to varying doses to find the lowest one at which an effect occurs. Also, for these other health problems, scientists believe that once exposure to the agent stops the effect usually goes away. But for cancer, since just a one-time mutation of a single gene can create permanent changes to the DNA that causes the disease, toxicologists use the MTD method, a more conservative and precautionary approach.)

A further imprecision arises in toxicology because in vivo tests in living lab animals, or in vitro tests of cells in a lab dish or beaker, isolate and test just one compound at a time. That's a smart way to find out with precision whether that particular agent is hazardous. But in the real world we're exposed to a stew of agents, and the mix can lead to different outcomes than exposure to any individual component. (Radon and smoking, for instance, apparently work synergistically and increase the risk of lung cancer more than the sum of one risk plus the other risk.) In addition, while the environment in the lab is stable and uniform, the real world is full of variables such as our environment, our health, our food, our emotional states, and our genetic makeup from one generation to the next and from one person to the next. These factors and many others affect how we react to a compound or circumstance.

In short, while toxicology can tell us a lot about the biological hazard of a particular chemical or element or compound, it can't tell us with absolute accuracy just what the substance being tested—at high doses to another species in a controlled lab—will do at lower doses to humans in the complicated real world.

Epidemiology

When we can't test a substance or hazard on people but we want to know whether it might be a threat to public health, we look around for circumstances in which people might already have been exposed. Studying what has happened, or is currently happening, to real populations in the real world, and trying to make sense of which hazards and exposures might be associated with which consequences, is the essence of epidemiology.

Like toxicologists, epidemiologists readily acknowledge that their science is imprecise. Epidemiology can usually provide only associations, not absolute proof, that some particular exposure *may be* what's causing some particular consequence. For example, in one kind of study epidemiologists investigate a specific small group of people who get sick. The book and movie *A Civil Action,* for instance, made famous the polluted drinking water in Woburn, Massachusetts.

A higher-than-expected number of cases of childhood leukemia showed up in just a few years in a small neighborhood. Epidemiologists investigated to find out what sources of exposure to potential hazards the neighbors shared. They discovered that one thing the neighbors had in common was that those who drank from a certain water supply had a higher rate of illness. Therefore, something about the water was the likely cause of the leukemia. They tested the water for chemicals suspected to cause that illness and estimated how much of the water people drank, for how long, and how polluted it was when people drank it. In the end, a peer-reviewed epidemiological study showed an association between how much of the well water pregnant mothers drank and the frequency of childhood leukemia in their offspring. The more they drank, the more likely it was that their children developed leukemia.

But that's not proof. Perhaps a couple of the neighbors were exposed to something else the researchers didn't ask about. Maybe the researchers never detected something else in the well water. These other factors are known as "confounders," hidden clues that can muddy the epidemiological waters and lead to an inaccurate assumption that A caused B. Hidden confounders can never be completely ruled out.

Epidemiologists can also do a different type of study, not just looking back in time at a small local group of people over just a few years but tracking a much larger population forward over longer periods of time. The famous Framingham Heart Study is an example of this kind of epidemiology, following an entire community over decades. Again, the researchers examine these wider populations for patterns in illnesses and exposures that suggest an association between the two. For instance, many epidemiological studies show that there is a strong association between air pollution increasing one day and hospital admissions for respiratory problems going up over the following several days, which suggests that the pollution is *probably* causing the respiratory problems. But again, that apparent association is not the smoking gun of absolute proof. Only when many long-term studies of different large populations repeatedly show the same thing, as with tobacco smoking and lung cancer, can epidemiology confidently say A causes B.

This isn't to suggest that the findings of epidemiology are weak or of little use in judging risks. In good epidemiological studies, researchers give the research subjects in-depth questionnaires about their health, their lifestyle, their diet, their social and economic characteristics, even their residential history (where they have lived and when), trying to rule out all confounders. They compare a group of people suffering some kind of health problem, like those families in Woburn, with other "control" groups, populations of similar size and socioeconomic status somewhere else, who presumably were not exposed to the same things. For the bigger long-term population studies, epidemiologists carry out multiple research programs in different places at different times to see if their results agree. With such techniques, epidemiologists can rule out every other possibility they can think of. They can become more and more certain of the associations they find.

But, like toxicologists, they can rarely be completely sure.

Statistical Analysis

In addition to the findings of toxicology and epidemiology, risk analysts also look for their clues among large sets of statistics. Those data collections are compilations of real-world information, on either morbidity (nonfatal health problems) or mortality (deaths). These databases can offer rich details, like how many people were injured or killed in motor vehicle accidents, categorized by speed, vehicle size, whether the victim was male, female, old, young, wearing a seat belt or not, and so on. There are data sets on hundreds of risks that offer information on the age, gender, and race of the affected population and the circumstances that led to the death or illness, such as the number of food poisoning cases connected with restaurants, or the number of workers murdered on the job. Other data collections provide risk analysts with information about hazardous materials emissions, local water or air pollution levels, or the presence of harmful chemicals in our blood or the food we eat. These details all offer insights about the hazard, exposure, consequence, and probability of various risks.

But the numbers in these data collections usually suffer from some imprecision. Not everybody who suffers food poisoning after dining at a restaurant, for example, actually goes to a doctor to report his illness. Not every police officer fills out every last detail on every accident report. Not every factory keeps accurate, or honest, records of its emissions. And not every government information collection system gathers the information and enters it into its database accurately.

Numbers are also subject to interpretation. Here's an example. According to national motor vehicle crash statistics, drivers 75 years old or over are involved in four times as many fatal crashes as the average of all other age groups. But does that mean that elderly drivers are killing other people, or just that because of frail health they're more likely to die themselves whenever they're in a crash? You can't tell by that statistic. The numbers don't tell you everything you need to know. As Mark Twain said, "There are three kinds of lies—lies, damned lies, and statistics."

Finally, no matter how precise and narrow statistical categories are, they lump everybody in that category together. For example, federal motor vehicle crash statistics group data by age, gender, the day and time of crashes, and the kind of vehicle involved. So you can determine how many 15- to 24-year-old males were involved in crashes on Sundays at 5 P.M. in pickup trucks. As narrow as that seems, that's still a large group of people and not everyone in it is the same. Individuals within that group have all sorts of differences in health, lifestyle, education, genetics, body size and shape, and on and on.

Risk statistics are generalities, and by definition cannot specifically answer the question we all want answered: "What is the risk to *me*?"

You will read a lot of numbers in this book. As we've stated, because you are unique none of those numbers will accurately and precisely answer your question. Risk numbers can be only a general guide. They give you a sense of which risks are bigger and which ones are smaller, and sometimes they can tell you which risks are higher or lower for the demographic groups to which you belong. But even risk numbers that define the categories as narrowly as possible still can't calculate the risk for each unique individual.

In sum, the sciences that supply the facts about risk, while growing more and more powerful, are still imprecise. They can provide us with valuable insights. But their results are uncertain and open to interpretation. There are very few unequivocal answers when it comes to defining and quantifying the risks we face. That's why in this guide our approach is to offer information in ranges: the *range* of exposures, the *range* of consequences, and so forth.

~

In addition to this scientific imprecision, sometimes we can't tell whether a risk is big or small, or real at all, simply because it's too new and hasn't been studied enough. Our modern world presents us with many new technologies (cell phones) or processes (genetic modification of food) or compounds (statin drugs to reduce cholesterol) that have profound benefits, but which also come with risks. Sometimes we are exposed to these technologies or processes or compounds before the risks have been adequately studied.

In the professional and policymaking world of people who deal with risk, how we should handle this uncertainty is a hotly debated issue. Some people argue that we should thoroughly study anything that might pose a risk *before* we start to use it.

The people on this side of the argument heed the advice of the eighteenth-century British politician Edmund Burke, who said, "Early and provident fear is the mother of safety." They suggest that we should adopt as a matter of law the "Precautionary Principle," the academic term for what most people think of as "Better Safe Than Sorry." These advocates argue that the best way to protect human and environmental health is to treat new compounds or technologies as guilty until proven innocent. They say that while we do this with some things, like new drugs, we don't do it with others, like new industrial chemicals. Advocates of the Precautionary Principle say that we must apply this careful approach across the board.

Others might subscribe to the advice of the American essayist Randolph Bourne, who wrote in his 1913 book *Youth and Life,* "We can easily become as much slaves to precaution as we can to fear. Although we can never rivet our fortune so tight as to make it impregnable, we may by our excessive prudence squeeze out of the life that we are guarding so anxiously all the adventurous quality that makes it worth living." These opponents of a sweeping Precautionary Principle argue that it would deny society many of the benefits of new technologies for years, even decades, until thorough scientific study can be completed. Those that argue against the Precautionary Principle also point out that almost anything carries *some* risk. Under the most rigorous application of the Precautionary Principle, these people claim, it would be hard to approve such things as motor vehicles or prescription drugs or vaccines. They argue that while it makes common sense to err on the side of caution, we should assess risks on a case-by-case basis, rationally weighing them against benefits. They say a blanket Precautionary Principle might deny society a public health advance that could save lives before all the scientific answers are in.

There are also times when we think science has come up with "the" answer, and we're reasonably certain about just how precautionary to be. And then things change. Even for risks that have been well studied, the facts are always evolving. We learn more and more every day. Between the time this book goes to press and the time you read it, our knowledge of the effects of hormone-disrupting chemicals will change. We'll certainly know more about human genetics and the risk of some diseases. We'll probably know more about the actual levels of particle pollution in the air we breathe. The statistical trends on established risks will have almost certainly shifted.

Further, our world of rapid technological development means that new risks are being created, and new solutions are being found, at an accelerating rate. Many of the risks in this book were largely unknown just a few years ago. Researchers are constantly developing new technologies or drugs that reduce some risks while potentially creating others. And science itself changes and grows more powerful. We can detect chemicals in our blood or in the air at levels much lower than we could just a decade ago. Risks that were always out there are just now being revealed. And all of these discoveries and new risks and new solutions interact in highly unpredictable ways. Methyl tertiary-butyl ether, or MTBE, was added to gasoline to improve air quality. But that policy led to the pollution of drinking water. The certainty we want in order to know how to judge risks is a tough ideal to achieve in our modern, dynamic world.

In short, our effort in this book to give you an accurate, reliable explanation of many of the risks you face is tantamount to shooting at a moving target. Which makes trying to score a bull's-eye with everything we present here a very risky proposition.

In an additional effort to ensure that what we present is accurate, thorough, and balanced, we asked experts to review each chapter. These reviewers were academics, scientists, doctors, government officials, engineers, risk assessors, and members of advocacy groups like the American Cancer Society and the Union of Concerned Scientists. We asked them to check our facts, to correct mistakes, to point out omissions, to clarify—and for feedback on whether we had provided a thorough and balanced overview of the issue. Their input was immensely valuable. But we take full responsibility for the final product. And the reviewers had nothing at all to do with the risk ratings in each chapter. We offer the names and backgrounds of the reviewers at the end of each chapter.

WHERE DO OUR FEARS ABOUT RISK COME FROM?

> "People are disturbed, not by things, but by the view
> they take of them."
>
> — Epictetus

As we wrote earlier, the facts about risk are only part of the matter. Ultimately we react to risk with more emotion than reason. We take the information about a risk, combine it with the general information we have about the world, and then filter those facts through the psychological prism of risk perception. What often results are judgments about risk far more informed by fear than by facts.

The terrorist attacks on the World Trade Center in New York and on the Pentagon and the subsequent anthrax attacks in the fall of 2001 are an example. Many of us were afraid, and rightly so. But some people responded by driving to a distant destination rather than flying, even though the facts clearly showed that flying remained the far safer mode of transportation, even after September 11. Some people bought guns, raising their risks from firearms accidents far more than reducing their risk of being attacked by a terrorist. Many people took broad-spectrum antibiotics even though they had no evidence that they had been exposed to anthrax—but they didn't get an annual flu shot.

Do these judgments make sense? Are they rational? Not based simply on the facts. But this is how humans respond to risk . . . with our hearts as well as our heads. The psychological study of this phenomenon, known as "risk perception," explains why our fears often don't match the facts. It is perhaps the biggest reason why writing this book is a risky affair.

We're confident that as you read this guide, your interpretation of what we say about various risks will differ from ours, and from some of your friends or family or neighbors. Same facts. Different interpretations. Despite our efforts to be neutral, many of the issues we write about are highly emotional and trigger powerful risk perception responses that all but guarantee that you might not like, or agree with, all of what you're about to read. We think it's valuable to understand what researchers have learned about risk perception because it might help you understand your own reactions to risk a little better.

Risk Perception

Humans tend to fear similar things, for similar reasons. Scientists studying human behavior have discovered psychological patterns in the subconscious ways we "decide" what to be afraid of and how afraid we should be. Essentially, any given risk has a set of identifiable characteristics that help predict what emotional responses that risk will trigger. Here are a few examples of what are sometimes called "risk perception factors."

- Most people are more afraid of risks that are new than those they've lived with for a while. In the summer of 1999, New Yorkers were extremely afraid of West Nile virus, a mosquito-borne infection that killed several people and that had never been seen in the United States. By the summer of 2001, though the virus continued to show up and make a few people sick, the fear had abated. The risk was still there, but New Yorkers had lived with it for a while. Their familiarity with it helped them see it differently.
- Most people are less afraid of risks that are natural than those that are human-made. Many people are more afraid of radiation from nuclear waste, or cell phones, than they are of radiation from the sun, a far greater risk.
- Most people are less afraid of a risk they choose to take than of a risk imposed on them. Smokers are less afraid of smoking than they are of asbestos and other indoor air pollution in their workplace, which is something over which they have little choice.

- Most people are less afraid of risks if the risk also confers some benefits they want. People risk injury or death in an earthquake by living in San Francisco or Los Angeles because they like those areas, or they can find work there.
- Most people are more afraid of risks that can kill them in particularly awful ways, like being eaten by a shark, than they are of the risk of dying in less awful ways, like heart disease—the leading killer in America.
- Most people are less afraid of a risk they feel they have some control over, like driving, and more afraid of a risk they don't control, like flying, or sitting in the passenger seat while somebody else drives.
- Most people are less afraid of risks that come from places, people, corporations, or governments they trust, and more afraid if the risk comes from a source they don't trust. Imagine being offered two glasses of clear liquid. You have to drink one. One comes from Oprah Winfrey. The other comes from a chemical company. Most people would choose Oprah's, even though they have no facts at all about what's in either glass.
- We are more afraid of risks that we are more aware of and less afraid of risks that we are less aware of. In the fall of 2001, awareness of terrorism was so high that fear was rampant, while fear of street crime and global climate change and other risks was low, not because those risks were gone, but because awareness was down.
- We are much more afraid of risks when uncertainty is high, and less afraid when we know more, which explains why we meet many new technologies with high initial concern.
- Adults are much more afraid of risks to their children than risks to themselves. Most people are more afraid of asbestos in their kids' school than asbestos in their own workplace.
- You will generally be more afraid of a risk that could directly affect you than a risk that threatens others. U.S. citizens were less afraid of terrorism before September 11, 2001, because up till then the Americans who had been the targets of terrorist attacks were almost always overseas. But suddenly on September 11, the risk became personal. When that happens, fear goes up, even though the statistical reality of the risk may still be very low.

People who first learn about these risk perception patterns often re-mark on how much sense they seem to make. It's little wonder. These are deeply ingrained patterns, probably ancient behaviors imprinted in us over millions of years of evolution. Long before we had our modern thinking brain, long before humans or primates even developed, only organisms that could recognize and successfully respond to danger survived and evolved. In Darwinian terms, these affective, "irrational" ways of protecting ourselves are adaptive. They help us preserve the species. Evolution selects for this type of behavior. That belief is supported by the fact that these patterns of risk perception cross cultures, age groups, genders, and other demographic groupings. There are some variations among individuals. Those variations make sense too because different people have different lives, different jobs, different family circumstances, different sets of experiences, different sets of values, and so on. Fearing a risk more if it involves children, for example, means parents will react differently from, say, teenagers. What is frightening to you might not be to your friend. Neither of you is right or wrong. You just each have a unique perspective on the same statistics and facts. But risk perception research shows that underneath our individual differences, we share certain patterns of risk response.

As we've written, this way of protecting yourself can be dangerous. What *feels* safe might actually be dangerous: driving instead of flying, antibiotics against anthrax instead of flu shots, arming yourself against a phantom risk. We explain risk perception, therefore, to help you understand the psychological roots of how we all respond to risk. That might help you understand your own concerns and put the risk issues in your life into clearer perspective.

THE RISKS WE INCLUDE, AND THOSE WE DON'T

This book does not include or omit risks based on whether they are "real." As we have mentioned, whether a risk is real is ultimately something you will decide for yourself. Rather, what follows in this book are many of the major risks you might want to learn about. In selecting what to include, we made no distinction among those that have been largely debunked (radiation from microwave ovens), those that are new (cancer from cell phones), those that remain poorly understood (hormone-disrupting chemicals), or those that have been well studied (electrical and magnetic fields from electricity lines and appliances). We include risks with high likelihoods, such as food poisoning, and low likelihoods, including cancer from pesticides on food.

That said, we do not include other risks. For example, we do not explore risks to the environment that don't have direct implications for human health. While risks like climate change, acid rain, and destruction of wetlands or forests all impact the biosphere on which we depend, they do not have a direct and immediate connection to human health.

We also do not include many of the risks that arise in detailed medical care. We do discuss medical errors in general as a category of risk, and we also explore some broad medical issues, such as vaccines or antibiotic resistance. We write about some health risks, like heart disease, cancer, and obesity. But we do not explain risks from drug reactions, the relative risk of one form of medical treatment over another, or many other specific medical risks. These risks are so unique to each individual that to discuss them in a book about risk in general might in fact be dangerous to the reader.

Nor do we discuss the risks of developing health problems because of genetic predisposition, which has more to do with susceptibility than cause. We don't include many things that you might think of as risks, but that are actually outcomes. Stroke and diabetes, for instance, are end results, the outcomes of natural biological processes or, in some cases, of exposure to a hazard. This book deals with the *hazards* to which we are *exposed,* because our actions have bearing on these parts of the risk equation. By the time we're faced with an outcome, it's too late. Still, we do include a few outcomes—heart disease, cancer, obesity—because they are such major killers and such common outcomes to many of the hazards we write about that we felt a general explanation of these issues would help.

We also don't go into detail about the risks of crime, a complex and unique set of issues. Some basic crime statistics are included in Appendix 1 in the back of the book. We leave for this appendix those risks that, while of interest, really don't need much of an explanation, like the risk of being hit by lightning, being killed in a plane crash, of snakebite or bee sting or shark attack. We also list in that appendix the statistics for some common causes of death and injury, such as stroke or homicide, diabetes or drowning, asthma or Alzheimer's disease. This appendix is simply a numerical listing of how many Americans suffer these outcomes each year. It is not a chart of your individual risk. As the historian Edward Gibbon wrote, "The laws of probability. So true in general. So fallacious in particular." Your individual risk depends on dozens of factors unique to your lifestyle, genetics, socioeconomic characteristics, and so on.

~

In the end, we hope we came up with a list, and an approach, that offers a review of what is known, and what is not known, about most of the risks that most people care about. We have tried to offer information in a neutral way, keeping matters simple by culling the essentials from the mountains of information on each risk. At the same time, we have rigorously pursued accuracy, and we've tried to offer some context and richness of detail. We hope this guide provides useful information that will help put the risks in your life in perspective. We hope it helps you lead a healthier, safer, less worried life.

· I ·

HOME, TRANSPORTATION, WORK

1. ACCIDENTS

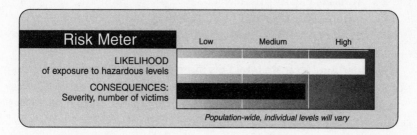

IT IS FITTING that "Accidents" is the subject of the first chapter of this book because it represents one of the greatest exposures to hazards that we write about. By the broadest definition, it is life itself that exposes us to hazards that may lead to the negative consequence of an accident. No matter how careful we are, accidents happen. Unintentional injuries, including motor vehicle crashes, are the fifth leading cause of death in the United States and the leading cause of death for people from ages 1 to 38. Not counting those caused by motor vehicles, accidents killed 54,334 U.S. residents in the year 2000, the eighth biggest category of deaths in the United States.

Accidents are of particular public health concern because unlike the four larger causes of death—heart disease, cancer, stroke, and chronic obstructive pulmonary disease—accidents kill people at an earlier average age. The average age for the four leading causes of death ranges between 70.3 for cancer and 78.9 for stroke. But the average age of someone killed in an accident is 50. Those who die from the top four killers in the United States have average remaining life expectancies of from 9.6 to 14.3 years. But accident victims have an average remaining life expectancy of 29.8 years. So this cause of death has a disproportionately large impact on life years lost to U.S. residents.

THE HAZARDS

This chapter will deal with the most common types of accidental death and injury, except motor vehicle accidents. There are so many important specifics to discuss regarding motor vehicle risks that we devote an entire chapter to those issues, Chapter 16.

We discuss these hazards, and their consequences, by category.

Falls

Falls are the most common accidental cause of death in the United States. They kill about 6 people per 100,000, and the largest number of those victims are elderly. In the year 2000, 16,200 people died from falls, 1,700 of whom were between 65 and 75 years old and 11,300 of whom were 75 or older. Each year in the United States, one out of every three older adults falls and suffers some sort of injury.

The elderly suffer from fatal falls more than any other age group for a number of reasons. Risk factors include having weak muscles, weak bones, or balance problems either from natural deterioration of the vestibular (balance) system or from taking medications that impair balance and visual acuity; suffering from a neurological disease or having previously had a stroke; and wearing shoes with thick, soft soles, which aren't good for balance. Sometimes the trauma from the fall leads to death either immediately or in a matter of days. More often, the fall leads to serious damage, such as hip fracture, from which the elderly person cannot recover, leading to a further deterioration of health and ultimately death, often within a matter of months.

Falling is not limited to the elderly, of course, and not all falls are fatal. An annual public health survey by the National Center for Health Statistics, a federal agency within the Centers for Disease Control and Prevention, has found that an estimated 11 million fall-related episodes led to some form of injury in 1997–98. The most likely place for falls may not be what you'd expect. The survey found that a third of falls were on the floor or level ground. Only about 12 percent occurred on stairs. The other falls occurred in lower percentages in a variety of other circumstances.

Poisoning

The next most common form of accidental death is from poisoning, with about 4 deaths per 100,000. This statistic includes deaths from drugs, inappropriately prescribed or administered medicines, alcohol, and household chemicals, such as cosmetics and cleaners, as well as commonly recognized poisons. It does not include suicides.

The people most at risk of death from poisoning, based on data from the year 2000, are those between 35 and 54 years of age. Of the estimated 11,700 U.S. residents who died of accidental poisoning that year, 6,800 of them were in that age group. A big portion of these deaths is explained by the statistics for a narrower part of that age group, those between 25 and 44. Almost half of the 6,800 poisoning deaths in this age group were from the use of illegal drugs or alcohol. In 1968, about 2 out of every 100,000 U.S. residents aged 25 to 44 died from accidental poisoning. By 1998 that rate was up to 7.5 per 100,000. By contrast, the death rate for the 0 to 4 age group dropped in that period from 1.6 to 0.1.

The gender statistics for poisoning reflect an interesting fact about all accidents. Men are far more likely to be accident victims than women. Nearly three times as many males died from poisoning accidents as females.

Fortunately, most poisonings are not fatal. The American Association of Poison Control Centers estimates that there were 2.2 million poisonings in the United States in 1998, resulting in about 50,000 hospitalizations and 400,000 doctor visits. About 9 out of 10 of these poisonings occurred in the home and two thirds of the victims were children six years old or younger.

Drowning

Between 1 and 2 U.S. residents per 100,000 drown each year. An estimated 3,900 U.S. residents drowned in the year 2000 in boating or swimming accidents. (This category does not include drowning from storms or floods.) Drowning was the leading cause of accidental death for children aged one to four. Most child drowning victims die in swimming pool accidents, with more than half occurring in residential pools at the child's own home. Roughly five times more males drown than females. Figures from the National Center for Injury Prevention and Control (NCIPC) suggest that drowning may not always be purely accidental. They show that alcohol is involved in 25 to 50 percent of drowning deaths for adolescents and adults.

Fire

After accidental death by water comes death by fire, which kills about 1.5 people per 100,000 each year. In the year 2000, 3,600 U.S. residents died from fires, either from burns, asphyxiation, falls caused by a fire, or by being struck by falling debris in connection with a fire. That's down roughly one third from 20 years ago. The National Fire Protection Association (NFPA) says that fires injured approximately 22,350 civilians. In 1999, the most recent year for which figures are available, the NFPA says 112 firefighters were killed and 88,500 were injured. Those numbers will surely rise when the data for 2001 are compiled, reflecting the 343 firefighters killed at the World Trade Center.

The NCIPC says that the United States has the highest death rate from fires of any industrialized country. By age group, death rates from fire are highest for children up to age 4, and for adults aged 65 and above.

The setting in which most fire deaths occur is the home. Three quarters of fire-related deaths, and two thirds of fire-related injuries, were caused by house fires. House fire deaths occur disproportionately in the southeastern United States, and more frequently in the colder months from November through February. The NFPA says that between 1994 and 1998, cooking and heating equipment caused roughly one third of all house fires, with about a quarter of house fire–related deaths caused by smoking materials, mostly cigarettes. The NFPA attributes 16 of every 100 fire-related deaths to arson or other suspicious causes.

Choking

Not far behind fires as an accidental cause of death is suffocation from unintentional inhalation of food or other objects that block the trachea—in other words, choking to death, which kills 1.2 U.S. residents per 100,000 annually. Though much attention is paid to this risk for children, it is actually the elderly who are most often the victims. Of the 3,400 U.S. residents who died this way in the year 2000, 2,600 were 65 or older. Experts don't have a good explanation for this phenomenon. Some think it may be caused by problems some elderly people have with chewing their food completely, the effects of dentures on chewing and swallowing, the effects of previous neurological events like stroke, or weaker chest muscles with which to expel something that begins to obstruct the trachea.

Firearms

According to the National Safety Council, approximately 600 U.S. residents died in the year 2000 from accidents involving firearms. That figure is down 43 percent from its peak in 1993. About 400 of those deaths occurred in the home. People ages 15 to 24 were twice as likely to die from a firearms accident as the average for the whole population. But the rate for 15- to 24-year-olds is still low, only 1 person per 200,000. Like fires, poisonings, and falls, not all accidents in this category are fatal. In 1998, the last year for which specific figures are available, hospital emergency room databases indicate that 13,698 nonfatal unintentional firearm-related injuries occurred.

Other Types of Accidental Deaths

There are all sorts of accidents that don't fit neatly into the major types. But grouped together, this "other" category is substantial. An estimated 14,500 U.S. residents died in 2000 from a variety of accidental causes, more than 5 people per 100,000. The principal causes include:

- Some medical errors, including accidental cuts, puncture, or perforation during surgical or other invasive treatments; foreign materials left inside the body after surgery; improper sterilization; improper radiation therapy; or failure to give a necessary drug or medication. (As you will read in Chapter 44, epidemiologists estimate that medical error is a much more significant cause of death than most people realize.)
- Natural and environmental causes, like excessive cold or heat
- Machinery accidents
- Falling objects
- Electrocution
- Water transport accidents (in which the death resulted not from drowning but a fall, fire, or collision)
- Injuries caused by animals

REDUCING YOUR RISK

The "A" word, *accidents,* is somewhat controversial in the field of injury prevention. Many experts are unhappy that the word *accidents* implies that these events are a matter of fate, that they are nobody's fault, and that there's nothing we can do about them. They worry that such an attitude leaves us resigned to accidents, instead of looking for ways to prevent them. As you'll learn in Chapter 16, "Motor Vehicles," many police organizations no longer use the word *accident,* preferring the word *crash.* They contend that no crash occurs that might not have been somehow avoided, and that therefore the word *accident* is inaccurate.

We take a middle view. Sometimes things happen over which we really don't have any control. True accidents do happen. But there are a number of things we can do to reduce the risk of accidents, of all types. For example:

There are lots of things the high-risk group of the elderly can do to reduce the risk of falls. Maintaining healthy muscles and bones is important. Activities to achieve these goals include progressive resistance training with weights, elastic bands, exercise equipment, or tai chi. Doctors also advise calcium supplements to help strengthen bones.

In addition, you can make environmental modifications: Install sturdy handrails on all stairways. Install ramps to replace outdoor stairs. Improve lighting at the top and bottom of stairways, indoors and out. Put a light near enough to your bed so you can flip the light on without walking around in the dark at night. Remove tripping hazards such as electric cords and loose rugs. Avoid climbing on ladders or chairs. Keep things that have to be accessed from cupboards or storage closets within easy reach. Put skid-resistant mats in bathtubs and showers. Install grip rails near baths and toilets. Wear sturdy, nonslip shoes. Use a cane or walker for support if necessary. You might also want to consider wearing hip pads, which reduce the risk of hip fracture in case of a fall by 50 percent.

It also helps simply to be aware of the factors that increase your risk of falling. Your eyesight is important, so keep eyeglass prescriptions up to date. Work with your doctor to adjust your medications, when possible, to avoid drugs that cause dizziness and other balance problems. Often this effect is caused by the combination of medication that seniors take, so talk with your medical provider about the side effect of dizziness if you are taking a combination of medicines. People who have suffered a stroke or other neurological disease should be aware that they are at a higher risk of falling. And people who have been injured in a fall are at a higher risk than the average population for falling again.

~

As we mentioned in the first part of this chapter, most of the poisoning fatalities in the United States each year occur among those taking illegal substances. Substance abuse programs are the most effective risk reduction for individuals in the 25 to 44 age group. For the age group that experiences the most nonfatal poisonings, children under 6, there are several steps that adults can take to reduce the risk.

First, it's useful to understand what poisons children are consuming most. The leading categories of poisoning for children 6 years old and younger in 1998 were 60 percent nonpharmaceuticals (731,407 incidents) and 40 percent pharmaceuticals (477,452 incidents). The top categories of nonpharmaceuticals are cosmetics (160,000), cleansers (126,000), foreign objects (74,000), and arts and crafts supplies (30,000). The top categories of pharmaceuticals that poison children 6 and under include pain relievers (90,000), cough medicines (64,000), topical medications (64,000), and antimicrobials/antibiotics (37,000).

The basics of childhood poison prevention include:

- Keep household products and medicines out of the reach of children, and locked up when possible.
- While you're using these products, never let them out of your sight when kids are around, even when you answer the phone or the doorbell or tend to an errand.
- Remember that nonprescription medications can be just as poisonous to small children as prescription drugs. Treat nonprescription drugs with the same precaution you take with prescription medication.
- Supervise kids closely when they are using arts and crafts materials or other products that might be poisonous.
- Keep products in their original containers, with the original labels.
- Don't store anything poisonous in a recycled food or beverage container.
- Avoid taking medications in front of small children, who tend to imitate what grown-ups do.
- Keep the phone number for your local poison control center handy, along with ipecac syrup, which induces vomiting. But never use it unless advised to by a poison control expert or your doctor. Some poisons, such as drain openers or oven cleaners, do even more harm as they are being expelled.

~

As with the other accidental risks we face, advice to reduce the risk of drowning differs depending on age group.

For young children, safety starts with supervision. Youngsters should never be left in the water alone, even if they've successfully taken swimming lessons. One quarter of the children who drown each year have taken such lessons. The rules for pool use should be hard and fast: no kids in or even near the water without adult supervision. And don't forget that children in the house also need supervision if there's a pool outside. In one study, half the kids who drowned or nearly drowned were last seen inside the house. Supervision must be vigilant. A child can slip under water quickly, without much noise. When you head out to the pool to watch youngsters, take everything you need—snacks, drinks, reading material, portable phone — so you don't have to leave the pool to get something inside.

Home pools should be fenced in, with self-closing and self-latching gates. Fences should be at least five feet high, and tables and chairs should be away from the edge so kids can't climb them. These steps can cut the risk of fatality in a residential pool by roughly two thirds. Keep a flotation or reaching device near both sides of a pool. And learn CPR.

For older kids and adults, or in open-water swimming conditions, always try to swim with a buddy, or with somebody watching from shore or in a boat nearby. Changing conditions (currents, temperatures, rip tides, weather) can overcome even the most accomplished swimmer. And remember that mixing swimming or boating with alcohol dramatically raises the risk of drowning.

Check water conditions, especially the depth if you're going to be diving. The minimum considered safe is 10 to 12 feet. Many swimming pools offer less than that as you dive off the diving board and enter the water where the deep end is rising up to the shallower part of the pool. When swimming in the ocean, watch out for rip tides, which are visible as columns of dirtier water rushing out to sea. If caught in one, swim *parallel* to the shore until you leave the column of the rip tide, then head toward shore. And don't forget that really cold water will quickly drop your body temperature, and your ability to swim.

Finally, boaters should always wear life jackets, especially in smaller or tippier boats, regardless of the distance they're going to travel or how well they swim.

∼

One of the easiest and most effective steps you can take to reduce your risk of injury or death from fire is to equip your home with smoke detectors on each floor, ideally just outside sleeping areas. (Most residential fire fatalities occur between 10 P.M. and 6 A.M.) But don't just put them there and forget them. More than 9 homes in 10 in the United States have smoke alarms, but one third of them have smoke alarms that don't work. Periodically check the batteries or the wiring. Replace batteries once a year. Wipe the detectors' outer surface periodically, to keep dust from building up and interfering with the devices. Homes with working smoke detectors reduce the chance of fatality from a fire by as much as 50 percent.

Keep a fire extinguisher in a handy spot in your kitchen. Know how to use it.

Have an evacuation plan for your house with two exits from each room. The plan should include routes out of each room (make sure windows aren't painted or nailed shut or blocked by air conditioners or fans), a place to meet outside the house, and a designated person to take care of family members who aren't mobile, such as infants or the elderly. Practice the plan a couple of times a year.

Of course, the best way to reduce your risk of injury or death from fire is to prevent fires from starting in the first place. Tips for reducing your risk of a fire in the home, which is where three quarters of fire-related deaths and two thirds of fire-related injuries occur, focus on the three leading causes of home fires: smoking materials, cooking, and heating equipment.

Smoking is the leading cause of home fire deaths. The NFPA recommends:

- Never smoke in bed.
- Always use a deep ashtray when smoking. Don't prop an ashtray on a piece of furniture that can catch fire.
- Don't leave burning cigarettes, cigars, or pipes unattended.
- Make sure smoking materials are *fully* extinguished before emptying an ashtray.
- Keep matches and lighters away from children.

Cooking is the leading cause of house fires. Roughly one quarter of house fires in the United States start in the kitchen. To reduce your risk of a cooking fire, or to deal with one if you have one:

- Never leave cooking food unattended. And monitor food cooking in the oven.
- Keep cooking areas free of things that can catch fire, like towels, paper, and wooden utensils.
- Make a three-foot area around your stove a "kid-free zone."
- Turn pot handles inward so you can't bump into them while walking past the stove.
- Be careful of loose clothing when working at the stove.
- Keep a potholder or mitt handy, so if there is a grease or oil fire in a pan, you can drop the lid on and snuff it out.
- If there is an oven fire, keep the oven door closed to suffocate it. Same thing for a microwave fire. And turn off the oven!

During the winter months, the leading cause of house fires in the United States is heating equipment, including space heaters like portable electric or kerosene heaters, woodstoves, fireplaces with inserts, and room gas heaters.

To reduce these risks:

- Remember that space heaters need space, at least a few feet between them and anything that can burn.
- Anything in which solid fuel burns—fireplaces, woodstoves, coal stoves, chimneys—should be checked annually for creosote buildup. This tarlike substance can ignite under certain conditions and create uncontrolled burning in your stove, fireplace, or chimney.
- Use a sturdy fireplace screen to keep sparks from shooting into the room.

~

We know we'll sound like your mother—but the best way to reduce the risk of choking is to "Finish chewing your food before you swallow it." For younger children, cutting food into smaller pieces and supervising your kids as they eat are important, especially if they're eating things like peanuts, hard candy, grapes, or thick foods like peanut butter.

When young kids play, anything small enough to fit through a circle an inch and a quarter in diameter—about as wide as two pennies next to each other—presents a possible choking hazard. Marbles, small batteries, coins, and pen or marker caps are examples. Supervise kids closely when they're playing with such objects, and make sure no one leaves them around for a child to pick up and mouth when they're not supervised. Remember that young children instinctively explore objects by mouthing them, so stay generally alert to early signs of choking.

Learn the Heimlich maneuver. Starting from behind the victim, make a fist with one hand, and put it, thumb side in, on the victim's abdomen, about midway between the waist and the rib cage. Grab that fist with your other hand and thrust it sharply in and up. You can do the same thing to yourself if you're choking—one fist into the abdomen, the other one to drive it sharply in and up. If that doesn't work, press your upper abdomen over the back of a chair or the edge of a table and use that extra pressure to do the Heimlich maneuver to yourself. It may take a few tries.

~

The best way to reduce the risk of firearms accidents in the home is to keep guns out of the hands of children, particularly teenagers, since the people most at risk for unintentional gun fatality are between 15 and 24 years old. That age group is three times more likely to die in gun accidents in the home. Keep guns locked up, unloaded, with the ammunition stored somewhere else, also locked up, with the keys hidden. A survey by the group Common Sense about Kids and Guns says that almost one third of the handguns legally owned in 40 million homes with children are stored loaded and unlocked.

FOR MORE INFORMATION

National Safety Council
www.nsc.org
1121 Spring Lake Drive
Itasca, IL 60143-3201
(630) 285-1121
Fax: (630) 285-1315

National Center for Injury Prevention and Control
www.cdc.gov/ncipc
Mailstop K65
4770 Buford Highway NE
Atlanta, GA 30341-3724
(770) 488-1506
Fax: (770) 488-1667

Consumer Product Safety Commission
www.cpsc.gov
4330 East-West Highway
Bethesda, MD 20814-4408
(301) 504-0990
Toll-free Consumer Hotline: (800) 638-2772
Fax: (301) 504-0124 and (301) 504-0025

This chapter was reviewed by Alan Hoskin, Manager of the Statistics Department at the National Safety Council; and by Lois Fingerhut, Special Assistant for Injury Epidemiology, Office of Analysis, Epidemiology and Health Promotion, National Center for Health Statistics.

2. AIR BAGS

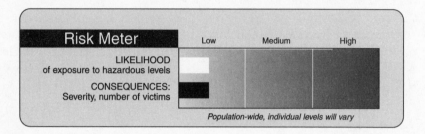

YOU MAY REMEMBER the early public service campaign "Seat Belts Save Lives." They do, but only if people wear them. When research made clear that a lot of people don't wear their seat belts, the federal government mandated a passenger protection system that would work automatically under certain crash conditions. After a multiyear battle with motor vehicle manufacturers, federal regulators required the installation of air bags. They have saved thousands of lives in the past several years, but they have also created some risk.

THE HAZARD

By 1999, all new cars and trucks, including vans and SUVs, were required to have air bags. A little more than half the cars and light trucks on the road today are equipped with some kind of air bag, and that percentage rises each year as newer cars replace older ones that are taken out of service. Some 50 to 60 million of these air bags are so-called first-generation devices, the kind that was installed in vehicles prior to 1996. First-generation air bags are the kind found to cause injury and death. They are designed to protect a 160-pound, 5 foot 10 inch driver in a head-on crash.

Newer vehicles, from about 1996 on, have so-called second- or third-generation devices, which have been engineered so they deploy with less force, or in different ways in different kinds of crashes. The newer devices also do a more accurate job at detecting crash forces and adjusting deployment accordingly. They can adjust inflation speed based on how close or far the seat is from the dashboard. They can also adjust based on whether the occupant is wearing a safety belt. These newer devices significantly reduce the risk of injury and death from air bags.

The risks we discuss as we talk about consequences deal only with the risks of the older, first-generation front air bags, since the data indicate that they pose the most risk. We also don't consider side air bags, which have not been demonstrated to pose a risk.

~

A typical air bag system is controlled by sensors that detect the rate at which a vehicle is slowing down. If these sensors detect that the vehicle is decelerating so quickly that a crash must be occurring, a chemical propellant rapidly inflates a large nylon cushion at a peak rate of up to 200 miles per hour. The cushion inflates within milliseconds, optimally before the driver or front seat passenger can be thrown into the dashboard, steering wheel, or windshield. The air bag has vents so it can deflate just a few seconds after deployment.

Air bags are designed so that normal operation of the vehicle, including braking and hitting bumps in the road, won't cause the device to deploy. The most common air bag systems are designed to deploy only in head-on crashes or in crashes with a significant frontal-force component since these crashes account for more than half of the severe and fatal injuries to motorists.

Most frontal air bag–induced injuries occur when the head, chest, arms, or hands of a driver or passenger are too close to the air bag housing when it deploys. The highest risk zone is 2 to 3 inches from the air bag housing, where the nylon material expands fastest. Some motor vehicle safety experts say being hit by an inflating first-generation air bag when you're that close is like taking a punch from a professional fighter who has been given a free shot.

It's also possible to be burned by a deploying air bag, though this is rare. The chemical propellant that causes inflation uses hot gases. They're gone in an instant, but if you're too close to the device when it deploys, those gases can burn.

THE RANGE OF CONSEQUENCES

The federal government reports that as of the end of 2001, air bags had deployed more than 3 million times and saved approximately 8,000 lives since they were first introduced. The National Highway Traffic Safety Administration (NHTSA) says air bags saved 1,584 people in the year 2000 alone. The Insurance Institute for Highway Safety says air bags reduce driver deaths by one quarter among drivers using safety belts, and one third among unbelted drivers, in direct frontal crashes.

The institute estimates that air bags reduce the risk of death for passengers by 14 percent if they're wearing safety belts, and 23 percent if they're not strapped in. NHTSA estimates that air bags reduce the risk of fatality in all crashes by 11 percent.

In some circumstances, however, first-generation frontal air bags cause injuries ranging from minor to fatal. While the fatal injuries have fortunately been rare, nonfatal air bag–related injuries are not. About 40 percent of air bag deployments result in at least one injury to a driver or passenger. The vast majority of these injuries are minor. The most frequent air bag injuries are bruises or abrasions to the face, neck, chest, or arms. Some researchers call this the "bag slap effect." There are also occasional fractures of the fingers, wrists, or arms of the driver because as she holds the steering wheel her hands and arms are closest to the air bag when it deploys. Minor injuries to the face, particularly bruises and abrasions, are also common among passengers. Burns from the propellant gas occur mostly to the hands and fingers as the gas releases from the deflating bag. Bruises to the chest are frequently reported among both drivers and passengers.

Rarely, deployment of first-generation air bags kills people. As of October 2001, the U.S. government had documented 195 cases of deaths caused by air bags in settings where the crash itself was judged unlikely to have caused a fatality if the air bag hadn't been there. In almost all the *fatal* injuries, the victims were not properly restrained by safety belts, and in most cases they weren't restrained at all and as a result they got too close to the air bag as it deployed.

Tragically, a majority of the victims have been children, and 20 of them were infants, strapped into rear-facing child car seats that were supposed to make them safe. But as the infant sat facing the rear, the car seat, and therefore the infant's head, were just inches from the air bag when it deployed. Nearly 100 air bag victims were children either in forward-facing child safety seats or who were simply too big for such seats. And 89 of these kids weren't strapped in at all.

Of the 76 adults killed, 68 were drivers. Of these, 45 weren't restrained at all, 20 were wearing their safety belts, and 3 were strapped in but not wearing their belts properly. Of the 8 adult passengers killed by air bags, 6 weren't strapped in.

If you add up those numbers, you find that of the 195 confirmed victims, 143 were either not strapped in at all or not strapped in properly. So it's pretty clear that proper safety belt use dramatically lowers your risk of air bag fatality.

Besides drivers who don't wear their safety belts, the risk of air bag injury is higher for drivers who are short, for females, and for those over the age of 55. People in these groups often choose to sit, or sometimes have to sit, closer to the steering wheel in order to safely operate the vehicle, so they're hit harder by the air bag when it deploys.

For front seat passengers, those most at risk are children 12 and under who aren't strapped in or who aren't using their seat belts properly (putting the shoulder strap behind their neck), or infants sitting in the front seat, especially in rear-facing child safety seats that end up within inches of the zone in which the air bag deploys with the most force.

THE RANGE OF EXPOSURES

Among first-generation devices, there are differences in air bag designs, and these differences have an effect on the degree of risk each type of bag poses. Some deploy horizontally, inflating directly toward the driver or passenger. Some deploy vertically, up toward the windshield first, then out toward the passenger. There are differences in how crash sensors are programmed and what kinds of crash forces trigger the device, in how fast air bags deploy and in the fold and shape of inflated cushions. Passenger-side air bags vary more in their features than driver-side devices. Because manufacturers keep much of this information private, there are no reliable studies that can say which designs serve best to protect and also minimize air bag–related risks.

First-generation front air bags were designed in the early 1980s, when less information was available about what happened during motor vehicle crashes. As we have learned more, manufacturers have switched to the newer, less dangerous devices and the rate of injuries seems to be going down. In 1996 there was 1 driver death per 7 million air bags. By 2000, even though Americans drove millions more total miles, there was only 1 documented death per 17.6 million air bags, a 60 percent decline. One review finds that between 1996 and 2000, the number of vehicles with air bags tripled, but the rate of fatality to children dropped 90 percent. This great improvement is due to the less forceful newer designs and to several public safety campaigns that educated the public about the greater risks of putting kids in the front seat.

REDUCING YOUR RISK

The best thing you can do to reduce your risk of air bag injury is to use your seat belt, including proper use of the shoulder strap. This precaution improves the likelihood that you won't be too close to an inflating air bag cushion.

The next most important thing is to sit kids 12 and under in the back whenever possible. If your vehicle has no back seat, or you can't use it for some reason, a child in the front should sit with the seat pushed all the way back and his seat belt properly positioned. *Never* put rear-facing infant seats in the front seat near a live air bag. If you have to seat your child in the front, you can get an on-off switch installed. You have to fill out a federal form to get authorization for your mechanic to install such a switch. You can get the form from state motor vehicle registries, some car dealers and repair shops, or from NHTSA. (See "For More Information.") If you have to transport multiple children and they don't all fit in the back, the largest child should sit in the front, again with the seat pushed all the way back and the seat belt properly strapped. Sitting kids in the back reduces their risk of death in the event of a crash by about a third.

Adults should maintain at least 10 inches between the air bag housing (in the steering wheel, in the case of the driver) and the chest. If you can't maintain this distance from the steering wheel because of a back problem or your physical stature, you also may qualify for an on-off switch. People who qualify include:

- People who must transport infants in rear-facing child safety seats in the front seat.
- People who must transport children 12 and under in the front seat.
- Drivers who can't change their customary driving position in order to keep 10 inches between their chest and the steering wheel.

If you have an adjustable steering wheel, another way to minimize injury from air bags is to point the steering wheel down toward your chest rather than up at your face. Also, if it's comfortable and doesn't interfere with your driving, tilt your seat back a bit, which moves your head another inch or two away from the air bag housing.

You can reduce the risk of hand and arm injuries from air bags by gripping the steering wheel on the sides, to the sides of the air bag, instead of at the top of the steering wheel, which puts your arms in front of the device.

FOR MORE INFORMATION

National Highway Traffic Safety Administration
www.nhtsa.dot.gov/airbags
400 7th Street SW
Washington, DC 20590
Toll-free Air Bags Hotline: (800) 424-9393

Insurance Institute for Highway Safety
www.hwysafety.org
1005 North Glebe Road, Suite 800
Arlington, VA 22201
(703) 247-1500

This chapter was reviewed by Maria Segui-Gomez, Ph.D., Associate Professor, Johns Hopkins School of Public Health, who has published several important research papers about air bags; by Susan Ferguson, Vice President for Research of the Insurance Institute for Highway Safety; and by James Simons, Director, Office of Regulatory Analysis and Evaluation Plans and Policy at the National Highway Traffic Safety Administration.

3. ALCOHOL

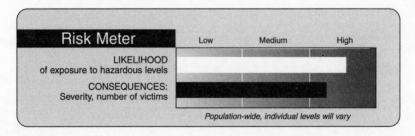

Risk Meter	Low	Medium	High
LIKELIHOOD of exposure to hazardous levels			
CONSEQUENCES: Severity, number of victims			

Population-wide, individual levels will vary

AMERICANS consumed 7 billion gallons of alcoholic beverages in the year 2000: 6.2 billion gallons of beer, 550 million gallons of wine, and more than 350 million gallons of distilled spirits. That comes to about

25 gallons for each man, woman, and child in the country. But of course not every man, woman, and certainly not every child, drinks alcohol, so the per capita consumption among those that do is even higher.

Many Americans drink only moderately, with no harm to themselves or others, and in some cases alcohol can actually have some health benefits. But the number of people who abuse alcohol is, pardon the pun, staggering. An estimated 18,380,000 U.S. residents consume alcohol in excess, according to medical definitions we'll discuss later in the chapter. Combine this group with the millions who consume alcohol moderately, but in dangerous circumstances, like driving, and it is easy to see why alcohol is one of our nation's top health threats.

THE HAZARD

Many people drink because they find that alcohol lifts their mood. But inside the body, alcohol depresses the activities of the central nervous system. And as the blood alcohol concentration (BAC—the percentage of alcohol per unit of blood) goes up, the effects of alcohol escalate.

A can of beer or glass of wine will create different blood alcohol concentrations in different people, based on body weight, percentage of

THE EFFECTS OF BLOOD ALCOHOL LEVELS

Blood Alcohol Level	Effects
.01–.05	No observable effects
.03–.12	Mild euphoria, decreased inhibitions, increased self-confidence, reduced capacity to pay attention and make judgments, beginning of sensory-motor impairment
.09–.25	Loss of critical judgment; impaired perception, memory, and comprehension; increased reaction times; reduced visual acuity; impaired balance; drowsiness
.18–.30	Disorientation; dizziness; exaggerated emotions; disturbed visual perception of color, form, and motion; decreased muscular coordination; slurred speech; staggering gait
.25–.40	Near total loss of motor function, inability to stand, vomiting, incontinence, marked drop in response to stimuli, stupor
.35–.50	Unconsciousness, reduced body temperature, incontinence, impaired circulation and breathing, possible death
>.45	Death from respiratory arrest

body mass that is fat or muscle, age, frequency of alcohol consumption, and how much food they've recently eaten. Body weight matters because alcohol has a strong affinity to bond with water. Since we're all made of mostly water, the more we weigh, the greater our water content. In essence, the bigger we are, the more water there is in our body to dilute the alcohol we drink.

The issue of water in the body also explains why body fat or muscle, gender, and age all matter. Fat cells hold less water. So the higher the percentages of fat per unit of body mass, the lower the percentage of water in the body. People who are overweight or out of shape have higher percentages of body fat and therefore lower water content. Women have higher percentages of fat per unit of body mass too. And elderly people tend to have lower percentages of body water. The same amount of alcohol will raise their blood alcohol content higher than it would in a person with lower fat levels, like someone who is well muscled.

~

Below is a chart of how different numbers of drinks might affect your BAC, based on your weight and gender. (We use the standard definition of "drink"—12 ounces beer, 5 ounces wine, 1.5 ounces 80-proof liquor. Numbers refer to the several minutes right after having the drink. This table is only a guide. Your actual BAC content, depending on how much you drink, will vary.)

BAC levels also depend on how quickly people eliminate alcohol from their body. Most of us metabolize alcohol at a steady rate of half an ounce per hour. But we eliminate alcohol from our system

BLOOD ALCOHOL LEVELS BY QUANTITY OF CONSUMPTION (BY WEIGHT AND GENDER)								
Drinks	100 lbs. M/F	120 lbs. M/F	140 lbs. M/F	160 lbs. M/F	180 lbs. M/F	200 lbs. M/F	220 lbs. M/F	240 lbs. M/F
1	.04/.05	.03/.04	.03/.03	.02/.03	.02/.03	.02/.02	.02/.02	.02/.02
2	.08/.09	.06/.08	.05/.07	.05/.06	.06/.05	.04/.05	.03/.04	.03/.04
3	.11/.14	.09/.11	.08/.10	.07/.09	.06/.08	.06/.07	.05/.06	.05/.06
4	.15/.18	.12/.15	.11/.13	.09/.11	.08/.10	.08/.09	.07/.08	.06/.08
5	.19/.23	.16/.19	.13/.16	.12/.14	.11/.13	.09/.11	.09/.10	.08/.09
10	.38/.45	.31/.38	.27/.32	.23/.28	.21/.25	.19/.23	.17/.21	.16/.19

faster when the blood alcohol content is either very low or very high. Chronic alcoholics usually metabolize alcohol faster than nonalcoholics. Older people usually don't get rid of alcohol as quickly as younger people.

If you drink, you may know that you tend to feel the effects of alcohol more quickly on an empty stomach. Alcohol is absorbed into the bloodstream much more efficiently from the small intestine than from the stomach. When your stomach is empty, the valve at the bottom that lets food or liquid pass down into the small intestine stays open. When you've recently eaten, that valve is sometimes closed, holding food in the stomach so it can be digested. So food in the stomach slows the passage of alcohol down into the small intestine and into your bloodstream. And because we eliminate alcohol faster when BAC levels are either low or high, by keeping levels lower food also speeds up that elimination. People who haven't eaten in several hours reach peak BAC levels in half an hour to two hours. People who have eaten something within several hours of drinking take one to six hours to reach peak BAC levels.

~

Alcohol can create physical dependence and addiction because the more your brain cells are in contact with alcohol, the more they adjust to its sedating effects. Then, when the alcohol is withdrawn, the central nervous system is suddenly released from this "sedating blanket," as one expert calls it, and becomes hyperexcitable. It's as though a system that was slowly shut off is suddenly turned back on. This phenomenon creates the symptoms of withdrawal that include nervousness, agitation, sweating, vomiting, and muscular convulsion, among others.

The clinical definition of the condition known as "alcohol abuse," according to the National Institute on Alcohol Abuse and Alcoholism (NIAAA), is a pattern of drinking that results in one or more of the following situations within a year:

• Failure to fulfill major work, school, or home responsibilities
• Drinking in dangerous situations
• Recurring alcohol-related legal problems
• Continuing alcohol-related problems in relationships

The NIAAA defines alcoholism as a disease that includes four symptoms:

- Craving or the strong need or compulsion to drink
- Loss of control or the inability to limit one's drinking on any given occasion
- Physical dependence, identified by nausea, sweating, shakiness, or anxiety during withdrawal
- Tolerance, or the need to drink ever greater amounts of alcohol to "get high"

Recent research has discovered an apparent explanation for why the sensations of alcohol exposure are more pleasurable for some people than others. The presence of alcohol stimulates production of the neurotransmitter dopamine, which is associated with pleasurable sensations. But it does so much more in some people than in others, particularly in alcohol abusers and alcoholics.

THE RANGE OF CONSEQUENCES

Alcohol is associated with approximately 100,000 deaths in the United States each year. Alcohol's ability to impair drivers' judgment and reaction time is perhaps the most well known of these fatal effects. Drunk driving killed more than 15,000 U.S. residents in 1999. There is an alcohol-related motor vehicle crash in the United States on average every two minutes. The federal standard for impaired driving is a BAC of 0.08, but traffic laws are adopted and enforced at the state level. As of the end of 2001, 16 states and the District of Columbia had made 0.08 BAC their standard for drunk driving: Alabama, California, Florida, Hawaii, Illinois, Kansas, Maine, New Hampshire, New Mexico, North Carolina, Oregon, Texas, Utah, Vermont, Virginia, and Washington.

In the other states, the drunk driving BAC level is 0.1. Studies suggest that reducing the drunk driving standard from 0.1 to 0.08 can reduce fatalities from 6 to 10 percent.

Beyond alcohol's effects on drivers, alcohol may be responsible for as many as 3 percent of the annual cancer deaths in the United States (about 16,000 of the 541,532 U.S. cancer deaths in 1998). Alcohol has been associated with various cancers, including cancer of the esophagus, larynx, liver, mouth, and pharynx. It is also associated with increased risk of breast and colorectal cancer. Alcohol causes cirrhosis of the liver in an estimated 900,000 U.S. residents. Cirrhosis killed 25,192 Americans in 1998. Together, cirrhosis and chronic liver disease were the tenth leading cause of death in the United States in 1998. Heavy alcohol use increases the risk of hypertension and may increase the risk of stroke. In 1997, there were approximately 421,000 hospital admissions for various problems for which alcohol was listed as the primary cause.

Alcohol creates unique dangers for pregnant women. It can affect fetal development and cause children to be born with smaller body size, reduced cognitive and learning capacity, and difficulty paying attention. Infants born to women who drink even moderately are at risk for fetal alcohol syndrome (FAS), a lifelong condition that includes symptoms such as a disfigured face and head; trouble sleeping and eating, seeing and hearing, following directions; and special medical needs. It is estimated that 4,000 to 8,000 U.S. children are born with FAS each year. One medical study found FAS to be the leading cause of mental retardation in newborns. No amount of alcohol has been found safe for pregnant women and their developing fetuses, but the more a pregnant woman drinks, the greater the likelihood of FAS in her child.

Beyond alcohol's physical effects, studies also show a relationship between alcohol abuse and behavioral problems. Of 11.1 million victims of violent crime each year, almost 1 in 4, or 2.7 million, tell police that the offender had been drinking before committing the crime. And children who are raised by parents who abuse alcohol, or who are otherwise exposed to alcohol abuse or dependence in their family, are more likely to become moderate or heavy drinkers when they grow up. Of the 72 million children in the United States under the age of 18, 1 in 4 of them is exposed to such abuse or dependence, so potentially more than one quarter of American kids could one day become problem drinkers.

The health consequences of alcohol consumption are not all negative. Studies indicate that low to moderate drinking can lower the risk of coronary heart disease, largely by reducing the concentration of fats in the blood. (Federal health officials define moderate drinking as no more than two standard drinks per day for men, and no more than one per day for women. As we've stated, a drink is defined as containing 0.5 ounces or 15 grams of alcohol, or the equivalent of 12 ounces of beer, 5 ounces of wine, or 1.5 ounces of 80-proof distilled spirits.) One study, which tracked the health of 490,000 U.S. residents for 9 years, found that both men and women who drank had a 30 to 40 percent lower risk of death from all cardiovascular diseases than abstainers, regardless of how much they drank. Alcohol can have other positive effects as well. Because it acts as a mild depressant in the body, alcohol can ease stress and generally relax the drinker, thereby elevating his mood and increasing sociability.

THE RANGE OF EXPOSURES

Alcohol is available everywhere, from liquor stores and bars to restaurants, supermarkets, clubs, and convenience stores. The use of this potentially harmful substance permeates our culture, and is frequently associated with glamorous and socially positive messages. With that pervasiveness come the hazards of excessive use.

The Centers for Disease Control and Prevention defines excessive drinkers as those who have had at least 12 drinks of any kind of alcoholic beverage in their lifetime *and* 5 or more drinks on one occasion at least 12 times during the past 12 months. By that definition, roughly 9 percent of U.S. residents are excessive drinkers. Males significantly outnumber female heavy drinkers in all age groups. More than half of the adults in the United States have a close family member who suffers or has suffered from alcoholism.

Alcohol exposure among the young is a particular concern. Approximately one third of eighth-graders and half of tenth-graders have been drunk at least once. One fifth of ninth-graders report binge drinking (consuming five or more drinks in a row) in the past month. More than 40 percent of the kids who begin drinking before age 13 will develop alcohol abuse or alcohol dependency at some point in their lives.

The children of alcohol abusers or alcoholics appear to have a genetically higher risk of becoming abusers or alcoholics themselves. Studies of the 11-year-old sons of men who died of alcoholism have discovered that from their very first exposure to alcohol, these children have a higher level of dopamine production in the presence of alcohol than children of nonalcoholics. They are apparently born with brain chemistry that predisposes them to alcohol abuse or alcoholism because their higher dopamine production means they are more likely to find the sensation of alcohol exposure pleasurable.

REDUCING YOUR RISK

The recommendations for dealing with alcohol abuse and alcoholism are straightforward. The NIAAA says you can tell whether you have an alcohol problem by checking the symptoms by which it classifies either alcohol abuse or alcoholism, which we mentioned earlier. In addition, if you've ever felt you should cut down on your drinking, been annoyed by people who criticize your drinking, felt bad or guilty about your drinking, or had a drink first thing in the morning to shake off a hangover, you should seek treatment through your doctor. Treatment may involve detoxification (removing your exposure to the hazard in a measured way and readjusting your body to the absence of an addictive substance), medication, counseling, or a combination of the three.

Reducing the risk from alcohol for those who don't meet the definition of an abuser or an alcoholic means minimizing your consumption, keeping it at or below the public health recommendations. While moderation in the amount you consume is important, there are several other actions you can take to reduce your risk from alcohol consumption.

- Try not to drink on an empty stomach. Eating a little as you drink, between sips, also helps slow the flow of alcohol to the blood.
- Sip your drink. Guzzling will raise your BAC faster.
- Become familiar with how much alcohol it takes to affect your behavior or physical control. Remember, the same amount of alcohol affects different people differently. Know your limit.
- Fight the social pressure to drink. Alcohol abuse programs teach that you should feel comfortable declining alcohol. Have a nonalcoholic drink in social settings, or have one alcoholic drink and make the others nonalcoholic.
- Beware of unfamiliar mixed drinks. Fruit drinks and some others can taste as though they have little alcohol in them, when in fact they're pretty strong.

FOR MORE INFORMATION

You can calculate your blood alcohol content, as measured by a breath test, at a website called "The Drink Wheel": www.intox.com/wheel/drinkwheel.asp.

National Institute on Alcohol Abuse and Alcoholism
www.niaaa.nih.gov
Willco Building
6000 Executive Boulevard, Suite 409
Bethesda, MD 20892-7003
(301) 443-3860

Al-Anon Family Group Headquarters, Inc.
www.al-anon.alateen.org
1600 Corporate Landing Parkway

Virginia Beach, VA 23454-5617
(757) 563-1600

Alcoholics Anonymous World Services, Inc.
www.alcoholics-anonymous.org
475 Riverside Drive, 11th Floor
New York, NY 10115
(212) 870-3400

National Council on Alcoholism and Drug Dependence, Inc.
www.ncadd.org
20 Exchange Place, Suite 2902
New York, NY 10005
(212) 269-7797

This chapter was reviewed by Henry Wechsler, Lecturer in Social Psychology at the Harvard School of Public Health, who has studied alcohol, especially its use on college campuses, for years; and by Dr. Joseph Pursch, worldwide lecturer on addiction and specialist in alcohol use by pilots, athletes, and public officials.

4. ARTIFICIAL SWEETENERS

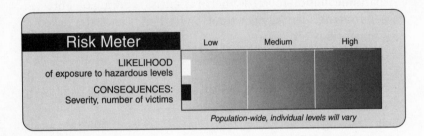

Risk Meter	Low	Medium	High
LIKELIHOOD of exposure to hazardous levels			
CONSEQUENCES: Severity, number of victims			

Population-wide, individual levels will vary

IT MAY TASTE GREAT in your whipped cream or on a bowl of strawberries, but natural sugar, high in calories, contributes to health risks big and small for hundreds of millions of people worldwide. More than half the adults in the United States are overweight, which increases the risk of heart disease, stroke, hypertension, and other life-threatening conditions. About 16 million Americans have diabetes, a blood sugar–associated condition. Sugar continues to be a contributor to tooth decay.

So for decades we have been turning to chemistry for substitutes, something that will give us the sweet taste without all the calories and associated risks. Initial efforts to create artificial sweeteners by reproducing the molecular characteristics that let sugar molecules trigger the sensation of sweetness failed. But within the past few decades, several artificial sweeteners have been discovered quite by accident. They have practically nothing in common with sugar molecules, but they can bond with the same receptors on the taste buds to initiate our sense of sweet. Most of these modern artificial sweeteners are byproducts of chemical experiments that were looking for other things. And the simple fact that these food supplements are industrial chemicals makes some people afraid.

THE HAZARD

The five products currently approved by the Food and Drug Administration (FDA) are:

- Saccharin: 200 to 700 times as sweet as sugar. Saccharin has been used in the United States since 1970. It's also known as Sweet 'n Low. It was discovered when a lab worker spilled some material on his hand, and noticed a sweet taste later while he was having dinner.

- Aspartame: 160 to 200 times as sweet as sugar. Used since 1981 in the United States, it's also known as NutraSweet or Equal. Aspartame was discovered when a scientist working on anti-ulcer drugs accidentally got some on his hands and later licked his finger.
- Acesulfame: 200 times as sweet as sugar. It's been in use in the United States since 1988. It's also known as Sunette or Sweet One. Acesulfame was discovered by a chemist when he licked his finger to help pick up a piece of paper.
- Sucralose: 600 times as sweet as sugar. Used since 1998 in the United States, sucralose is also known as Splenda. It was discovered in a lab testing sugars as intermediate chemicals in reactions. A foreign graduate student misunderstood the instruction to "test" a substance and instead gave it a "taste."
- Neotame: 7,000 to 13,000 times sweeter than sugar. Like aspartame, it contains phenylalanine, which can threaten the health of people with a rare disease, phenylketonuria (see "The Range of Consequences").

Other artificial sweeteners, not approved for use in the United States, include:

- Cyclamate: 30 times sweeter than sugar. Cylcamate was used in the United States beginning in the 1940s. A lab assistant working on antifever drugs put his cigarette down on a workbench and noticed the sweet taste of what turned out to be cyclamate when he put the cigarette back in his mouth. Much of our continuing concern that artificial sweeteners might pose a risk originated with cyclamate, which was banned in 1969 based on a study suggesting an association with bladder tumors in rats. While 75 subsequent studies have failed to show cyclamate is carcinogenic, the sweetener has yet to be reapproved for use in this country. Cyclamate is currently used in Canada and Europe. Many scientists consider that it tastes and behaves more like natural sugars than other artificial sweeteners.
- Alitame: 2,000 times sweeter than sugar. No known hazards.

(There is also a class of low-calorie sweeteners, known as polyols, which are derived from naturally occurring sugars and which contribute just a few calories. These include sorbitol, mannitol, and xylitol. We mention them because you may have seen them on food labels. But we don't deal with them in this chapter because they are essentially versions of natural sugars rather than truly artificial substances, and none of them has been associated with any serious health concerns.)

THE RANGE OF CONSEQUENCES

The following are the consequences we know about for the four artificial sweeteners currently approved for human consumption in the United States.

Saccharin causes bladder cancer in extremely high doses in male rats. The doses are the human equivalent of drinking 750 cans of soft drinks or using 10,000 tablets of saccharin a day, every day, for your entire life. After the early results linking saccharin to bladder cancer were released, in 1977 the FDA proposed a ban on saccharin. Congress put a moratorium on the ban and it never took effect. Since 1977, numerous studies have shown that saccharin does not cause cancer in mice, monkeys, or humans with the normal range of exposure, and the proposed ban has been withdrawn by the FDA. The National Toxicology Program has taken saccharin off the list of potentially carcinogenic substances. Still, the FDA has not removed the requirement that products containing saccharin must carry a warning label that it has been shown to cause cancer in laboratory animals.

Aspartame and neotame produce phenylalanine, a significant hazard to individuals with phenylketonuria (PKU), a rare genetic disorder in which the body can't break down the amino acid phenylalanine, which comes from food proteins. When phenylalanine accumulates in the bloodstream and body tissues, it can lead to impaired brain and physical development. Individuals with PKU already suffer from dangerous levels of phenylalanine in the blood, so they are particularly sensitive to aspartame. PKU affects males and females equally. Every state in the country has mandatory PKU testing within the first 3 days of birth, so most people who have PKU know they have the disease.

Some people have complained that aspartame causes a long litany of problems, including headaches, seizures, nausea, numbness, muscle spasms, weight gain, rashes, fatigue, irritability, increased heart rate, insomnia, vision problems, hearing loss, breathing difficulties, slurred speech, loss of taste, tinnitus (ringing in the ears), vertigo, memory loss, and joint pain. Extensive scientific testing has failed to confirm any of these symptoms. These effects may show up in individuals with unique sensitivities but remain too rare to show up in larger studies. The FDA approved neotame in July 2002.

Acesulfame: No known adverse effects.

Sucralose: No known adverse effects.

~

All the approved artificial sweeteners can be consumed by pregnant women and children, two population subgroups with unique health concerns.

Numerous studies have failed to confirm the suspicion that artificial sweeteners change how we eat, that they make us eat more, or less.

THE RANGE OF EXPOSURES

Exposure to artificial sweeteners is widespread. According to a 1998 survey by the Calorie Control Council, 144 million U.S. residents regularly consume low-calorie, sugar-free products. Artificial sweeteners are found in foods labeled "sugar-free," "low-calorie," "light," or "lite," among others. These include diet sodas and fruit drinks, sugar-free gums and candies, sugar-free gelatin desserts, sauces, "light" yogurts and ice creams.

The four approved artificial sweeteners are considered safe according to FDA standards, which are based on certain maximum acceptable daily intake (ADI) levels. The FDA defines an ADI as the amount of a compound that can be safely consumed on average every day over an entire lifetime. Usually ADIs are developed by exposing lab animals to varying amounts to find the lowest level that causes any effect. Then regulators reduce that level by a factor of 100, to reflect the uncertain-

ACCEPTABLE DAILY INTAKE AND ESTIMATED ACTUAL INTAKE OF ARTIFICIAL SWEETENERS		
Product	Acceptable Daily Intake (for a 132-lb. person)	Estimated Daily Intake (per person)
Saccharin	300 mg	50 mg
Aspartame	3000 mg	<300 mg
Sucralose	300 mg	<60 mg
Acesulfame	900 mg	<180 mg

ties in extrapolating data from animal studies to humans and to account for differences in sensitivity among people. If human studies are available, those data are used to calculate ADI too.

REDUCING YOUR RISK

The risk to your health from the consumption of any of these artificial sweeteners is considered negligible at levels of intake up to the ADI. Moderation of intake is key. Some health care professionals suggest that children and pregnant women should consult a physician before adopting a low-calorie, low-sugar diet involving foods containing artificial sweeteners, in case they have unique health conditions.

FOR MORE INFORMATION

International Food Information Council
www.ific.org
1100 Connecticut Avenue NW, Suite 430
Washington, DC 20036
(202) 296-6540
Fax: (202) 296-6547

Food and Drug Administration
www.fda.gov
5600 Fishers Lane
Rockville, MD 20857-0001
(888) INFO-FDA or (888) 463-6332

National Cancer Institute
www.nci.nih.gov
NCI Public Inquiries Office
Building 31, Room 10A31
31 Center Drive, MSC 2580
Bethesda, MD 20892-2580
Cancer Information Service: (800) 4-CANCER or (800) 422-6237

This chapter was reviewed by Dr. Eric Walters, Assistant Professor in the Department of Biochemistry and Molecular Biology at the University of Chicago (Dr. Walters has done extensive research in the field of artificial sweeteners), and by Jo-

seph Brand, Professor of Biochemistry, School of Dental Medicine, University of Pennsylvania, and Associate Director of the Monell Chemical Senses Center, one of the leading institutes in the world studying taste, smell, and chemosensory irritation.

5. BAD BACKS, CARPAL TUNNEL SYNDROME, AND OTHER REPETITIVE TASK INJURIES

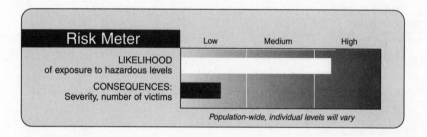

Risk Meter	Low	Medium	High
LIKELIHOOD of exposure to hazardous levels			
CONSEQUENCES: Severity, number of victims			

Population-wide, individual levels will vary

THE HUMAN BODY has 206 bones, more than 600 muscles, and thousands of tendons and ligaments that connect those muscles and bones and help them work together. We are all familiar with what happens if we suddenly put too much pressure on our musculoskeletal system. Things break, strain, or tear. But many of us suffer musculoskeletal injuries not from too much sudden pressure on bones, muscles, or tendons, but from chronic, repetitive motions in positions and under loads that our bodies are not designed to sustain.

While musculoskeletal disorders (MSDs) can occur anywhere, they most frequently occur on the job, because that is where we usually perform repetitive physical tasks for hours each day. A report by the National Academy of Sciences, Institute of Medicine (IOM), reviewing the scientific research on the issue of musculoskeletal disorders, says that about 1 million workers suffer these disorders each year to the point of needing time off from work. The Occupational Safety and

Health Administration says that 1.8 million workers suffer MSDs annually, and 600,000 lose time on the job as a result. The IOM report estimates that Americans make 70 million doctor visits each year for treatment of MSDs (not all of which are necessarily work-related.)

THE HAZARD

Most broadly, MSDs occur when the musculoskeletal system is subjected to repeated pressures, positions, and motions that it's not designed to handle. The degree of hazard depends on four basic components: the intensity of the physical activity (how hard we're working), the duration of that activity (how long it lasts), the frequency of the activity (how often we do it), and posture (the position we're in as we do the activity). These conditions most often combine when we're on the job, which explains why most MSDs are work-related.

Typing is a good example. An estimated 45 million Americans work at a keyboard. Anybody can sit and type for a little while without any harm. But holding the wrist in a position in which the hand is pulled back and held slightly above the arm, as most keyboards require, is not something the wrist can sustain, especially if one repeatedly moves the fingers while the wrist is in this position. This unnatural arrangement can cause inflammation of the tendons on the back of the wrist and inside a tunnel in the wrist—the carpal tunnel—which carries a key nerve that controls hand movement and sensation. The intensity of this activity is not great, but duration can be hours and the frequency is daily.

Any of us can pound a nail with a hammer or play a musical instrument for a while, with little or no discomfort. But consider a professional carpenter, or a violinist, or a surgeon, all of whom can suffer MSDs because they have to grip tools for long durations, frequently and tightly. Anyone can drive a truck, operate a power tool, or occasionally lift a heavy object. But workers who drive vehicles that vibrate, or use vibrating tools like chain saws or jackhammers, or who have to lift and carry heavy objects, all the time, are experiencing stresses, positions, or motions of higher-than-normal intensity for extended durations and with high frequency.

Our bodies are designed to deal with a certain amount of wear and tear. Everyday activities cause damage to our bones, tendons, ligaments, and muscles, imperceptible damage that we might not even be aware of or that we describe as just a little stiffness or soreness. These aches and pains and inflammations are not MSDs, because given time and rest, these injuries will heal. MSDs occur when the intensity, duration, or frequency of our activities creates more damage to the musculoskeletal system than natural healing can repair. We are exposed to the pressures, motions, and positions that lead to MSDs in many ways, resulting in more than 100 health consequences, many of which we will discuss in this chapter.

THE RANGE OF CONSEQUENCES

The main symptoms of MSDs are pain, tingling, numbness, swelling and inflammation, burning sensation, or stiffness. These can range from minor annoyances to disabling pain or dysfunction. They are usually temporary if caught early, but can become chronic and disabling if allowed to persist with no intervention. The symptoms can be experienced in most parts of the body, depending on the specific MSD from which you're suffering. The major ones fall into four categories: low back pain, disorders of the nerves, of the tendons, or of the muscles themselves.

Low Back Pain

Low back pain occurs when something irritates the soft tissue disks and muscles that cushion the vertebrae, the bony sections of the spine. According to the American College of Neurological Surgeons, more than 65 million Americans suffer from low back pain every year. Backaches are the most common reason for doctor visits, after cold and flu symptoms. But despite the fact that as many as four Americans in five will experience low back pain at least once in their lifetime, the causes are not well understood.

In general, low back pain is caused when something creates irritation to any of the nerves or muscles, where they run between the spinal cord and the rest of the body. That irritation, and the resulting tension in the surrounding muscles, contributes to low back pain. Sources of that irritation include obvious things like the strain of lifting objects improperly, or twisting the torso and the lower back muscles. But another major cause of low back pain is simply sitting. Sitting for long periods of time is not something our bodies evolved to do. Sitting causes the knees to bend up toward the chest, which stretches muscles in the lower back, which in turn compresses the vertebrae, and that puts pressure on the disks between the vertebrae. And lifting something while sitting is a double whammy, putting even more strain on the back while it's already in a stressed position.

Nerve Problems

Like biological wires, nerves carry electrical messages to and from the brain, signaling sensation and controlling locomotion. Anything that damages nerve tissue or causes inflammation of the nerves or puts pressure on the nerves can produce pain, tingling or numbness, and loss of motor control. Perhaps the most well known of the nerve-entrapment MSDs is *carpal tunnel syndrome.*

A series of bones and tissues in the wrist form the carpal tunnel, through which runs the vital median nerve and nine tendons that control the movement of your thumb and fingers. When either the tendons or the protective synovial tissue around the nerve and tendons are irritated, the synovium swells, squeezes on the median nerve, and causes the familiar symptoms of carpal tunnel syndrome: tingling, numbness, or pain in your hand, particularly on the palm side of your thumb, index, and middle finger. Symptoms are most frequent at night. Causes

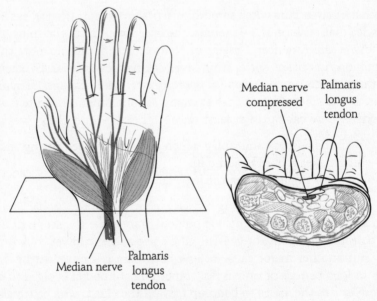

The carpal tunnel

include repetitive and forceful grasping, or repetitive and forceful bending of the wrist.

An estimated 850,000 new cases of carpal tunnel syndrome are reported each year, and 260,000 people a year undergo surgery to open the carpal tunnel and relieve the pressure on the nerve (although there is controversy about how many of these surgeries are necessary). About half of these operations are work-related. Sufferers include people in the food production industry, such as poultry processors, who spend much of their workday cutting meat in a cold refrigerator; assembly line workers; grocery store clerks; and some keyboard operators. While accounting for less than 2 percent of all workplace-related health problems, carpal tunnel syndrome results in the highest number of days of work lost—a median of 30 days off the job (which means half the cases involve more than that, and half involve less).

Nonoccupational sufferers include pregnant women, because they retain water and undergo hormonal changes that can compress the nerve. (The carpal tunnel problems usually go away after giving birth.) Elderly people who do a lot of gardening with hand tools are also frequent sufferers of carpal tunnel syndrome.

∼

Another commonly known nerve-entrapment MSD is *sciatica*. The sciatic nerve runs from the lower spine down through the buttocks to the leg. It's the largest nerve in the body. Sciatica is usually caused by pressure on the sciatic nerve from a herniated disk, a condition in which something causes a disk to stick out farther than it should. That puts pressure on the sciatic nerve. Depending on which disk is herniated, and how far it sticks out (and thus how much pressure it puts on the nerve), symptoms can include:

- Pain in the buttock and/or leg which gets worse when sitting and which makes it hard to stand up
- Burning or tingling down the leg
- Either a constant or an occasional shooting pain, usually in one buttock and down the leg
- Weakness or numbness in one leg or foot

Sciatica pain can be mild or severe, occasional or constant. You're most likely to experience sciatica between ages 30 and 50. Most people who suffer sciatica get better with treatment over a few weeks, without surgery. Anyone who places a strain on the lower back either

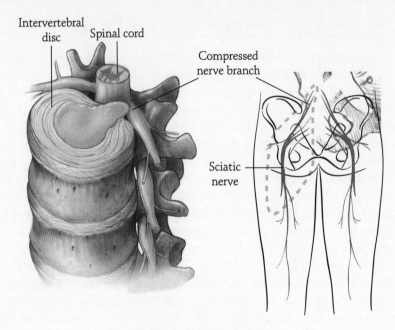

The anatomy of sciatica

through chronic or sudden pressure, motion, or an unusual position is at risk for a herniated disk and sciatica. One in 50 people suffer a herniated disk over the course of their lifetime.

If you've ever had a sharp burning or tingling sensation run through your shoulder, you may have experienced another common nerve-related MSD called *thoracic outlet syndrome (TOS)*. Most people call TOS a pinched nerve in the neck, and they're right. TOS results from pressure on the nerves (and, rarely, the blood vessels) that run out of the upper part of the spine and from the chest, through an opening behind the collarbone called the thoracic outlet, through the shoulder, and down the arm. Symptoms include pain in the shoulder, arm, and/or hand, and a general feeling of weakness or fatigue in the arm. The hand pain is often worst in the fourth and fifth fingers.

TOS is also treatable with therapy. Workers vulnerable to thoracic outlet syndrome include people who work with elevated shoulders, such as computer operators without arm supports set in the proper position to relax their shoulders, painters, or car mechanics (who hold their hands and arms over their heads while making repairs on a car that's up on a lift).

Tendon Problems

Tendons are the elastic tissues that connect muscle to bone. One well-known and easily identified tendon is the Achilles tendon in the back of your ankle. Many MSDs involve tendon problems. Pain, burning, and inflammation are common symptoms of tendonitis.

One common form is *lateral epicondylitis,* or *tennis elbow,* an inflammation of the outside ligaments in the elbow. Inflammation of the inside tendons is called *medial epicondylitis,* or *golfer's elbow.* Both conditions affect anyone who puts stress on these tendons with too much intensity, for too great a duration, and too frequently. Besides tennis players and golfers, anyone who uses his or her forearms to do work, including carpenters, people in the lumber industry, and computer operators, is often among the afflicted.

Another common occupational tendonitis is *de Quervain's disease,* an inflammation of the tendons running between the thumb and the wrist. It is also called *washerwoman's syndrome,* since it was first identified in the late 1800s among women wringing out laundry. It can affect anyone who uses hand tools or who works on the finishing or assembling of small parts.

Baseball players are among the workers who suffer *rotator cuff* tendonitis, an inflammation of the tendons that encapsulate the shoulder joint. But this MSD also afflicts other professions, including plasterers or painters, because it's not only throwing that causes the problem, but the abnormal posture of holding the arm over the shoulder so often and for so long.

Another tendon MSD may sound like something from the Wild West. *Trigger finger* is actually a condition in which repetitive motion of the fingers causes calcium deposits to build up on some of the tendons in the hand that help extend and curl the fingers. These tendons have to slide back and forth smoothly, but the calcium deposits interrupt this function and make it painful, sometimes impossible, to flex or extend the fingers. Besides gunslingers, people who get trigger finger include workers who do a lot of power gripping, including jackhammer operators, butchers, carpenters, or steelworkers. Trigger finger becomes more common as people age.

Muscle Disorders

Besides the muscle sprains, strains, and tears that occur from sudden traumatic stress, muscle disorders also arise from repetitive injury.

You may have heard of *writer's cramp,* a form of *focal dystonia.* Another is *musician's cramp.* These conditions are more serious than their names imply. Repeated use of certain muscles appears to reprogram the section of the brain that sends and receives signals in a way that makes those muscles more difficult to control and causes them to activate involuntarily. People who suffer focal dystonia continue to have normal use of the hand and finger muscles, except for the activity they have engaged in repeatedly—writing or playing an instrument—that caused the problem in the first place.

Writer's cramp was first reported in the 1830s. One victim called it "scrivener's palsy" and wrote: "The paralyzed scrivener, though he cannot write, can amuse himself in his garden, can shoot, and cut his meat . . . at the dinner table, indeed he can do almost anything he likes, except earn his daily bread as a scribbler." One study suggests that this muscular-neurological disorder occurs in 60 people per million.

THE RANGE OF EXPOSURES

Remember, MSDs can be caused by activity anywhere. But they are principally work-related since that's where we spend long periods of time in the same position repeating the same range of motion.

There are some general factors that affect your risk for MSDs. The prevalence of hand- and wrist-related disorders is higher among women than men. Carpal tunnel syndrome, for instance, is more common in women. The incidence of back problems is highest among men aged 20 to 24 and among women 30 to 34. And here's yet another reason not to smoke. Several studies have found that smoking increases the likelihood of low back pain and sciatica.

Body size and shape also play a role in MSDs. Obesity increases the risk of carpal tunnel syndrome. Greater height and weight seem to increase the risk of low back pain. These body sizes and types may be indicators of greater likelihood of a sedentary lifestyle, which contributes to elevated risk for some MSDs.

~

Here are some common exposures to musculoskeletal disorders and some of the occupations with which they're associated. The federal government deems most of these activities excessive if done for a total of more than two to four hours a day. (That doesn't necessarily mean two hours straight, just two hours total.)

Awkward Posture

Working with your hands above the head or your elbow above the shoulder (painters), working with your back or neck bent (masons, or anyone working at a computer monitor who has to lean forward or back or sit in some awkward position in order to read the screen), squatting or kneeling (airplane baggage loaders), or holding the wrists in a bent position (anyone who works at a keyboard).

High Hand Force

Gripping something tightly, or holding onto something that's heavy, or pinching something (anyone working with hand tools, such as dentists, small parts assemblers, musicians, surgeons, food packers). A "pinch grip" by the tips of the fingers causes much more force on the tendons than a "power grip" by the whole hand.

Highly Repetitive Motion

Similar movements with the neck, shoulders, elbows, wrists, hands, or fingers, repeatedly, for two hours or more. Intensive work at a keyboard for four hours or more (packers, inspectors, small parts handlers or assemblers, cashiers, telephone operators, sewer workers, data processors).

Repeated Impact

Using the hand or knee as a hammer or pounding device more than 10 times per hour (packers, assemblers, carpenters, construction workers).

Heavy, Frequent, or Awkward Lifting

Lifting objects that weigh more than 75 pounds once per day, or objects that weigh more than 55 pounds but less than 75 pounds more than 10 times a day, or objects that weigh more than 10 pounds more than twice per minute. Lifting objects that weigh more than 25 pounds above your shoulders, below your knees, or at arm's length more than 25 times a day (construction workers, laborers, assemblers, shippers, nurses, emergency medical technicians, janitors, warehouse workers, package deliverers, sanitation workers).

Vibration

Using tools that vibrate or operating equipment with constant vibration (plumbers, carpenters, electricians, truck/bus/train/heavy equipment operators, demolition workers, steelworkers). Sitting in a seat that vibrates (truck drivers) also affects the spine.

Cold

Working in a meat packing plant or frozen food processing area, or in construction outdoors: not only are you doing repetitive motions and/ or using high hand force, but you lose feeling because of the cold.

Direct Pressure

Resting arms or legs on a sharp edge for any length of time, compressing nerves and reducing blood flow to the part of the arm or leg beyond the pressure (precision workers, anyone leaning on a desk or working at a keyboard).

REDUCING YOUR RISK

The science and practice of designing jobs, workplaces, and working conditions to match the abilities and limitations of the human body is called ergonomics. The study of ergonomics has produced many suggestions about how you can reduce your risk of musculoskeletal disorders.

Most importantly, pay more attention to how you feel physically while you're on the job, and afterward. Little aches and pains that you might dismiss can become debilitating conditions that are not good for you or your employer. Are you comfortable as you sit at your desk? Do your chair and workstation allow for comfortable posture? Is your computer monitor too close or too far away, too high or low, forcing you to lean forward or back or hunch your shoulders as you work? Does your equipment require you to perform in an awkward position, or put an odd strain on your muscles? Is your body not supported properly? Do you experience a lot of vibration, jerky motion, and/or bumps at work?

Specific things you can do to reduce the risk of some of the common MSDs include: For problems of the forearms and the wrist, like tendonitis and carpal tunnel syndrome, make sure that you can keep your wrists in a neutral position as much as possible while you work. Your hand and arm should be basically flat and your wrist should not bend up or down. Also, take little breaks frequently, just 10 to 15 seconds, when you relax your hands and wrists: try clasping your fingers behind your head and stretching your elbows back.

For shoulder problems, such as TOS—a pinched nerve in your

neck—make sure your posture is good when you sit at your desk. Set your chair and computer monitor so you can keep your eyes at a 15-degree downward angle, your shoulders relaxed, your elbows flexed about 90 degrees, and so you don't have to lean forward or backward to read the screen. The keyboard should ideally be no more than an inch or two above your lap, at the same level as your elbow when your shoulders are relaxed. The armrests of your chair should be adjusted so you don't have to hunch your shoulders. If you talk on the phone a lot, get a headset so you don't have to tense your muscles to squeeze the phone between your head and shoulder.

Don't force any specific posture. Your body should be supported so it can be as relaxed as possible. Also, make sure your glasses are prescribed for the type of work you do. The body follows the eyes, and if you don't have the correct prescription, you may experience not only eyestrain, but neck, shoulder, and back pain from unsupported postures. Vary your work habits to include different motions. For example, use the phone instead of relying solely on e-mail. As one expert in ergonomics put it, "The best posture is the next one."

For low back pain, avoid lifting heavy objects alone. Work in lifting teams. Position your work properly to avoid excessive twisting or lifting while also reaching out. Keep the weight of the object as close to your body as possible. Bend your legs at the knees and use your legs, not just your back, to lift. Use lifting devices, such as height-adjustable scissors tables, manually operated forklifts, cherry pickers, and small cranes and winches. Do not carry heavy items long distances. Use carts when possible. Avoid excessive pushing and pulling of loads.

FOR MORE INFORMATION

National Institute for Occupational Safety and Health
www.cdc.gov/niosh/ergopage.html
200 Independence Avenue SW
Washington, DC 20201
(800) 35-NIOSH or (800) 356-4674

Occupational Safety and Health Administration
www.osha-slc.gov/SLTC/ergonomics
U.S. Department of Labor
200 Constitution Avenue

Washington, DC 20210
(800) 321-OSHA or (800) 321-6742

This chapter was reviewed by Jack Dennerlein, Assistant Professor of Ergonomics and Safety, Department of Environmental Health, Harvard School of Public Health; and by Colin Drury, Professor of Industrial Engineering at the University of Buffalo, a Fellow of the Institute of Industrial Engineers, of the Ergonomics Society, and of the Human Factors Ergonomics Society.

6. CAFFEINE

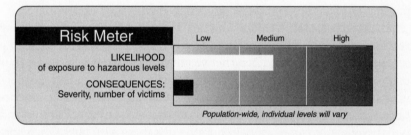

TENS OF MILLIONS of people start their day with it. Students pop it in pill form to stay awake through all-night study sessions. Parents keep it away from their kids so they won't "get hyper" or be unable to fall asleep at night. It's caffeine, the most widely consumed psychoactive substance in the world. In the United States, roughly 80 percent of adults consume caffeine daily in their coffee. Nearly 60 percent of children consume caffeine in soft drinks, chocolate, tea, and some cold relief and pain relief medications.

THE HAZARD

Caffeine works principally by interfering with a chemical in the body called adenosine. Adenosine is a compound involved in nerve signaling that acts as a modulator. It affects different body systems in different ways, but in each case adenosine essentially turns activity down rather than up. It does so by locking onto specialized receptors on the surface of certain cells. Caffeine binds to the same receptors. So when those receptors are "occupied" by the caffeine, the adenosine can't latch on and trigger its modulating effect. The activities of those cells aren't modulated—aren't turned down—because of the presence of caffeine. That's why we identify caffeine as a stimulant.

Caffeine's effect on us is moderate because it doesn't outcompete every adenosine molecule for every available receptor. Plus, caffeine is quickly metabolized and eliminated, in approximately three to six hours. Sensitivity to caffeine varies among individuals.

THE RANGE OF CONSEQUENCES

Many people commonly think of caffeine as an addictive drug, but the truth is closer to the way Ralph Waldo Emerson put it: "Tobacco, coffee, alcohol, hashish, prussic acid, strychnine, are weak dilutions: the surest poison is time." Regular consumption of caffeine does produce physical dependence, which means that regular use causes physical and psychological symptoms upon withdrawal. But caffeine is not clinically defined as addictive. Clinical dependence and addiction have more rigorous definitions, which, according to most experts, caffeine does not meet. *The Diagnostic and Statistical Manual of Mental Disorders* of the American Psychiatric Association states, "A diagnosis of substance dependence can be applied to every class of substances except caffeine." Beyond the question of addiction, in the moderate amounts consumed by most people, caffeine's effects are not considered harmful and of course many caffeine consumers describe its effects as positive.

Remember that caffeine counteracts the modulating effects of adenosine. In the central nervous system, adenosine reduces spontaneous electrical activity, decreasing signals between neurons. Caffeine allows for more signaling, which is why caffeine use is associated with alertness. Adenosine contributes to a slowing of heart function, which is why caffeine use is associated with increased heart rate and irregular heart rhythms. Adenosine is involved in the modulating events that help us fall asleep, which is why caffeine is associated with keeping us awake. Adenosine reduces kidney activity and urine output. That's why caffeine has the opposite, or diuretic, effect. In the stomach, adenosine turns down production of gastric acids. Caffeine allows for more production of these acids, which is why it is associated with heartburn.

Adenosine dilates the diameter of blood vessels. Caffeine constricts blood vessels, raising blood pressure. Adenosine constricts the diameter of the air passageways. Caffeine dilates them, which is why caffeine (or theophylline, a similar drug found in tea) is part of some treatments for asthma. Adenosine encourages the storage of fat. Caffeine helps release fat into the bloodstream, where it can be consumed by the body's metabolism. That's why caffeine used to be included in some diet pills.

~

Tolerance to caffeine's effects can build up in a matter of days. One study found that people ingesting the same moderate daily amount of caffeine no longer experienced an increase in blood pressure after just three days, and no longer experienced interference with sleep after seven days. These internal adjustments to caffeine play a major role in the body's reaction to the sudden absence of caffeine, a condition called caffeine withdrawal, which is the rebound back to "normal."

One of the most common symptoms of caffeine withdrawal is headache, believed to come from a "rebound effect" when adenosine again starts to widen blood vessels, including those that supply the brain. After being consistently constricted, the sudden change in diameter of the blood vessels in the brain is associated with headache. Caffeine withdrawal headaches generally start 12 to 24 hours after the last ingestion of caffeine. They are usually diffuse and general throughout the head, and get worse with exercise. Like the other symptoms of caffeine withdrawal, these headaches usually go away after two to four days, though they can recur sporadically for up to two weeks.

Other symptoms of caffeine withdrawal include sleepiness or drowsiness, impaired concentration, depression, anxiety, irritability, and in more severe and rare cases, nausea or vomiting and muscle aches and stiffness. The severity of the symptoms depends on the amount of caffeine that has been used, the length of time it's been used, the suddenness of the withdrawal, and the individual's sensitivity to caffeine. Most of these symptoms are also probably rebound effects, as adenosine again starts to modulate and the body's systems readjust.

One controlled experimental test of caffeine withdrawal found that 52 percent of the participants had moderate or severe headache, 8 percent had symptoms of fatigue, and 8 to 11 percent had anxiety or depression. Other studies have found that between one third and one half of caffeine users who quit suffer some withdrawal symptoms. Caffeine withdrawal can be caused by ingestion of as little as 100 milligrams of caffeine per day. That's about one cup of brewed coffee or two to three 12-ounce caffeinated sodas.

The Diagnostic and Statistical Manual of Mental Disorders calls excessive use of caffeine "caffeine intoxication" and defines it as follows: "After consuming at least 250 mg of caffeine—about 2–3 cups of brewed coffee—five or more of the following symptoms occur, strong enough to cause distress or impairment in social, occupational, or other important areas of functioning: restlessness, nervousness, excitement, insomnia, flushed face, diuresis, gastrointestinal disturbance, twitching muscles, rambling thought or speech, heart rhythm changes, periods of inexhaustibility, psychomotor agitation."

Little hard information exists on how widespread such use of caffeine might be. One study found that of 166 people who said they used caffeine regularly, 12 percent met the criteria for caffeine intoxication. A 1981 study of college students found only about 1 percent met the criteria. Nine years later a study of college students found that 19 percent reported a history of caffeine intoxication at some point.

Some people have raised the concern that caffeine consumption can cause cancer. The American Cancer Society finds no link. Investigations into possible connections between caffeine and heart disease, elevated cholesterol levels, birth defects, miscarriages, fertility problems, benign growths in the breast (fibrocystic disease), osteoporosis, and clinical hyperactivity in children have also found no confirmed associations. The Food and Drug Administration (FDA) lists caffeine as "Generally Recognized as Safe" but recommends moderate consumption for pregnant women because of some studies suggesting an association between very high caffeine intake (five cups of coffee a day or more) and increased risk of miscarriage.

THE RANGE OF EXPOSURES

The average daily intake of caffeine among adults is 200 milligrams. The average daily consumption by kids, mostly from soda, candy, and tea, is 35 to 40 milligrams. See the table "Caffeine Content in Common Sources" for more specific measurements.

A lot of products, like chocolate, contain caffeine because it occurs naturally in the plants from which those products are made, like coffee beans, tea leaves, and cocoa beans. In its pure form, it has a bitter taste. Sometimes caffeine is added to products such as soft drinks as a flavoring ingredient. It is most commonly found in colas, rather

CAFFEINE CONTENT IN COMMON SOURCES

COFFEE / TEA (7 oz.)

(Coffee is given in ranges, since precise amount depends on the coffee beans used and the preparation, neither of which matters with tea)	Milligrams of Caffeine
Drip	115–175
Espresso (1.5–2 oz.)	90–110
Brewed	80–135
Instant	65–100
Decaf coffee	2–4
Brewed tea, imported	60
Brewed tea, domestic	40
Iced tea	40
Instant Tea	30

SOFT DRINKS (12 oz.)

Jolt Cola	71
Mountain Dew	59
Surge	51
Tab	47
Coca-Cola	46
Diet Coke	46
Shasta Cola	44
Dr Pepper	40
Pepsi Cola	37
Diet Pepsi	35
RC Cola	36
Diet RC	36
Barq's Root Beer	23

CHOCOLATE

Hershey's Special Dark bar (1.5 oz.)	31
Hershey's Milk Chocolate bar (1.5 oz.)	10
Cocoa—Hot Chocolate (8 oz.)	5

OVER-THE-COUNTER MEDICATIONS

NoDoz, maximum strength; Vivarin (1 tablet)	200
NoDoz, regular strength (1 tablet)	100
Excedrin (2 tablets)	130
Anacin (2 tablets)	64

than sweeter-tasting soft drinks. Because it constricts blood vessels, caffeine is added to some pain relief medication. And because it widens breathing passageways, it's used in medications for asthma. So is theopylline. Since the FDA found that it had no measurable positive effect on weight loss, it is no longer an ingredient in diet pills. Caffeine is the principal active ingredient in over-the-counter medications to help with alertness.

REDUCING YOUR RISK

In a survey of medical specialists, three out of four doctors recommended that patients suffering certain preexisting conditions should reduce or eliminate their caffeine intake. These conditions include anxiety, insomnia, arrhythmia, heart palpitations, esophageal or hiatal hernia, and GERD (gastroesophageal reflux disease, or excessive heartburn).

Remember that aside from the low-level, short-term effects of caffeine and the effects of caffeine withdrawal, moderate caffeine consumption poses no health risks. Still, if you want to reduce your caffeine consumption without suffering the symptoms of withdrawal, make a daily diary of your caffeine intake over a few weeks. Then taper off by reducing your intake by roughly 10 percent every few days.

FOR MORE INFORMATION

MEDLINEplus
www.nlm.nih.gov/medlineplus/caffeine.html

This chapter was reviewed by Dr. John Daly, former Chief of the Laboratory of Bioorganic Chemistry at the National Institutes of Health; and by Roland R. Griffiths, Ph.D., Professor of Behavioral Biology and Neuroscience, Johns Hopkins University School of Medicine.

7. CELLULAR TELEPHONES AND DRIVING

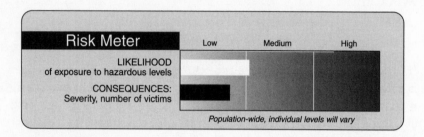

IF IT SEEMS to you that cell phones are everywhere now, you're right. At the beginning of 2002, 123 million Americans, almost one in two, was a mobile phone subscriber. And if it seems to you that drivers using cell phones are everywhere, you're also right. Surveys have found that 80 to 90 percent of mobile phone owners use their phone while driving at least some of the time, and about 30 percent of them regularly use their phone and drive at the same time. Depending on the region of the country, between 40 percent and 70 percent of calls from mobile phones are made by people who are driving. These people are DWP— driving while phoning—and their habit is creating a risk.

THE HAZARD

Like any form of distraction while you're driving, operating a mobile phone can be dangerous. The National Highway Traffic Safety Administration (NHTSA) reports that driver distraction, from various causes, plays a role in 20 to 30 percent of all motor vehicle crashes. Research has identified three different ways in which the use of a mobile phone can distract a motorist:

- Vision: While you're searching for the phone in your vehicle, reaching for the phone to initiate or receive a call, looking up a phone number, or looking at the phone while dialing or receiving a call, you're not looking at the road.
- Biomechanics: When your hands are holding the phone (as of the end of 2001, about 75 percent of cell phones in the United States

were hand-held), or dialing or receiving a call, they're not holding the steering wheel or helping to control the vehicle.

- Cognition: Several studies show that a driver paying attention to a mobile phone conversation is distracted simply because he is paying attention to the conversation, reducing the attention he is paying to driving. Studies show that a driver is more distracted by talking on a mobile phone than by talking to another person in the vehicle. Several studies have found that cognitive distraction affects motorists whether the phone they are using is hand-held or hands-free. Other studies suggest that cognitive impairment may distract drivers for up to two minutes after the phone conversation has ended. Still other studies show that cognitive distraction leads to a narrowing of the visual field.

No statistics exist to prove which of these three types of distraction is most important in causing mobile phone–related crashes. Research at the University of Utah suggests that the cognitive distraction may be the most important, since test subjects did equally as poorly when using hands-free or hand-held devices. A report by NHTSA finds that the time when a motorist is dialing a mobile phone is the most dangerous, since it involves all three levels of distraction.

But there is no question that using a mobile phone while driving is a distraction. Studies of driver performance, observing drivers both on simulators and in the field, have shown that mobile phone use while driving can adversely affect reaction time, swerving ability, ability to execute difficult driving tasks, and other indicators of safe driving. University of Utah research found that test subjects took longer to react to traffic signals and completely missed twice as many of those signals when they were talking on their mobile phone.

THE RANGE OF CONSEQUENCES

Though drivers using mobile phones have caused a handful of high-profile fatal crashes, there is very little reliable overall information on how many crashes or how many deaths have been caused by the use of mobile phones by drivers. As of the end of 2001, 16 states included mobile phone use as a factor on their accident reports. But since most of these crash investigations rely on voluntary reporting by the driver who has caused the crash, information about cell phone use by these drivers is not considered reliable.

Based on the limited information available, drivers use their mobile phones anywhere from half an hour to two hours per month. Assuming usage of one hour per month, the additional probability of death in crashes due to mobile phone use in 2001 was about 12 chances in a million per year. For perspective, that's about one third the fatality risk for someone who drives with a blood alcohol concentration of 0.10 percent for the same total amount of time, one hour each month. The risk of death to another motorist, a bicyclist, or a pedestrian due to a driver using a mobile phone is four in a million per year. That is about one fourth the annual risk of being killed by a driver with a nonzero blood alcohol content.

Mobile phone use by drivers probably contributes to more nonfatal injuries than deaths because mobile phones are often used in traffic circumstances, like rush hour, when vehicle spacing and travel speeds are reduced. However, due to the lack of data collected on the causes of nonfatal crashes, the probability of mobile phone use causing nonfatal crashes can't be calculated.

One study suggests that mobile phone use by drivers plays less of a role in crashes than other forms of distraction. The following table is based on the University of North Carolina Highway Safety Research program analysis of information from the National Crashworthiness Data System, a sample of 5,000 incident reports from state and local police at crash scenes where at least one vehicle was towed away. The categories are those that appear on police reports. These data are based on self-reporting by people involved in the crashes, creating a wide range of uncertainty.

SOURCES AND FREQUENCY OF DRIVER DISTRACTION

Distraction	Percent of Drivers Affected
Outside person, object, or event	29.4
"Other"	25.6
Adjusting radio/cassette/CD player	11.4
Other occupant	10.9
Source unknown	8.6
Moving object in vehicle	4.3
Other device/object	2.9
Adjusting vehicle controls	2.8
Eating or drinking	1.7
Using mobile phone	1.5
Smoking-related	0.9

THE RANGE OF EXPOSURES

Sketchy data are starting to give us an idea of who uses mobile phones, when, and where. According to NHTSA, at the beginning of the year 2000 an average of 3 percent of American motorists—500,000 people—were on their mobile phone at any given time. About half of drivers say they have mobile phones with them, and of that group, three quarters say they use their phone while driving.

Women used mobile phones more, particularly drivers of vans and SUVs. Mobile phone usage rates by female drivers of SUVs and vans were twice as high as for men. Use by passenger vehicle drivers was twice as high during non–rush hours as during rush hours. More drivers of vans and SUVs used mobile phones in non–rush hour driving than any other group by vehicle type and time of day. The lowest group by vehicle type was drivers of pickup trucks.

Rates of usage across age groups were similar, except for those 70 and over, who used mobile phones half as frequently as younger drivers.

Usage was higher during the week than on weekends.

Other hints about circumstances in which mobile phone use exposes motorists to risk comes from an online (and highly *un*scientific) poll of customers of an Ohio insurance company. Eight hundred people responded.

- 43 percent said they had accelerated on at least one occasion while using their mobile phone.
- 23 percent said they had tailgated.
- 18 percent said they had cut someone off.
- 10 percent had run a red light.
- 41 percent said they had accelerated to get away from someone else in another vehicle using a mobile phone.

But very little reliable data exist about cell phone use by drivers. In 2001 several states passed laws amending police accident reports so officers on the scene of a crash would have to record whether mobile phone use had been involved in the crash.

REDUCING YOUR RISK

In 2001, legislation to restrict mobile phone use by drivers was proposed in 43 states. It passed only in New York. The Rhode Island legislature passed a bill, but Rhode Island's governor vetoed it, saying that the benefits of mobile phones to public safety and to personal and professional productivity far outweigh the risks. It's estimated that nearly 144,000 emergency calls are made by motorists from their mobile phones in the United States each day, though many of these are probably multiple calls about the same event.

Opponents of mobile phone use by drivers say the passage of the state law in New York should give momentum to similar efforts in dozens of states in the coming years. But many legislators say the conflicting and limited research makes it tough to know what kind of law to pass in order to improve public health and safety. Until research answers the key questions about this issue, the following steps by drivers can make things safer:

- Try not to dial new calls until you're stopped or in a safe low-traffic environment.
- Keep calls short, less than 1 or 2 minutes.
- When you're on the phone, try to avoid difficult maneuvers such as sharp acceleration or passing, which require a lot of attention.
- Try to avoid stressful or anxiety-ridden calls. They're more distracting.
- Have a passenger dial, send, or receive calls.
- Familiarize yourself with the features on your phone, like speed dialing, that let you use it more efficiently.
- Avoid using hand-held phones in vehicles with manual transmissions.
- Limit additional distractions while using the phone, such as drinking, eating, or changing the radio station or a CD or tape. Any single distraction is bad enough.
- Invest in a hands-free model, especially if you use your phone regularly. While the cognitive distraction may be the most dangerous part of using a mobile phone while you're driving, it helps to avoid the mechanical distraction when you can. Most hands-free designs still require some manual operation, but they come with an earpiece and microphone so you can listen and talk without holding the phone. Some newer devices are voice-activated, completely eliminating the need to manually operate the phone. Many new cars have options that allow for hands-free mobile phones.
- If it's safe, you might want to consider pulling over and stopping before using your mobile phone. Do so only in rest areas or a parking space, not on the shoulder of the road, since pulling over and

stopping creates its own risks and isn't feasible or safe in many conditions.

FOR MORE INFORMATION

National Highway Traffic Safety Administration
www-nrd.nhtsa.dot.gov/departments/nrd-13/
    DriverDistraction.html
400 7th Street SW
Washington, DC 20590
(800) 424-9393

Cellular Telephone Industry Association
www.wow-com.com/consumer
1250 Connecticut Avenue NW, Suite 800
Washington, DC 20036
(202) 785-0081

National Conference of State Legislatures
www.ncsl.org/programs/esnr/2000cell.htm
1560 Broadway, Suite 700
Denver, CO 80202
(303) 830-2200

This chapter was reviewed by Michael Goodman, one of the leading experts on driver distraction at the National Highway Traffic Safety Administration and principal author of NHTSA's 1997 wireless communications report, "An Investigation of the Safety Implications of Wireless Communications in Vehicles"; and by Matt Sundeen, Program Principal and Investigator on mobile phone issues for the National Conference of State Legislatures.

8. CELLULAR TELEPHONES
AND RADIATION

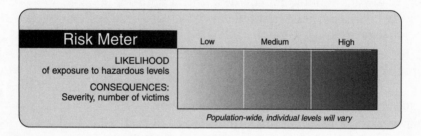

A CELL PHONE may seem like a modern technological device. But it is just a variation on an old and familiar technology. It's really just a sophisticated two-way radio, sending and receiving information via electromagnetic energy waves. But the mobile phone represents a pervasive new use of this form of communication. When Ronald Reagan was sworn in for his second term as president, almost nobody in the United States had a mobile phone. At the beginning of 2002, there were more than 123 million subscribers. Since cell phones transmit and receive radiation, there is concern that they might pose a health risk.

THE HAZARD

Mobile phones work the same way that radios do, and the same way that televisions, portable phones at home, even infant monitors, work: by sending an electromagnetic energy signal from one place to another. With cell phones, there are two sources that send and receive these signals, the handset and the base antenna. Each base antenna carries conversations within a several-square-mile area (which is that antenna's "cell").

The energy levels generated by mobile phone handsets and base antennas are very low. The first cell phones typically operated at a maximum power of 0.6 watts, 5 times less than a typical flashlight. Newer phones typically operate at 0.2 watts or less, 500 times less than the energy emitted by a household light bulb. Mobile phone power levels are similar to the newer 900-megahertz cordless home phones.

Base station antennas emit more radio wave energy, about the equivalent of between one and several household light bulbs. But since base antennas point most of their energy at the horizon, and since the intensity of the radiation exposure drops off quickly as the distance from the source increases, radiation exposure to people is very low, probably even lower than from the phones themselves. Base station radiation, however, is continuous, whereas handset emissions are intermittent.

It's not just the strength of the signal that matters. The potential health risk associated with any device that emits electromagnetic energy also depends on the nature of the radiation, specifically its frequency. Frequency is a measure of how many waves of electromagnetic energy pass a fixed point in a given period of time—essentially a measure of how wide or narrow the waves are. (For more information, see Chapter 34.) The frequency of the electromagnetic energy used by cell phones is similar to that used by microwave ovens (see Chapter 15) and TV channels above 13. Energy at this frequency is called nonionizing. It isn't strong enough to break chemical bonds and damage tissue or DNA molecules. That's what higher-frequency ionizing radiation does. X rays are a form of ionizing radiation. (See Chapter 48.) Since the radiation from mobile phones can't break chemical bonds, it can't cause mutation to the DNA molecules in our cells, and therefore most scientists agree it can't cause the damage that leads to cancer.

However, like microwave ovens, nonionizing radiation at frequencies like those used by mobile phones can cause water molecules to move in a way that creates heat in objects that contain water—like the human body. If the energy levels aren't too high, your body can easily dissipate the heat generated by microwave energy. At higher levels, though, microwave energy can cause burns. (The heat you may occasionally feel from a mobile phone is from the battery and electric circuits, not the radiation the phone is emitting to send or receive your conversation.)

Even though cell phone frequencies are not believed to be harmful unless they heat things up, some people fear that cell phone use may contribute to brain cancer because mobile phones are held close to the head. Public concern about this theory accelerated in the early 1990s when a Florida man sued a mobile phone manufacturer and several service providers, claiming that his wife's cell phone was responsible for her brain tumor. That suit was dismissed for lack of evidence. In the late 1990s a Baltimore neurologist filed a similar suit claiming that mobile phone use caused a brain tumor on the same side of his head that he usually held his cell phone to. As of the end of 2001, that lawsuit was still pending.

THE RANGE OF CONSEQUENCES

There is no scientific evidence to support the concern that mobile phones, either the handsets or the base antennas, cause cancer. An extensive analysis of the issue done for the European Union concluded: "In particular, in humans, no evidence of carcinogenicity in either children or adults has resulted from epidemiological studies (the size of some of which was very large, although the period of observation was not long enough for a definitive statement). A relatively large series of laboratory studies has not provided evidence of genotoxicity (poisonous effects on DNA)."

Scientists have conducted a substantial amount of research investigating the potential health effects from exposure to radio frequency energy. The Food and Drug Administration (FDA), the World Health Organization, and expert panels commissioned by the United Kingdom and Canada have concluded that the studies conducted so far do not demonstrate adverse health effects. (One exception is the very low but recognized potential for cell phones to interfere with electronic medical devices, like pacemakers, defibrillators, and hearing aids, when they are placed near the device. However, the FDA has adopted standards that should eliminate this type of interference in the future.)

But much of the evidence is not precisely applicable to mobile phones because many studies investigated exposure conditions that differed from normal cell phone use, and those studies looked at frequencies that did not apply to cell phones. So while the scientific consensus is overwhelming that cell phone radiation exposure is not a health risk, most of the scientific groups that have reviewed these studies have concluded that, as is the case with almost any risk, adverse health effects cannot be absolutely ruled out.

A massive review in Great Britain of mobile phone research found no convincing evidence that cell phones were harmful to anyone. The report, issued at a time when the British government was being fiercely criticized for its handling of other public health risks, such as mad cow disease, recommended that regardless of the evidence, to err on the side of caution, children should not use cell phones.

THE RANGE OF EXPOSURES

If cell phones can be used where you live, work, or travel, you are exposed to cell phone radiation, regardless of whether you use a phone yourself. Most residents of the United States living or working in cities or suburbs are constantly exposed to cell phone radiation. Base antennas are spaced just a few miles apart and, as we've stated, emit radiation constantly. In addition, people who use mobile phones are exposed to energy emitted by the handset. Even just being near someone on a mobile phone exposes you to some radiation. But the frequencies and power of this energy are so low that scientists consider them harmless.

Because the only demonstrated biological effects of radio frequency energy are thermal—heating—the federal government and various international organizations set limits to ensure that both handset exposures and base antenna exposures are safely in the subthermal range. They set those standards by starting with the lowest energy level shown in any experiment to affect animals. They then divide this energy level by 50 to arrive at the standard for mobile phone handsets and base antennas.

Regulatory limits cap whole-body exposure at five to ten watts (the energy emitted by a typical night-light). However, even at their strongest point, exposures from base antennas are roughly 100 times smaller than that. And the handsets themselves are much weaker still.

REDUCING YOUR RISK

There does not appear to be any risk from cell phone radiation. Remember, a risk requires both hazard and exposure. We're exposed to lots of cell phone radiation, but not to levels that are hazardous. Still, if you want to reduce your exposure to cell phone radiation, increase the distance between the phone and your body. You can get headsets or earpieces that connect to the cell phone, so the transmitting antenna is kept away from your head. While this may reassure you, there is no evidence that it actually makes you safer.

In response to public concern, cell phone manufacturers in 2001 began labeling their devices with an SAR value—specific absorbed radiation—which tells you how much radiation you are likely to absorb if you use a mobile phone held to your ear. This information lets you compare which devices subject you to what levels of radiation. It appears inside the package of the phone and on the websites of phone manufacturers. All the phones on the market meet the federal standard for SAR, and levels from newer digital models are usually significantly below this standard. To give you some perspective on the level the government has set: if you turn on a cell phone that emits the maximum federally accepted level of radiation and put it on a 2.2-pound block of ice, it would take two and a half days to melt the ice.

Some companies sell devices that they claim will shield you from cell phone energy. According to the Federal Communications Commission, such devices don't work, and may even *increase* exposure to cell phone energy by concentrating the radiation in certain ways.

Cell phone users who have medical devices like a pacemaker or implanted defibrillator should check with their doctor and/or the phone manufacturer to see what precautions, if any, they should take. The FDA says cell phones do not pose a risk for most pacemaker and defibrillator users.

Some hospitals have rules that prevent cell phone users from turning the devices on while in the hospital, not to protect patients from radiation, but to reduce the very small risk of electronic interference with life support equipment.

FOR MORE INFORMATION

Medical College of Wisconsin
www.mcw.edu/gcrc/cop/cell-phone-health-FAQ/toc.html
8701 Watertown Plank Road
Milwaukee, WI 53226
(414) 456-8296

European Union
europa.eu.int/comm/health/ph/programmes/pollution/
    ph_fields_index.html

Wireless Information Resource Center
www.wirc.org
407 Laurier Avenue

P.O. Box 56066
Ottawa, Ontario
Canada K1R 7Z0

This chapter was reviewed by Kenneth R. Foster, Professor, Department of Bioengineering, University of Pennsylvania, who has studied and written extensively on electrical and magnetic fields and nonionizing radiation; and by Peter Valberg, who has a Ph.D. in physics and for 12 years was a member of the faculty of the Harvard School of Public Health. Valberg is now an environmental consultant for the Gradient Corporation. He has studied and published on electrical and magnetic fields, microwaves, and other energy issues.

9. ELECTRICAL
AND MAGNETIC FIELDS

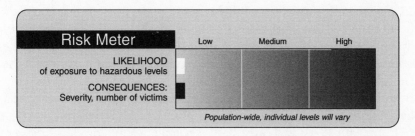

AS FAR BACK as ancient Greek civilization, humans have recognized the power of electricity. But not until 1790 did Luigi Galvani discover that electricity could flow. Thirty years later, a Danish scientist discovered a magnetic field around a flowing electrical current. And 160 years later still, in the 1980s, epidemiologists and journalists brought to the public's attention the possibility that the magnetic and electrical fields around electricity could damage human health. If this theory proved true, the implications were staggering for a modern industrialized world so absolutely dependent on this form of energy. Fortunately, intense worldwide research has largely disproved the idea.

THE HAZARD

Electricity does more than provide direct power. As it flows through wires in an open circuit or waits, as voltage, to flow when a switch is flipped on, electricity generates electrical or magnetic fields (EMFs). These zones of energy exist within a few inches, or a few feet, of the wire or appliance or motor that is carrying the electricity.

These fields differ from radiant energy such as light from a light bulb or a star. Radiant energy travels in waves away from its source, and those waves continue to exist and travel even if the source is turned off. When you see a star, for instance, you're seeing energy waves and light particles—photons—radiated out from that star millions of years ago. Radios and cell phones emit radiant energy too. EMFs, however, are nonradiant energy. They do not travel across space. They cease to exist as soon as the electricity generating them is gone. And though these fields are often lumped together in discussions of risk, EMFs are actually different. Electrical fields are generated around anything that contains electricity, even when that electricity is not flowing through an open circuit. Magnetic fields, on the other hand, exist only if the electricity is flowing.

Think of a cord running to a lamp and imagine that the lamp is off. But the wire connecting the lamp to the plug still contains electricity, so it's generating an electrical field. But it doesn't generate a magnetic field until you turn on the lamp, completing the circuit. The electricity flows down one strand of wire, through the lamp, and back down the other strand of wire to the plug. If you turn off the lamp and break the circuit, you stop the flow, and the magnetic field is gone. Pull the cord from the plug and it no longer contains electricity and the electrical field is gone too.

In the late 1970s and early 1980s several epidemiological studies reported an association between living near power lines and some forms of disease, particularly childhood leukemia. This finding prompted widespread public concern, and years of intense research into whether EMFs might be harmful to human health. The research was done on many levels, from lab tests on molecules, cells, and animals, to dozens of epidemiological studies analyzing the experience of human populations. A panel of experts convened by the National Academy of Sciences (NAS) in 1996 reviewed the research and found that "the current body of evidence does not show that exposure to these fields presents a human health hazard." The experts also wrote, "No conclusive and consistent evidence shows that exposures to residential electric and magnetic fields produce cancer, adverse neurobehavioral effects, or reproductive and developmental effects." Reviews by other government agencies in the United States, and government health authorities in other countries, came to similar conclusions.

Those reports acknowledge that several epidemiological studies observed a correlation between childhood leukemia and electrical wiring. The 1996 NAS report finds "living in homes classified as being in the high wire-code category is associated with about a 1.5-fold excess of childhood leukemia, a rare disease." But the expert reviews went on to say that since all the tests on molecules, cells, and lab animals were negative, and since many other epidemiological studies found no such association, it is unlikely that EMFs from the wiring were causing the childhood leukemia.

The National Institute for Environmental Health and Safety (NIEHS) wrote in 1999 that "associations reported for childhood leukemia and adult chronic lymphocytic leukemia cannot be dismissed easily as random or negative findings. The lack of positive findings in animals or in mechanistic studies weakens the belief that this association is actually due to electrical and magnetic fields but cannot completely discount the findings."

Following up on its 1996 report, the NAS reported in 1999 that "The results of [further research] do not support the contention that the use of electricity poses a major unrecognized public health danger."

The largest epidemiological studies done so far, including one by the National Cancer Institute in 1997 and a major study in the United Kingdom, found no association between EMFs and childhood cancer.

The scientific reviews by the NAS, the NIEHS, and international agencies suggested that the association between these power lines and childhood leukemia is not from the EMFs themselves, but from some other as yet unknown factor. This missing link could range from the socioeconomic class of those who live near power lines to how the studies were done and interpreted.

However, in the summer of 2001, the World Health Organization's International Agency for Radiation Control listed EMFs as a class 2B or possible carcinogen, their lowest category of possible carcinogenicity, based on the several studies showing a small but statistically significant association between EMFs and childhood leukemia.

THE RANGE OF CONSEQUENCES

Aside from the questionable association between childhood leukemia and wiring codes, nearly a decade of research has failed to establish any health consequences from exposure to EMFs. Some studies have suggested effects including miscarriage, birth defects, Alzheimer's disease, multiple sclerosis, or sleep disorders, but attempts to reproduce these findings have failed. Most of the original studies, when reviewed by experts, were judged either too small, defective in their methodology, or inconsistent in their findings.

EMFs are nonionizing radiation, with wavelengths too long to break apart molecules, the way more powerful ionizing radiation such as X rays can. Nonionizing radiation can still do damage, by vibrating molecules and thus heating biological tissue. A microwave, for example, is a form of nonionizing radiation. But if the energy waves get too long, they can't even vibrate water molecules enough to heat biological tissue. EMFs are generated by electricity with wavelengths too long to produce heat in our bodies.

Some observers have suggested that another way EMFs might cause harm is by interfering with the internal electrical impulses our body uses for nervous system signaling. Research has confirmed that this interference doesn't occur with most exposures, probably because the EMFs are very weak compared with the more powerful electrical impulses in our body. Very large exposures to EMFs *can* produce stimulation of our nerves and muscles, producing an effect like a twitch or jolt from an electric shock. Such exposures are rare, confined to occupational settings.

One consequence that research did establish was the effect that strong EMFs can have on cardiac pacemakers. Fortunately only very strong fields, much greater than those normally found in residential environments, interfere with pacemakers, and these fields occur only within several feet of high-voltage transmission lines of 230 kilovolts and above. People with implanted electrical medical devices should receive warnings about this hazard from their doctors.

THE RANGE OF EXPOSURES

Since electrical fields are too weak to penetrate skin, scientists believe that only exposure to magnetic fields could have any biological effect, though none has been found so far. Magnetic fields are measured, in the

United States, in units called gauss. (Overseas the unit used is the Tesla.) Magnetic field levels can be 1500 milliGauss within an inch of a refrigerator motor, or a ceiling fan, but the power of both electrical and magnetic fields drops off dramatically the farther you are from the source. Just a foot away from that refrigerator or ceiling fan, magnetic field levels would drop to just 2 milliGauss.

In residential settings, magnetic fields are strongest near anything with a motor, since motors use magnetic fields to operate. MilliGauss fields from some common appliances (at six inches distance) include:

- Washing machines: 100
- Electric shavers: 600
- Mixers: 600
- Hair dryers: 700
- Vacuum cleaners: 700
- Can openers: 1500

Research indicates that even these relatively high levels are far too low to cause any harm, at least not any *physical* harm. When public concern about EMFs was at its peak a few years ago, people were so afraid of living near power lines that some courts upheld homeowners' claims that their property value was lessened because there was a high-voltage transmission line nearby.

In late 2001, researchers proposed a new theory to explain the connection between EMFs and childhood leukemia. The idea is that the "contact currents" that occur when we touch something, like a faucet or pipe or other metal object that causes a current to flow through us, might have something to do with it. These contact currents are too small for us to notice in most cases, but they're big enough to have a biological effect. But serious questions remain about how these exposures could cause childhood leukemia. Since childhood leukemia is a disease that occurs only after multiple events cause a series of DNA mutations, and since it peaks in kids ages two to five, the question remains whether children or pregnant women are exposed to enough of these currents to trigger the multiple events that lead to disease at those young ages. Research on the idea of contact currents was just getting under way at the beginning of 2002.

REDUCING YOUR RISK

No safety standards have been established for EMF exposure in the United States. No recommendations regarding precautions exist, other than those pertaining to people using electrical medical devices or to some electrical workers. The NIEHS has reported that "the level and strength of evidence supporting EMF exposure as a human health hazard are insufficient to warrant regulatory actions." They recommended "passive measures" aimed at reducing exposures in future wiring of buildings and in siting of power transmission lines.

In 1996 the Swedish government, while finding "no basis for any limit values or other compulsory restrictions on low-frequency electrical and magnetic fields," recommended, "If measures generally reducing exposure can be taken at reasonable expense and with reasonable consequences in all other respects, an effort should be made to reduce fields radically deviating from what could be deemed normal in the environment concerned."

Such passive measures rely mostly on putting sources of EMFs farther away from people. Higher towers for transmission lines are sometimes suggested. Burying power lines, or shielding sources of electricity, reduce EMF exposure, but are very expensive.

FOR MORE INFORMATION

National Institute for Environmental Health and Safety
www.niehs.nih.gov/emfrapid
P.O. Box 12233
Research Triangle Park, NC 27709
(919) 541-3345

Medical College of Wisconsin
www.mcw.edu/gcrc/cop.html
8701 Watertown Plank Road
Milwaukee, WI 53226
(414) 456-8296

This chapter was reviewed by Peter Valberg, who has a Ph.D. in physics and for 12 years was a member of the faculty of the Harvard School of Public Health. He is now an environmental consultant for the Gradient Corporation. He has studied

EMFs and published several articles reviewing the issue. Dr. Robert Kavet, one of the leading researchers in the field and Area Manager for EMF Health Assessment for the Electric Power Research Institute, also reviewed this chapter.

10. FIREARMS

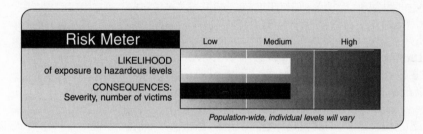

GUNS are inanimate objects. By themselves, they do nothing. Yet firearms are one of the most contentious and emotional public health issues in the United States today. In 1999, according to the Centers for Disease Control and Prevention (CDC), guns killed 28,874 people nationwide: 16,599 killed themselves, 10,828 were murder victims, 824 died from unintentional firearms use, 324 were shot to death but the specific cause was never determined, and 299 were killed by firearms in legal intervention. For every person killed by a firearm, approximately three more are injured seriously enough to seek medical attention. In the year 2000, firearms were used in approximately 331,000 robberies and aggravated assaults.

Regardless of one's position on gun ownership and gun control, it is inarguable that the potentially lethal nature of firearms makes them a significant public health risk. As a public health expert observed, "Guns may not cause violence, but they do make violence more lethal and allow traditionally nonfatal incidents to become fatal."

THE HAZARD AND THE CONSEQUENCES

Despite the significant numbers of deaths caused by firearms, the slowing trend in America is encouraging. Deaths from firearms in 1999 were lower than they had been in 20 years, down 27 percent since the high point in the last two decades, 1993, when guns killed nearly 40,000 people. Since that peak, the largest drop has come in the category of homicides, which are down 40 percent. Gun use in suicides is also down since 1993, but by much less, only 12 percent. Approximately 40 percent fewer U.S. residents died in the year 2000 from accidents involving firearms than in the peak year of 1993.

~

In 1977, a group of physicians in the U.S. Public Health Service met to draw up a list of the 12 most important steps to prevent deaths in the United States. In identifying the leading causes of death for people under age 65, they found that homicide and suicide were among the top five causes of death, and that a majority of people killing one another, or themselves, were using guns.

Seven years later, Surgeon General C. Everett Koop declared that violence was as much a public health issue for modern physicians as smallpox was for the medical community a few generations earlier. States, counties, and cities across the United States began establishing violence prevention programs.

But the trend that brought nationwide attention to the public health implications of firearms was the dramatic rise during the late 1980s and early 1990s in youth homicide, mostly in large cities. In California, for example, between 1985 and 1995, youth homicides more than doubled. During this period, deaths due to stabbing fell by nearly a third, but homicides by guns nearly tripled. Three quarters of the youth homicides in California in that ten-year span involved firearms, 90 percent of which involved handguns. These murders got widespread media attention, along with a string of mass murders in schoolyards, restaurants, and workplaces.

Some research on this health risk found a correlation between the number of new firearms manufactured and the rates for homicide and suicide since the late 1950s. From the early 1970s, the availability of new firearms and the number of homicides using firearms often paralleled each other. The number of new firearms made every year in the United States increased from about 2.4 million in 1950 to a peak of 5.7 million in 1975, with an average of more than 5 million produced per year from 1979 to 1982. By the mid-1990s, epidemiologists estimated that there were 192 to 216 million firearms in the nation. Across regions, states, and cities, higher levels of firearm prevalence are statistically associated with higher levels of homicide, suicide, and unintentional gun deaths, for both males and females, and across all age groups.

But sociologists, criminologists, and other researchers noted other trends and suggested that they had more to do with rising gun crime than merely the availability of firearms. Rising unemployment, population demographics (particularly a bump in the teens-and-early-20s age group most associated with criminal behavior), the emergence of gang cultures in many large cities, and more illegal drug use (including the introduction of crack cocaine) all contributed to violent crime, as did general issues of poverty, child abuse, and continued poor education in urban schools. Still, the lethality of firearms certainly raised the risk that the violence resulting from these other trends was more likely to be deadly.

~

The most hazardous type of firearm is the handgun, based on the number of victims of gun death in all categories. Epidemiologists have found that although handguns make up only about one third of firearms manufactured every year, they are involved in most of the firearm deaths. An estimated two thirds of the 1 million people killed by firearms in the United States between 1962 and 1997—in homicides, suicides, and unintentional shootings—were killed with handguns.

Small, inexpensive, easily concealed handguns, often called Saturday night specials or junk guns, began to be disproportionately involved in crime in the 1960s. In 1968, after Robert Kennedy was killed by an attacker using a concealed handgun, Congress began what has been an ongoing battle to deal with the public health issue of firearms, particularly handguns. They passed the Gun Control Act, and banned the importation of handguns not "suitable for sporting purposes." The Bureau of Alcohol, Tobacco and Firearms developed criteria regulating such guns, which included requiring safety features to prevent unintended shootings. At that time, most handguns were imported, and the legislation exempted domestic manufacturers. U.S. firearms manufacturers began producing inexpensive handguns, and a number of new companies specialized in them. By the end of the 1980s, most handguns manufactured and sold in the United States would not have been legal if they had simply been made outside the country.

~

While the use of firearms in crimes receives most of the media attention, the largest number of gun deaths remain suicides. From 1981 through 1999, approximately 271,000 people have been murdered by guns, but about 338,000 have died from a self-inflicted gunshot. The number of murders by firearm rises and falls in cycles like other categories of violent crime, but the number of suicides per year remains nearly constant. The highest rate of suicide by firearm, by age group, is among those over the age of 65. The greatest overall number of people killing themselves with firearms by age group is among those 15 to 24. More than 2,400 young people between those ages used a gun to commit suicide in 1997.

In addition to the question of firearms and suicide, there is also the issue of whether firearms in the home raise or lower risk. Studies of the risks of owning a firearm find that having a gun in the home, either for sport or self-protection, increases the risk for adult suicide, adolescent suicide, homicide, and accidental shooting. One study found that a firearm purchased for self-protection is 22 times more likely to be used in a murder, a suicide, or to harm someone by accident than it is to kill someone in self-defense. Another study using a national telephone survey found that guns in the home are probably more often used to frighten family members or friends than to thwart crime.

EFFORTS TO REDUCE THE RISK

The Second Amendment to the U.S. Constitution says:

"A well-regulated militia, being necessary to the security of a free state, the right of the people to keep and bear arms, shall not be infringed." This amendment was written in the days when a colonial militia had just used its guns to overthrow a repressive foreign government. With one recent exception, the federal courts have always interpreted the Second Amendment to mean that having a gun is a collective constitutional right, in order for states to be able to maintain a militia, not an individual right. However, many states specify an individual right to have firearms. It is in the context of that legal debate, and a general resentment among many about government interference with what they feel are their personal rights, that the fight over gun control and gun safety is being waged.

Many efforts to reduce the risk of firearms to public health take the same approach many gun supporters take. As a leader of the National Rifle Association, Charlton Heston, puts it, "A gun in the hands of a bad person is a bad thing." A number of efforts at firearms control attempt to keep guns out of the possession of people most likely to use them illegally.

Studies found that denying the purchase of handguns to people at high risk for violence—those with prior criminal records—reduced the risk for later criminal activity by 20 to 30 percent. Other research found that people with one prior misdemeanor conviction were more than seven times as likely as those with no prior criminal history to be charged with a firearm or violence-related offense after a handgun purchase. A subsequent study found that denying handgun purchases to people convicted of at least one violent misdemeanor was associated with a reduction in the risk of arrest for new gun and/or violent crimes.

While seven states have attacked the problem of gun violence by simply outlawing ownership of certain kinds of firearms, particularly Saturday night specials, many more states have acted to take guns out of the hands of the people most likely to use them to harm others or themselves. For instance, 22 states have established waiting periods for handgun purchases, to make it harder for people to buy guns at a time of high emotion. And 15 states have licensing or registration laws for some type of gun, while 18 states require a background check when a handgun is bought from an unlicensed seller, whether from a friend or at a gun show.

But 28 states have no waiting period for handgun purchases, and 32 states require no background checks when a handgun is purchased from an unlicensed private or gun show seller. Some 35 states require neither licensing nor registration for any type of gun, and 40 states prohibit or restrict communities from enacting local gun laws.

One particularly contentious aspect of the firearm safety debate is the issue of concealed handguns. States have taken different approaches to this matter. Some want to allow people to carry concealed weapons as a way to reduce crime and make the public safer. Others say it increases the risk to public health by encouraging criminals to carry guns to make sure they can commit their crimes against a possibly armed victim. These people also fear that if more people carry guns, it becomes more likely that guns will be used in the course of escalating arguments. One research study showed that states that loosened their laws to make it easier for people to carry concealed weapons had a 9 percent reduction in violent crime. However, a study of 1996–1997 crime data showed that states that rely on permissive concealed weapons laws as a crime-fighting strategy had a significantly smaller drop in the crime rate (2.1 percent) than states that used other methods to combat crime (4.4 percent).

~

The U.S. Congress has passed three firearms acts since 1993. The Brady Handgun Violence Prevention Act, which went into effect in 1994, is a response to the assassination attempt on President Ronald Reagan, which also seriously injured his press secretary, James Brady. It requires federally licensed firearms dealers to do a background check with law enforcement on a prospective purchaser before selling a firearm. Since the law took effect, more than 600,000 criminals and other people prohibited from owning guns have been stopped from purchasing firearms from licensed dealers. However, the law does not regulate the sales of firearms at the estimated 4,500 gun shows per year, or sales among private gun owners. Together these amount to a significant portion of all gun sales, 40 percent by one estimate.

The Brady Act also stiffened the requirements for licensing of firearms dealers. The dealer-licensing fee was increased, and licensees were required to provide photographs and fingerprints and to notify local law enforcement of their business. The number of federally licensed firearms retailers dropped from a peak of nearly a quarter of a million in 1992 (more than the number of gasoline stations in the United States at that time) to roughly 70,000 in the year 2000.

The second federal gun control law, the Violent Crime Control and Law Enforcement Act of 1994, banned the manufacture, sale, and possession of 19 types of semiautomatic assault weapons and some copycat models. It also banned some other semiautomatic guns with certain characteristics, particularly the capacity of the firearm to carry more ammunition than would reasonably be needed either for self-defense or sporting use. It also banned juvenile possession of a handgun or handgun ammunition, with limited exceptions, and made it a crime to sell or give a handgun to anyone 18 or younger.

Finally, the Domestic Violence Offender Gun Ban of 1996 prohibits anyone convicted of a misdemeanor domestic-violence offense from buying or owning a gun.

The campaign to increase firearm safety has also made its way into the court system. Dozens of local and regional governments across the United States began suing the firearms industry in 1998, taking the same approach as many did in demanding responsibility for product safety from auto manufacturers and tobacco companies. The suits against gun manufacturers claim that they have declined to incorporate safety devices and warnings that would help prevent accidental shootings, have misled the public with advertising about the consequences of having a gun in the home, and have used distribution practices resulting in a large, illegitimate secondary market for guns.

Many suits have been dismissed. Others are making their way through the system. In 2001, the Smith and Wesson Corporation entered a consent decree to make changes in design, marketing, and distribution of its firearms. Twelve cities agreed to drop their lawsuits against the company, but others decided to continue legal action. A number of individual suits were also pending as of the beginning of 2002.

Another tactic taken by governments to reduce firearms deaths and injuries is to institute gun buyback programs. Many of these have been run sporadically at the local level. The federal government funded a buyback program during the Clinton administration. The program purchased 20,000 guns in 12 cities before President Bush terminated it shortly after his inauguration. Some research questions the effectiveness of these programs.

$$\sim$$

Over the last five years, crime rates have dropped 20 percent. Between 1993 and 1999, domestic handgun production is down by roughly half, from 2,825,000 to about 1,332,000. Still, roughly two thirds of homicides in 2000 involved firearms, and roughly three quarters of those involved handguns. In 1998, 57 percent of the people who took their own lives used a gun.

Unfortunately, no comprehensive, neutral data collection systems exist for developing the facts on which to base more informed firearms policy. The CDC is working with the Harvard Injury Control Research Center at the Harvard School of Public Health and 15 states to establish the National Violent Death Reporting System (NVDRS) as a pilot program on which to base a national reporting system to guide violence prevention efforts. The system is modeled on the Fatality Analysis Reporting System (FARS), which was established in 1974 in response to the thousands of deaths and injuries occurring in motor vehicle crashes. The information derived from FARS has resulted in motor vehicle safety laws and regulations that have led to a 60 percent decrease in auto deaths—an estimated 24,000 lives saved annually.

Wisconsin, one of the states participating in the NVDRS pilot program, collects the following data on each violent death:

- For suicide, whether the victim was being treated for depression, had a history of substance abuse, or suffered a serious physical illness
- For crime, the victim's and perpetrator's or suspect's age, race, sex, education, occupation, marital status, criminal history, and the presence of drugs or alcohol
- Cause of death and the type of fatal wound
- If death by firearm, the make, model, ammunition, casings, owner, storage, place of purchase, and the time from purchase to incident
- Location of death (highway/alley/street, victim's residence, offender's residence, drug house, bar, convenience store, woods)
- Circumstances (drive-by, argument over money or property, during divorce proceedings, child playing with gun)

With that kind of detailed information, communities can make informed decisions about preventing firearms deaths. For example, Wisconsin knows that men in their early 20s and early 70s are most at risk for suicide by firearm. Milwaukee knows that gun buyback programs aren't likely to reduce homicides because most of the firearms murders in Milwaukee involved types of guns that rarely turned up during the city's buyback program.

REDUCING YOUR RISK

Owning a gun is a personal choice. Like many choices, it comes with risk. Learning about the risks that accompany gun ownership will help you make your choice more wisely.

If you do choose to own a firearm, public safety officials strongly advise that you take a course in how to use it and store it safely. You should take a refresher course periodically, too.

Avoiding the risk of being victimized by a crime involving a firearm is far too complex and personal a matter for us to discuss on your behalf. So is the issue of reducing the risk of using a gun in a suicide, beyond the obvious; if you don't have a gun, it's harder to use one to take your own life, as an average 17,000 Americans do each year.

We can suggest ways to reduce the risk of firearms accidents in the home. As we also mention in Chapter 1, "Accidents," keep guns out of the hands of children, particularly teenagers, since the people most at risk for unintentional gun death are those between ages 15 and 24. That age group is twice as likely to die from firearms accidents in general, but three times more likely to die in gun accidents in the home. So if you have guns at home, keep them locked up. Store them unloaded, with the ammunition stored somewhere else, also locked up and the keys hidden. A survey by the group Common Sense about Kids and Guns states that nearly 30 percent of the handguns legally owned in 40 million American homes with children are stored loaded and unlocked.

FOR MORE INFORMATION

Bureau of Alcohol, Tobacco and Firearms
www.atf.treas.gov
Office of Liaison and Public Information
650 Massachusetts Avenue NW, Room 8290
Washington, DC 20226
(202) 927-8500

National Center for Injury Prevention and Control
www.cdc.gov/ncipc
Mailstop K65
4770 Buford Highway NE
Atlanta, GA 30341-3724
(770) 488-1506
Fax: (770) 488-1667

Federal Bureau of Investigation Uniform Crime Reports
www.fbi.gov/ucr/ucr.htm
J. Edgar Hoover Building
935 Pennsylvania Avenue NW
Washington, DC 20535-0001
(202) 324-3000

Criminal Justice Information Services Division
www.fbi.gov
1000 Custer Hollow Road
Clarksburg, WV 26306
(304) 625-4995
Fax: (304) 625-5394

National Violent Injury Statistics System
www.hsph.harvard.edu/hicrc/nviss/
Harvard Injury Control Research Center
Harvard School of Public Health
677 Huntington Avenue
Boston, MA 02115
(617) 432-3353
Fax: (617) 432-4494

National Rifle Association
www.nra.org
11250 Waples Mill Road
Fairfax, VA 22030
(703) 267-1000

Brady Center to Prevent Gun Violence
The Brady Campaign to Prevent Gun Violence
www.bradycenter.org
1225 Eye Street NW, Suite 1100
Washington, DC 20005
(202) 898-0792
Fax: (202) 371-9615

Open Society Institute: Center on Crime, Communities, and
   Culture
Funders Collaborative for Gun Violence Prevention
www. soros.org/crime/guncontrol.htm
400 West 59th Street, 3rd Floor
New York, NY 10019
(212) 548-0135

This chapter was reviewed by Susan Sorenson, Professor at the UCLA School of Public Health, a violence prevention researcher who has extensively investigated firearms issues; by David Hemenway, Professor of Health Policy at the Harvard School of Public Health and Director of the Harvard Injury Control Research Center; and by Philip Cook, Professor of Public Policy at the Sanford Institute for Public Policy at Duke University. Cook has been studying firearms issues for more than 25 years and is coauthor with Jens Ludwig of Gun Violence: The Real Costs.

11. FOODBORNE ILLNESS

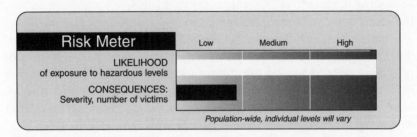

YOU ARE MORE LIKELY to be affected by foodborne illness than by almost any other risk in this book. Approximately 76 million Americans, roughly one in four, suffer food poisoning each year. The consequence for the vast majority of those victims is a few hours or a day or two of an uncomfortable stomach. But foodborne illness kills approximately 5,000 Americans a year.

You probably have heard of a few of these diseases—salmonella, *E. coli,* botulism—but more than 200 foodborne diseases have been identified. Most are caused by bacteria, viruses, or parasites. You may not have heard of the most common cause of foodborne illness, the *Caliciviruses* (also known as the Norwalk-like viruses, since they were first discovered through an outbreak in Norwalk, Ohio, in 1972). *Campylobacter, Salmonella,* and *Clostridium perfringens* are three common infectious bacteria. But the pathogens that cause the most infections are not necessarily the most deadly. Two of the top foodborne killers are the parasite *Toxoplasma gondii* and the bacteria *Listeria monocytogenes.*

THE HAZARD

No matter which of the various agents spread these diseases, they have several things in common. They are carried by food that isn't handled carefully, washed properly, or cooked thoroughly. Most of them cause similar symptoms—diarrhea (sometimes bloody), stomach cramps, vomiting, and fever. And the most severe effects from all these diseases are experienced by the very young and the very old or anyone else with a compromised immune system who is less able to fight off internal infection.

THE RANGE OF CONSEQUENCES
AND THE SOURCES OF EXPOSURE

The Centers for Disease Control and Prevention (CDC) report that foodborne illness sends 325,000 people to the hospital annually. But like the 76 million cases and 5,000 deaths a year, this number is an estimate. It's impossible to know exactly how many cases, hospitalizations, and deaths occur, for several reasons. Many people who suffer from food poisoning have only mild symptoms and don't seek medical attention. Also, many pathogens that spread through food also spread through water, so it's hard to know how many cases are strictly food-related. And most important, experience teaches us that at any given time germs are out there that we haven't yet identified. Many of the pathogens of most concern today, like *Campylobacter jejuni* and *E. coli* 0157:H, weren't even known as causes of foodborne illness just a couple of decades ago.

Plenty of uncertainty remains even within these estimates. Of these 76 million total estimated cases of food poisoning, only 14 million come from known causes. Of the 325,000 hospitalizations, only 60,000 come from known causes. And of the estimated 5,000 deaths, only 1,800 come from known causes. It is unsettling, but the causes of the rest are simply unknown.

The Main Culprits

Here are the main foodborne illnesses, their symptoms, and their main sources:

Norwalk-like viruses cause an estimated 9.2 million cases of foodborne illness each year, 20,000 hospitalizations, and 124 deaths. Outbreaks are usually associated with raw oysters and clams or poor sanitation among food handlers. Interestingly, the contaminated seafood, known as "filter feeders," apparently pick up the virus from human waste dumped into the ocean. The marine bivalves don't become sick themselves but pass the virus back to us in a form that makes us sick. Symptoms include diarrhea, cramps, vomiting, headache, and low-grade fever, and last one to five days.

Campylobacter is the most common bacterial cause of diarrheal illness in the United States. It's estimated to affect about 2 million people each year, send 10,500 of them to the hospital, and kill 1000. It causes diarrhea, cramping, abdominal pain, and fever two to five days after exposure and lasts about a week. Rare cases—about 1 in every 1,000—get into the bloodstream and lead to a temporary but severe paralysis called

Guillain-Barré syndrome. The main source of *Campylobacter* is poultry, since birds are a common carrier. Half of the raw chicken in the United States, in fact, carries *Campylobacter,* according to the CDC.

Salmonella affects 1.3 million Americans a year, sends nearly 15,600 to the hospital, and kills approximately 550 people. Victims suffer diarrhea, abdominal pain, and fever beginning 12 to 72 hours after eating the contaminated food. The disease usually lasts 4 to 7 days. One of the main sources is infected eggs. Even eggs that have been disinfected on the outside can harbor *Salmonella* inside the shell.

Clostridium perfringens bacteria sickens a quarter of a million Americans a year, but sends fewer than 50 to the hospital and causes fewer than 10 deaths. Symptoms include intense abdominal cramps and diarrhea, which begin 8 to 22 hours after consumption of contaminated food. The illness usually disappears within 24 hours but can persist for up to 2 weeks. *Clostridium p.* is thought to be the most common pathogenic bacterium known and most often causes illness in people who eat meat that either hasn't been heated long enough or becomes contaminated after it's been cooked.

The parasite *Giardia lamblia* is the fifth most common cause of foodborne illness in the United States, causing 200,000 cases a year and sending 500 people to the hospital. Fortunately, it is estimated to cause only 1 death per year. It is usually carried in water, but poor sanitation by food handlers can transfer it to any food. Vegetables eaten raw are commonly implicated in foodborne *Giardia* cases.

Escherichia coli bacteria in their various pathogenic forms cause roughly 173,000 cases of disease annually, 2,800 hospitalizations, and 80 deaths. The majority of these cases come from one strain, *E. coli* 0157:H. Undercooked ground beef is the usual culprit, since the germ lives in the intestines of cattle and spreads to meat in the human food chain through unsanitary slaughtering and processing procedures. *E. coli,* especially *E. coli* 0157:H, causes bloody diarrhea and cramps but usually no fever. The disease lasts five to ten days. In 2 to 7 percent of cases, exposure to *E. coli* can lead to kidney failure and hemolytic uremic syndrome, usually in children under five and the elderly. This syndrome is the cause of most *E. coli* deaths.

Listeria monocytogenes bacteria produce listeriosis, the symptoms of which include fever, muscle aches, and sometimes gastrointestinal cramps and diarrhea. Only 2,500 or so listeriosis cases occur each year, but this serious disease sends nearly everyone who gets it to the hospital and kills 1 patient in 5. People with AIDS are 300 times more likely to get it. Pregnant women are 20 times more likely to get listeriosis than other healthy adults. Most cases are treatable but a few can lead to meningitis, and the bacteria can pass to the fetus and cause severe damage or miscarriage. *Listeria* lives in soil and water. Vegetables fed with manure can carry it. So can animal and dairy products. It has also been found in soft cheeses and uncooked cold cuts, which is why pregnant women are warned to avoid these foods.

Toxoplasma gondii is a parasite that makes an estimated 113,000 Americans sick each year. About 2 in 10 get sick enough to require hospitalization, and *T. gondii* kills 375 people a year. It is present in cattle, sheep, chicken, and pigs, and can be carried by house cats. Most people are infected from eating undercooked meat, but inadequate sanitation after handling cat litter can also transfer the parasite to food. Consumed by pregnant mothers, *T. gondii* can be especially harmful to the fetus. It's also particularly dangerous to people with weakened immune systems. People with healthy immune systems rarely suffer any symptoms.

Shigella bacteria is spread by people who don't wash their hands thoroughly while handling food. Foodborne carriers have been known to include tuna, chicken, potato, egg, shrimp, and macaroni salads; raw vegetables; dairy products; and poultry. The CDC estimates that there are 90,000 illnesses, 1,250 hospitalizations, and 14 deaths from *Shigella* exposure each year.

REDUCING YOUR RISK

Foods can become contaminated with bacteria, viruses, or parasites in many ways. Most have to do with inadequate sanitation in processing or preparation. Fortunately, you can wipe out most of the microbes on your food with some simple steps. (All these precautions are especially important for people at high risk—the young, the elderly, pregnant women, and those with weakened immune systems, including people receiving chemotherapy or taking steroidal medication. A normal healthy immune system can keep most of these pathogens in check, which is why the vast majority of foodborne illness results in little more than a few hours of mild symptoms.)

Cook everything well. Heat kills nearly all these organisms. Use a meat thermometer to make sure your cooked meats achieve proper temperature.

- Cook roasts and steaks to at least 145°F. Cook ground beef to at least 160°F.

- Cook eggs until the yolks are firm.
- Cook chicken until there is no pink meat left and the juices run clear. Experts recommend 170°F for white meat, 180°F for dark meat.
- Reheat leftovers, including sauces and gravies, to at least 165°F to kill any germs that may have survived the first cooking, or gotten on the food after it was first cooked. Reheat leftovers only once. After that, the risk goes up that they can make you ill because there have been more chances for bacteria to grow in the food. Don't store leftovers in containers that hold large amounts. Divide them into containers that hold just the right portion for the next meal.
- In restaurants, where you can't be sure of the sanitary habits of the people preparing your food, order your food thoroughly cooked. Undercooked animal foods like tuna tartare, raw oysters, a rare T-bone steak, or eggs sunny-side up may all be tasty, but the more raw your food, the greater the risk.

Refrigerate meat, poultry, dairy, fruits, and vegetables. Heating kills pathogens, but cooling them slows their growth. That's important, because often it takes quite a lot of them to make you sick. Chicken with only a few *Campylobacter* bacteria on it or an egg with a few *Salmonella* aren't as likely to make you as ill as chicken or eggs that have been left at room temperature, where a single bacterium cell can reproduce into tens of millions of cells in just a few hours. The danger zone for growing bacteria is between 40° and 140°F. The Food and Drug Administration estimates that one quarter of all refrigerators aren't set to a cold enough temperature. Cooling tips include:

- Refrigerate raw foods, prepared foods, or leftovers within two hours. If you can't get restaurant leftovers home and refrigerated within that time, don't take them at all. "Doggy bags" are a common source of foodborne illness.
- Marinate foods in the refrigerator, not at room temperature.
- Don't defrost food in hot water or even on the counter at room temperature. While you're still trying to defrost the inside of a roast, the outside can be breeding bacteria. Defrost food in the refrigerator, in cold water, or in the microwave if you'll be cooking it right away.
- Refrigerate foods for a buffet, like tuna or chicken or egg salad, right up until the time you serve them. Place these foods on chilled trays or on a bed of ice cubes to keep them cold.
- Separate raw foods from cooked or ready-to-eat foods when you refrigerate them. Put things like raw meat or chicken in sealed

plastic bags on the lowest level of the refrigerator to keep juices from dripping. The drips can carry bacteria to other foods.

Wash your food. Wash your hands. Wash your food preparation surfaces and tools. Often. You can wash away many of the microbes on the outside of your food, but your hands can pick up and carry contamination from one food to another. If you handle contaminated chicken and then cook it, the chicken is safe, but your hands can carry the bacteria to the next food you handle. The same thing goes for food preparation surfaces and tools. If the germs from those hamburger patties remain on your cutting board or kitchen counter, it won't matter how well you cook the hamburger if you then slice the onions or tomatoes on the same cutting surface, or use the same knife.

- Wash your food preparation areas—cutting boards, countertops, dishes—with soap and hot water after you handle each food there.
- Harried parents of infants: remember to wash your hands after changing diapers.
- Everyone should be attentive to regular and thorough hand washing with soap and hot water before and after handling food, and especially if you are experiencing diarrhea.

Beware of certain foods. A few foods have been associated with outbreaks of foodborne illness with some frequency. In fact, the FDA and the CDC officially warn people about the risks from eating raw alfalfa, clover, and other sprouts, after repeated outbreaks of *E. coli* and *Salmonella* illness over the past several years were traced back to sprouts. Raw sprouts are a healthful food, but they are grown in a humid environment, ideal for promoting the growth of bacteria. Sprouts can become contaminated through exposure to untreated water, improperly cleaned harvest or processing machines, and inadequate sanitation by workers handling them. Even washing sprouts is sometimes not enough to remove bacteria lodged in the nooks and crannies of the folded shoots.

Other foods that carry higher risk include any beef, poultry, pork, or seafood dish served raw or nearly raw. And as we mentioned, pregnant women are advised by food safety experts to avoid soft cheeses like Brie, Camembert, and blue cheese. They should thoroughly heat ready-to-eat foods like cold cuts and hot dogs, as an extra precaution against listeriosis, which can travel to the fetus. Though we point out that the

FDA issues these warnings, it's important to keep in mind that the risk from these foods remains quite low.

FOR MORE INFORMATION

A good online gateway to food safety information is:
www.foodsafety.gov

Food Safety Initiative Program
Centers for Disease Control and Prevention
www.cdc.gov/ncidod/dbmd/diseaseinfo/
    foodborneinfections_g.htm
1600 Clifton Road
Atlanta, GA 30333
(404) 639-2213

Food and Drug Administration
www.nal.usda.gov/fnic/foodborne/fbindex/index.htm
5600 Fishers Lane
Rockville, MD 20857-0001
Toll-free Food Safety Hotline: (888) SAFEFOOD or (888) 723-
    3366

Department of Agriculture
Food and Drug Administration
Foodborne Illness Education Information Center
National Agricultural Library
www.nal.usda.gov/fnic/foodborne/fbindex/index.htm
Beltsville, MD 20705-2351
(301) 504-5719

This chapter was reviewed by Dr. Lester Crawford, Deputy Commissioner of the Food and Drug Administration; by Dr. Nancy Ridley, Assistant Commissioner of the Massachusetts Department of Public Health and Director of the state's Division of Food and Drugs; and by Priscilla Neves, Food Safety Epidemiologist, Erica Berl, Foodborne Illness Response Coordinator, and Frauke Argyros, Infectious Disease Microbiologist and Foodborne Illness Epidemiologist, all of the Massachusetts Department of Public Health.

12. FOOD IRRADIATION

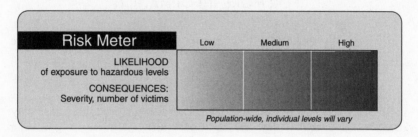

ANYONE who's had to toss out moldy bread, sour milk, or spoiled meat knows the effect of bacteria and microbes on food. So does anyone who has experienced food poisoning. Bacteria and other microbes in food create more than just the risk of spoilage. Pathogens in food are a significant risk to human health. An estimated 76 million Americans suffer food poisoning each year. Most cases are mild, but approximately 5,000 people die from foodborne illness annually (see Chapter 11).

To fight back against both spoilage and food-induced disease, we dry our food, pickle it, pasteurize it, refrigerate it, and, of course, we cook it. Another way to kill microorganisms in food is to expose it to radiation. The first patent to irradiate food was filed back in 1905, as a way to preserve the food itself. But since the process kills the same microorganisms that cause both spoilage and foodborne illness, it also works to make our food safer. Radiation has been used on some spices in the United States since 1983. Since 1971 some of the food that astronauts eat has been irradiated. (Diarrhea in space would be a problem.) Food irradiation has been used in dozens of countries around the world for years.

Irradiation is commonly used to sterilize medical equipment and health care and hygiene products. And since the anthrax attacks in late 2001, irradiation has been used to remove pathogens from the U.S. mail.

THE HAZARD

Irradiated foods are treated with gamma rays or X rays, or passed through an electron beam. The food travels on a conveyor belt through a shielded chamber, where it is exposed to carefully calculated doses of radiation. The radiation levels vary, depending on the food itself and the microorganisms the treatment is supposed to kill. The ionizing radiation has enough energy to break up DNA molecules. As a result, the living microorganisms in the food die off. Often, the byproducts of metabolism from these microorganisms, not the organisms themselves, make us sick. Once the microbes are dead, they can no longer produce these byproducts and therefore are no longer a hazard. And even if a few of them survive, there aren't enough to produce levels of metabolic byproducts that can be dangerous.

Irradiating food does not make the food radioactive. The easiest way to think of food irradiation is to compare it to exposing your hand to a light bulb. When the bulb is off, no radiation is emitted, and your hand feels nothing. Turn the bulb on and you will feel a little warmth, depending on how close your hand is to the bulb and the wattage. Turn the bulb off again, and the radiation is gone. Your hand has not become radioactive in the process. The same is true when food is irradiated. Exposure to ionizing radiation affects the microorganisms living in the food, and to a small degree the constituents of the food itself. But it doesn't make the atoms in the food break apart. So it doesn't create subatomic particles that continue to emit radioactive energy after the irradiating source is turned off. Exposure to radiation does not make food radioactive any more than exposure to X-ray radiation makes a person radioactive.

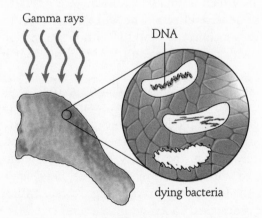

Gamma rays

DNA

dying bacteria

Food irradiation

Irradiation can alter enzymes, vitamins, and other substances in the cells of the food itself. The flavor of irradiated foods may be changed in the process. This phenomenon limits the foods that can be treated with irradiation as well as the radiation levels that can be used. Certain foods with high moisture, like fresh fruits and vegetables, are not good candidates for irradiation because the amount of energy required to kill the microbes changes their flavor or color to an unacceptable degree. However, irradiation at relatively low levels will slow down the ripening process and increase the shelf life of some fruits, like berries, by as much as 10 days to 2 weeks.

The levels of irradiation approved for food safety do not sterilize the food. Low levels of potentially pathogenic organisms may remain. Irradiated foods will have longer shelf lives than untreated foods under the same storage conditions, but they still need careful storage, handling, and preparation.

THE RANGE OF CONSEQUENCES

Irradiation can effectively control some of the most common microbial pathogens that cause foodborne illness, including *E. coli* 0157:H, *Salmonella, Campylobacter, Listeria, Cyclospora,* and *Shigella.* But some people have raised concerns that the radiolytic products of food irradiation—new molecules created when radiation interacts with the food— might be dangerous. Extremely low amounts of compounds like benzopyrenes and formaldehyde are formed during food irradiation. However, studies in which animals or humans ate irradiated foods show no evidence of adverse effects. The USDA, FDA, Centers for Disease Control and Prevention, the World Health Organization (WHO), and the American Medical Association (AMA) all say that irradiated food is safe.

Another concern about food irradiation is that its use will reduce the pressure on food companies to maintain other sanitary procedures at earlier stages in the production and processing system. Some fear that if food manufacturers cut corners elsewhere along the way, it could ultimately reduce the quality or safety of what we eat. This hasn't happened with milk or milk products, however, which are pasteurized near the end of their processing. Pasteurization is basically a heating process designed to kill germs. Several regulations still require proper hygiene and sanitation during dairy production and handling, before products are subjected to pasteurization. Food producers who are permitted to use irradiation must still meet all hygiene rules and regulations.

Some people also worry that food irradiation might cause a reduction in the nutritional value of the food. Irradiation leads to small losses of some nutrients, especially thiamine, but not at levels significantly different from other processing methods.

THE RANGE OF EXPOSURES

The U.S. government has approved the use of food irradiation for many foods, including red meat and poultry. But as of the end of 2001, food irradiation was not widely used in the United States. The U.S. General Accounting Office reported in August 2000 that about 95 million pounds of spices, herbs, and dry vegetable seasonings were irradiated in 1999. That may seem like a lot, but it's only about 10 percent of the total amount consumed in the United States. Some fruits and vegetables are irradiated to kill insect pests that might be carried from where they are grown to where they are sold.

In spite of government approval, as of the beginning of 2002 only a small fraction of poultry and red meat was being treated by irradiation. Some irradiated poultry and ground beef were being sold in Florida, where some elderly consumers are particularly concerned about foodborne pathogens. Some ground beef was being sold in Minnesota, and nationally by some specialty chains. Some irradiated beef and poultry were also being sold in Midwestern states.

Many in the food industry acknowledge that lack of consumer acceptance is slowing wider use of the technology. This may change as consumers learn more about irradiation through news reports about its use on the U.S. mail.

Even though the FDA and USDA have declared food irradiation safe, they require that food sold at retail (except spices when they are only a small part of the product) be labeled with the international radura symbol and one of two statements: "treated with radiation" or "treated by irradiation."

REDUCING YOUR RISK

According to the WHO, the AMA, and U.S. government agencies, irradiated food does not pose a health risk. But it bears repeating that though the risks of foodborne illness are significantly reduced by irradiation, they are not eliminated. Microorganisms left behind or picked up after treatment can still cause problems if food is not handled properly. All standard food-handling practices still apply.

FOR MORE INFORMATION

Centers for Disease Control and Prevention
www.cdc.gov/ncidod/dbmd/diseaseinfo/foodirradiation.htm

National Center for Infectious Diseases
Division of Bacterial and Mycotic Diseases
www.cdc.gov/ncidod/dbmd/diseaseinfo/
    foodborneinfections_g.htm
Mailstop C-14
1600 Clifton Road
Atlanta, GA 30333
(888) 232-3228

Food and Drug Administration
Center for Food Safety and Applied Nutrition
vm.cfsan.fda.gov/list.html
200 C Street SW (HFS-555)
Washington, DC 20204
(800) 532-4440

Another useful federal government site on food safety issues:
    www.foodsafety.gov

This chapter was reviewed by Dr. Harry Mussman, who was Administrator of the USDA's Animal Protection and Health Inspection Service from 1980 to 1983 and Deputy Assistant Secretary of Agriculture from 1989 to 1993; and by Dr. James Steele, former Chief Veterinary Officer of the U.S. Public Health Service, Assistant Surgeon General, and Professor of Environmental Health Science at the University of Texas. Steele is now a consultant to the World Health Organization and the Pan American Health Organization, among others.

13. GENETICALLY MODIFIED FOOD

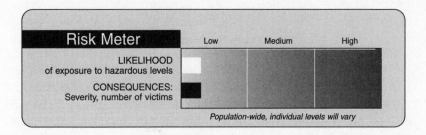

WE HAVE BEEN genetically modifying what we eat for a long time. In fact, many scientists believe that it was the domestication of wild plants and animals through genetic modification that first prompted humans to shift from being nomads to settlers. Today, most of the food we eat has been genetically modified. Ninety percent of the world's plant food comes from just 15 species, which have been selectively bred for centuries. And two thirds of the world's food comes from just 3 of these species, wheat, corn, and rice.

Of course, until the industrial age, genetic modification used slow natural processes. More recently, over the last several decades, we have genetically modified our plant food using techniques that expose the seeds to known chemical and radiological mutagens. These techniques alter an unknown number of genes in the plants' DNA, and if those changes produce favorable traits in their offspring, those offspring are bred, even though we don't know what other genetic changes this more technological "shotgun" breeding has produced.

Within the last decade, science has developed a much more powerful and precise way of producing new strains of plants and, potentially, farm animals and fish. Using the technology of genetic engineering, individual genes responsible for particular traits in one organism can be transplanted into the DNA of any other organism to give the recipient the desired trait. For example, a gene from a flounder can be inserted into a strawberry (to help increase the strawberry's resistance to cold), or from a human to a bacterium (so that the bacteria will produce large quantities of some necessary human hormone, like insulin). With genetic engineering, the old requirement that donor and recipient had to be closely related species is gone. This new ability creates a vast potential for crops and animals that could produce higher yields, require fewer chemicals, grow on land that currently cannot support them, offer greater nutritional value, or yield pharmaceuticals. It also, however, creates the possibility for unexpected new traits in the foods we eat, prompting concern that this promising technology might pose unpredictable risks.

THE HAZARD

Genetic engineering has a significant safety advantage over the existing form of industrial agricultural crossbreeding. New strains produced through genetic modification don't have unknown numbers of changed genes (as do those altered by the shotgun blast from a chemical or radiological mutagen) but rather just one specific changed gene. Scientists can study this gene beforehand and observe the changes in the offspring much more precisely. But while advocates cite the advantages of genetic engineering—its precision, the possibilities for increasing agricultural productivity, producing health-enhancing drugs, reducing the need for pesticides and herbicides—critics cite the dangers of creating new organisms by inserting foreign genes from one species into a completely different one. They say that while these new transgenic species may be benign in the lab, introducing them into natural ecosystems is dangerous because we can't predict how these new genes and traits will interact with other organisms. They fear that unpredictable outcomes could be bad not only for the environment, but also for human health.

~

There are several steps involved in creating a transgenic plant or animal. After a gene responsible for a desired trait is identified and isolated, it is inserted into the host species by one of several methods, typically by using a bacterium known for its ability to carry genetic material from one species to another. Then the resulting offspring are tested to see if the desired gene has been successfully inserted in the host's DNA. Successful genetic transfer is rare. Only once in a thousand times, sometimes only once in a million, do genes transferred into new organisms successfully show up in the DNA of their offspring. Then, a "promoter," a piece of DNA usually derived from a plant virus, is used to induce the inserted gene into expressing itself and producing the intended trait. Tests are then run to assure that the new trait is stable and that the new organism will produce consistent offspring over several generations. Scientists also do tests to make sure that the new gene isn't producing unexpected harmful effects.

But genetic modification of food is not quite as precise as proponents claim. The location at which the new gene splices into the target species genome is completely uncontrolled, and may produce unexpected results (for example, by inserting itself into the middle of an existing gene and disrupting *its* function).

Because of the possibility of these unexpected results, genetically modified (GM) foods have become one of the most contentious environmental issues of recent times. Critics, especially in Europe, call genetically engineered agricultural products and livestock "Frankenfoods" that threaten the environment and consumers' health. Advocates reply that such technology could dramatically ease global disease and hunger. To many of those who have studied the issue, such extreme positions may exaggerate both the potential risks and benefits. So far, none of the genetically engineered products on the market have shown signs of producing any negative human health effects. But confirmed benefits of GM food have been modest. Cotton bred to resist pests has contributed to reduced use of some pesticides. In Canada, canola bred to tolerate herbicides has contributed to reduced use of these chemicals and a 10 percent increased yield. One form of transgenic corn can be grown with less herbicide, but studies show that many farmers growing this type of corn still use the same amount of herbicide and view the GM species only as a reassuring backup. No GM crop has yet to demonstrate an ability to grow in areas where its natural counterpart cannot.

One heated controversy rages over how the government should control GM food. Initially, there was some disagreement as to whether, from a regulatory point of view, the genetic manipulation of food crops or animals constituted a truly new technology, or should be regarded simply as an extension of traditional breeding. The U.S. Food and Drug Administration (FDA) asked biotech developers to consult with the agency, on a voluntary basis, as they moved forward with their products. But an FDA report in 1992 concluded that there was nothing truly novel about genetic engineering and that no new legally binding testing procedures or regulations were needed.

The World Health Organization (WHO), however, came to the opposite conclusion, finding that genetic modification was substantially different from older ways of crossbreeding. In 2001, the FDA changed its earlier position and said that such methods do indeed present new potential health risks and require special scrutiny. As of January 2002, an FDA rule requiring companies to notify the agency 120 days before releasing any new bioengineered organism into the environment was in the "proposed" category and had not yet been formally adopted.

~

As far as judging safety for human consumption, the basic method used for assessing GM foods is based on the principle of "substantial equivalence," essentially testing the plant to detect only chemical compounds that may be new. Those compounds are subjected to further testing for health effects. The rest of the plant is judged to be substantially equivalent to the nontransgenic original, and therefore doesn't need further testing. An analysis by the WHO and the UN Food and Agriculture Program in 2000 concluded that the substantial equivalence method contributes to "a robust safety assessment framework." Their report concedes, however, that it is impossible through present testing techniques to assure that GM foods don't include new proteins or other compounds that remain undetected, or produce other unexpected risk factors. But these international health agencies deem that risk to be very small. And remember, existing methods of crossbreeding change many genes in unknown ways, while only one gene is changed in GM food. So the GM transgenic plants are closer equivalents to the original plant than are traditionally crossbred plants.

THE RANGE OF CONSEQUENCES

Not much hard data on possible human health effects is available. As of the end of 2001, there were no peer-reviewed clinical studies of the health effects of GM foods on humans. Regulators say such tests aren't necessary, at least for existing commercial crops. Tests on animals, however, are done routinely. Some animals are fed only the unique proteins from the transgenic plant, and other animals are fed the whole crop, for 20 or 90 days. These tests have not raised any potential concerns for human health so far.

GM foods could, at least in theory, present some risk. By crossing the traits of one species into a new one, foods could produce new and unexpected allergens, toxins, or proteins that might cause health problems in people. Allergens are a potential risk. One early experiment involved inserting a gene from a Brazil nut into a soybean to increase its protein content. But the company doing this work did tests that found that the GM soybeans produced allergic reactions in some volunteers who were allergic to Brazil nuts. Critics of GM foods cite this experiment as an example of what can go wrong and point to the fact that mandatory screening, which would have identified this problem, is not required. Testing is at the discretion of the companies. But supporters of GM technology point to the Brazil nut–soybean episode as a success story, noting that the company did in fact test its product and caught the problem before the general public was ever exposed.

Along with the prospect of spreading *known* allergens, genetic modification of food raises the possibility that an introduced gene from a species never before used as food might have *unknown* allergenic potential. And in some cases, entirely new proteins are produced by GM organisms, some of which might also prove allergenic. Allergies can produce a wide range of impacts, from respiratory-tract inflammation or skin rashes to potentially fatal anaphylactic shock. An estimated 2.5 percent of adults, and two to three times as many infants, have serious food allergies. The most common food allergens come from peanuts, soybeans, milk, eggs, fish, crustaceans, wheat, and nuts; together, they account for 90 percent of all food allergies, although a total of 160 foods have been associated with occasional allergic reactions. While most people know if they have strong allergies, GM foods could potentially contain allergens not normally associated with that food—or, for that matter, with any known food. At present, however, there are no GM products on the market that contain any substances that have been identified as allergenic.

In addition to the risk of allergens, critics worry about what genetic modification might do to toxic chemicals that occur naturally in plants. These are usually in amounts too small to do any harm, or occur on parts of the plant that aren't used as food or that can be easily removed through traditional preparation methods. For example, potatoes and tomatoes contain some toxic chemicals, especially in their leaves and stalks. Some critics contend that genetic manipulation might increase the amount of these toxins, or alter their location in the plant in a way that would increase human health risk. The risk remains theoretical, and no evidence for any such toxic compounds in GM foods has been reported.

As of the end of 2001, no GM food product on the market was known to contain any allergens or toxic agents not present in ordinary varieties of the same product.

THE RANGE OF EXPOSURES

Exposure to GM foods is already ubiquitous in this country, and the prevalence of GM food is growing fast. In the United States in 2001, 76 million acres of GM corn were planted, 75 million acres of soybeans, and 60 million acres of wheat. Recent studies have shown that an estimated 70 percent of all processed foods in the United States contains some genetically engineered ingredients. Major GM crops include corn, soybeans, squash, wheat, and canola.

Most of the genetic modifications to these plants were made to help reduce production costs. Some modifications made the plants resistant to parasites or diseases previously controlled by costly chemical treatments. Some modifications made plants resistant to specific herbicides so that farmers could use these chemicals to kill weeds—without hurting the plants—and thus get by with far less tilling.

Plants are not the only foods by which we are exposed to genetic modification. A genetically engineered hormone is injected into many cows to increase milk production or to accelerate growth. The first transgenic food ingredient, chymosin, an enzyme used in cheese production, has been used for more than 15 years with no known reactions. Transgenic salmon that grow faster have been developed but have not yet been approved for commercial use.

Sometimes GM foods show up where they're not supposed to, although such errors have never caused any health problems. Tests in the year 2000 found an unapproved variety of GM corn in a wide variety of food products because of inadvertent mixing of crops. No evidence was found that the corn could produce an allergic reaction or other harm, but the uncertainty about the risks of GM foods led to public concern. Thousands of tons of corn and corn products were recalled and destroyed, food retailers promised that their products would be free from GM corn, and some farmers decided to cut back on their plantings of GM crops.

More fear was raised late in 2001 when scientists found that genes in a GM species of corn in Mexico had spread to a native species, at least 60 miles away. Despite controversy over this finding, it raised the fear that any given gene could spread from the transgenic crops planted in a field to naturally occurring plants, and end up appearing throughout the environment—a risk that could have unpredictable consequences for natural ecosystems on which we rely for our food.

REDUCING YOUR RISK

As of the beginning of 2002, GM foods in the United States were not required to be identified as such on their labels. The European Union and Japan, both major markets for U.S. crops, do require labeling of GM products. While labeling would provide consumers with information, and choice, GM products are so widespread in the United States that they are hard to avoid and are mixed in with all sorts of consumer products. The existing storage and distribution systems for grains, soybeans, and oil seed are not set up for complete segregation of crops. In 2001 the *Wall Street Journal* tested food samples labeled "natural" and "GM-free": more than half contained some GM ingredients. In most cases, this contamination occurred despite active efforts by the producers to ensure that their supplies were pure. However, the "GM-free" foods are likely to contain lower amounts of modified genes than the foods labeled "natural." Eating organic produce and products from countries that require labeling of GM foods may minimize your exposure. But for U.S. consumers it is virtually impossible to eliminate all exposure. As of the end of 2001, with GM foods present in roughly 70 percent of the processed foods we eat, there was no evidence that exposure to GM foods would cause any health damage.

As of the end of 2001, the government did not require testing of GM foods once a product was released, nor did it require monitoring for potential health effects. As we stated, the FDA was considering a requirement that biotech and agricultural companies notify the agency before any new GM organism is released. But the proposed regulation contained a provision to protect corporate trade secrets, which means that while the government would know which plants and animals were transgenic, in order to test them for safety, the public would not. And the debate over labeling continues, with many companies privately admitting that it might be a good idea, but none of them wanting to go first and risk financial disadvantage against competitors.

FOR MORE INFORMATION

Food and Drug Administration
CFSAN Outreach and Information Center
Center for Food Safety and Applied Nutrition
www.cfsan.fda.gov/~lrd/biotechm.html

200 C Street SW (HFS-555)
Washington, DC 20204
(888) SAFEFOOD or (888) 723-3366

Biotech Info
A website cosponsored by several consumer and environmental
   organizations, including Consumers Union and the Science
   and Environmental Health Network
www.biotech-info.net

Industry Council for Biotechnology Information
www.whybiotech.com/en/default.asp

World Health Organization
UN Food and Agriculture Organization
"Safety Aspects of Genetically Modified Foods of Plant Origin,"
   June 2000
"Evaluation of Allergenicity of Genetically Modified Foods,"
   January 2001
"Safety Assessment of Foods Derived from Genetically Modified
   Microorganisms," September 2001
www.who.int/fsf/GMfood

Scientific American
130.94.24.217/2001/0401issue/0401hopkin.html

Institute of Food Technologists
www.ift.org/govtrelations/biotech

This chapter was reviewed by Professor Fergus Clydesdale, Head of the Depart-
ment of Food Science, University of Massachusetts at Amherst. He has served on
the FDA Advisory Board, the National Academy of Sciences, and the International
Life Sciences Institute; by Professor Ken Lee, Chair of the Ohio State University
Food Science and Technology Department, College of Food, Agricultural, and Envi-
ronmental Sciences; and by Ariane König, Fellow in the Belfer Center for Science
and International Affairs at Harvard University's Kennedy School for Govern-
ment and a member of an expert panel for the evaluation of biosafety research for
the German Federal Government and consultant on these issues to the European
Commission.

14. MAD COW DISEASE

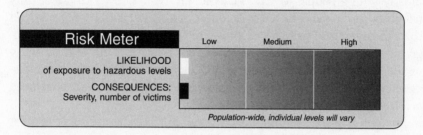

Risk Meter	Low	Medium	High
LIKELIHOOD of exposure to hazardous levels			
CONSEQUENCES: Severity, number of victims			

Population-wide, individual levels will vary

IN THE MID-1980S, farmers and health officials in the United Kingdom discovered a new disease that was killing dairy cows. The animals led healthy lives until, between three and six years of age, they began to suffer neurologic symptoms including unexplained aggression, abnormal posture, and the inability to walk or even stand up. Within a few months of developing these symptoms, the animals died. Because of the way the animals behaved, they were called "mad cows."

Autopsies revealed damage to the animals' brains similar to that seen in other known animal diseases like scrapie in sheep and chronic wasting disease in deer and elk. These conditions belong to a family of diseases called transmissible spongiform encephalopathies, or TSEs. But lab work showed that a new type of TSE was killing these cows. Scientists named it bovine spongiform encephalopathy, or BSE.

The disease appeared only in Great Britain and quickly began to affect hundreds, then thousands of cows. In tracing the epidemic, officials found that the most likely cause of the disease was the animals' feed. It had been supplemented with protein in the form of meat and bone meal from slaughtered cattle and was apparently passing the infectious agent from sick cows, through the animal food chain, back to healthy animals, thereby making more and more cattle sick.

THE HAZARD

Then, in 1995, the first human cases of a disease called variant Creutzfeldt-Jakob disease (vCJD) appeared in three British citizens. *Classic* CJD had been found in various populations around the world for years, apparently arising from spontaneous genetic mutation at a rate of about one case per million people. Autopsies of the brains of classic CJD victims showed the same kind of spongy tissue found in the brains of TSE-affected animals. But the newly discovered variant of Creutzfeldt-Jakob disease had different characteristics. It affected mostly young people, whereas classic CJD was known to affect only people 60 and older. The average course of vCJD was 14 months, longer than the average 4 to 6 months for classic CJD. And electroencephalograms of vCJD victims showed a different pattern of electrical activity in the brain from that of classic CJD victims. Unfortunately the diseases had many other things in common, including the fact that both are invariably fatal.

Epidemiologists studying the outbreak of vCJD determined the most likely cause was the BSE in cattle. Apparently, the infectious agent was being carried in some part of processed cattle products that people were eating, probably the brain or spinal cord. Research determined that the infectious agent was most likely a misshapen protein, called a prion, which for still unknown reasons can cause other proteins in the brain to change shape. This growing number of misshapen proteins destroys normal brain function and leads to death.

THE RANGE OF CONSEQUENCES

Variant Creutzfeldt-Jakob disease begins with predominantly psychiatric problems, including anxiety, depression, withdrawal, and progressive behavioral changes. The first physical symptoms are unpleasant abnormal sensations in the face and limbs. That progresses to problems with walking and muscle coordination, forgetfulness and memory disturbance, severe cognitive impairment, paralysis and inability to speak, involuntary muscle spasms, and ultimately death.

Autopsies reveal the same general kind of damage to the brain—spongy plaque buildup—as is seen in people who die of Alzheimer's and Parkinson's diseases.

THE RANGE OF EXPOSURES

As of the end of 2001, BSE had shown up in cattle born and raised in at least 18 European nations, as well as Japan. Of the nearly 190,000 cattle that had the disease, more than 98 percent were in the United Kingdom.

Most of the human vCJD cases have also been in Great Britain, where, as of the end of 2001, 113 cases had been confirmed or suspected. There were also 4 cases in France, and 1 in Ireland. A person in Hong Kong whose vCJD had been detected late in 2001 was thought to have been exposed in the United Kingdom.

Because the incubation period between infection and symptoms has not been confirmed, some health scientists believe there may be people in the United Kingdom and possibly in other countries with BSE-infected cattle who are still carrying the prions but who have yet to get sick. They think new cases of vCJD will continue to show up for at least several years. Estimates range from just a few hundred additional cases to tens of thousands. The wide range is explained by the remaining uncertainty about the mechanisms of the disease.

As of the end of April 2002, there were no known cases of BSE or vCJD that had originated in the United States. In April 2002, a woman in Florida was identified as having vCJD. She had lived in England and is believed to have been exposed there. A series of government rules designed to keep potentially infected animals or animal products out of the country had apparently been effective. A risk analysis by our center found that even if BSE found its way into this country, the chances of it spreading to a lot of animals, or causing a significant human health risk, is very low. The U.S. government's "feed ban" makes it illegal to render the leftover parts from slaughtered animals into protein that is mixed into animal feed. This precaution breaks the cycle by which a sick animal can spread disease to a healthy one. Our study found that even if some BSE-infected animals or infectious material gets into America, and even if some animal feed producers violate the feed ban, there is enough compliance to keep BSE from taking hold here.

(The feed ban applies only to some species, including cows, sheep, and goats. Protein from pigs, chicken, and fish can still be used to supplement cattle feed, since these species are not known to suffer any TSEs.)

In roughly 50 different hypothetical scenarios, introducing as many as 500 sick animals, we found that BSE disappears within several years, and in no case does it do more than make a small number of additional animals sick. We also found, in all our scenarios (and 1,000 variations of each one), that the amount of infectious material that people would be exposed to would be so low that it's unlikely (but not absolutely guaranteed) there would be any human cases of vCJD from exposure in the United States.

A lot of uncertainty remains about BSE and vCJD, but science has been able to narrow down the types of animal tissue most likely to carry the infectious agent and cause human infection. Since cattle showing signs of BSE have most of the suspicious prions in their brains, spinal cord, and other parts of the central nervous system, experts think these parts are the most likely sources of infection in the human food supply. Tests have shown that meat, dairy products, fat, and blood from infected cattle do not transmit the disease.

But some people eat brain or spinal cord tissue directly. Many others eat mixed meat products like ground beef and hot dogs that may contain these parts. Meat processors sometimes use what is called mechanical or advanced meat recovery systems, which harvest the remaining bits of tissue from the skeletons of cattle that have already been butchered. Even though the spinal cord is supposed to be removed from the animal before this process, sometimes that doesn't happen, and these mechanical systems occasionally capture spinal cord and/or other central nervous system tissue that gets added to various mixed meat products.

For Americans who travel to countries where indigenous cattle have been found with BSE, exposure is still extremely low. Perhaps most importantly, these countries have banned the practice of using meat and bone meal from slaughtered animals to supplement animal feed. European countries that have detected BSE test any cows that are older than two and a half before allowing them to be used in the human food supply. (The United Kingdom doesn't allow cows older than two and a half to be used for human food at all.) Scientists believe the disease doesn't really get going, and produce larger amounts of the suspected prions, until the animals are older than that. In addition, European countries where BSE has been found have slaughtered millions of animals to try to get rid of the disease. And they have regulations forbidding the use of potentially infected tissues from being used in human food. Because of all these steps, the likelihood of BSE-infected cows introducing infectious material into the human food supply is very low.

You may have heard that the American Red Cross and the U.S. Food and Drug Administration have restricted blood donations from Americans who lived in the United Kingdom going back to 1980. There is no evidence that human-to-human blood transmission can transmit vCJD. But those agencies feel that even the theoretical risk is enough to warrant this action, though it will reduce the blood supply for the 4.5 million Americans a year who need a transfusion.

REDUCING YOUR RISK

As of the end of 2001, there was no known risk of getting vCJD in the United States, and only a very low risk of getting it anywhere else.

FOR MORE INFORMATION

Department of Agriculture
Animal and Plant Health Inspection Service
www.aphis.usda.gov/oa/bse
Washington, DC 20250
(301) 734-7799

This chapter was reviewed by Silvia Kriendel, former researcher at the Harvard Center for Risk Analysis and coauthor of the center's study on mad cow disease; and by Linda Detwiler, Senior Staff Veterinarian for the U.S. Department of Agriculture's Animal and Plant Health Inspection Service and a leading government official working on BSE for 15 years.

15. MICROWAVE OVENS

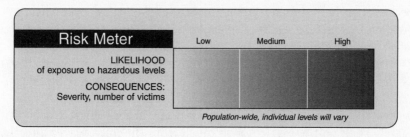

THE NAME of one of the most popular early microwave ovens, Radarange, helps to convey how this technology became such a common part of our everyday lives. Microwaves are electromagnetic energy waves with characteristics that make them ideal for radar. As radar was first developed and installed on American ships during World War II, sailors noticed that they would warm up if they stood in front of the devices, and some discovered that if they held their food in front of the radar transmitters it would cook. Percy Spencer, an engineer at the Raytheon Corporation who pioneered the development of radar, accidentally found that radiation from the magnetrons—the part of the radar device which generates the microwave—would melt a candy bar. He tried it on popcorn, which popped. He tried it on an egg, which exploded.

From those early days, microwave ovens have evolved into one of the most common household appliances in America. One producer of microwave ovens estimates they are now in 9 of every 10 homes in the United States. A public opinion survey in 2000 found that consumers rate the microwave oven as the number one technology that makes their lives easier. Yet some people remain concerned that these appliances pose a risk because they are a source of radiation.

THE HAZARD

Microwave energy causes only certain kinds of molecules, called polar molecules, to vibrate. H_2O, water, is one of these molecules. Microwave radiation makes the water molecules in food flip back and forth millions of times per second. All this flipping back and forth doesn't heat up the water molecule itself. But as millions of water molecules in a piece of food or a mug of coffee flip around, they rub against one another, and against other adjacent molecules, and that friction causes the heat by which microwave ovens work. (Glass, ceramic, or paper containers can be safely used inside a microwave oven because they're not made of polar molecules. And they don't reflect microwaves, as metal does. Some plastics are safe, some are not. Check the labels.)

Based on the size and liquid content of the "target," different foods of varying thicknesses heat at different rates. The amount of liquid content in the food also affects the heating. More solid foods absorb more waves and cook more on the outside than inside. These foods ultimately cook on the inside not so much from the microwaves directly, but from the heat created out on the surface, which then penetrates slowly down into the food. The more liquid a food contains, the more the waves can penetrate and cook the food all the way through.

Biologically, microwaves have the same effect on human tissue as they have on anything that contains water. But this heating effect occurs only if the tissue is exposed to enough microwaves, transmitted with enough power. It occurs inside a microwave oven because the metal walls, ceiling, and floor reflect microwaves and concentrate them inside the device. Even the semiclear window in the oven door contains a thin metal wire mesh, with holes big enough to see through but too small for the microwaves to escape through. That keeps the microwaves from being emitted out into the room. Door seals on microwave ovens are specially made to ensure that no radiation leaks through the seams. As a result, the level of microwave radiation outside the oven is too weak to produce heating, which is the only known way microwaves can harm people.

THE RANGE OF CONSEQUENCES

Properly functioning microwave ovens do not pose a risk to health. When they were new, a real technological marvel in their day, many people feared that leakage, microwave radiation getting out of the device through seams or inadequately shielded windows, could interfere with pacemakers or other implanted electronic medical devices. This was back when microwave ovens and pacemakers were both in their infancy. Now pacemakers and other implanted electrical medical devices are built with shields to protect them from most ambient radiation. And microwave ovens have to meet rigorous standards to protect against leakage. One of those standards is a system to make sure that as soon as the door opens, the components producing the radiation shut off.

Microwave ovens can still cause injury, for simple reasons unrelated to radiation. You can be burned if you do not properly handle the heated material inside the oven. Also, overheating an object too much, as with any oven, can cause a fire. And then there's the problem of superheated water. When a liquid in a cylindrical container with a diameter between one and three inches is heated for too long at too high a level in a microwave oven, it can lead to superheating. This is a condition in which some of the water in certain spots within the container, not near the surface, reaches temperatures above 212°F—the boiling point—without changing from liquid to gas. Then, when the container is moved as you take it out of the oven, the disturbance gives that heated water just enough extra energy to finally convert from liquid to gas. That sudden boiling causes the water to "explode" and burn anyone it splashes.

THE RANGE OF EXPOSURES

Animal tests have determined that the minimum amount of micro-wave radiation it takes to induce damage to the most sensitive tissues, the eyes and the testes, is hundreds of milliwatts per square centimeter. The amount of microwave energy permitted by federal standards just 2 inches from a microwave oven is just 5 milliwatts per square centimeter. And the energy drops off as a square of the distance between you and the source of the radiation. So 20 inches away from the oven, the level would be one four-hundredth the level 2 inches from the device.

REDUCING YOUR RISK

The only concern about radiation from a microwave oven is radiation "leakage." So don't operate a device with a warped door or that continues running with the door open. To be extra safe, avoid prolonged direct contact with the door to the oven while it's operating.

The main risk from microwave ovens is the same from all heating devices: being burned by the heated objects inside, or causing a fire by overheating. Remember to refer to recommended heating times to avoid a fire from overcooking. Don't heat liquids in small cylindrical containers for extended periods. And don't forget to handle objects that have been heated in the oven carefully.

FOR MORE INFORMATION

Washington State Department of Health
www.doh.wa.gov/ehp/rp/rp-oven.htm

This chapter was reviewed by Dr. John Osepchuk, longtime Raytheon engineer, now a consultant and nationally recognized expert on microwave radiation.

16. MOTOR VEHICLES

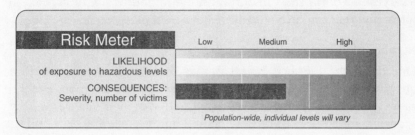

IF YOU ARE between 4 and 33, this may be the most important chapter in the book. People in your age group are more likely to die from motor vehicle crashes than any other cause of death. And for people of all ages, motor vehicles are the eighth leading cause of death in the United States. In the year 2000, motor vehicle crashes killed nearly 42,000 people and injured approximately 3,189,000. On average, one U.S. resident dies in a motor vehicle crash every 13 minutes.

But the news is not all bad. Compared with 1990, deaths from motor vehicle crashes in 2000 were down 8 percent. The number of people killed in alcohol-related crashes was down 25 percent. Pedestrian deaths were down 27 percent. Safety belt use is up 50 percent. From 1975 through 2000, lap and shoulder belts saved more than 135,000 people. In 2000 alone, child safety seats saved 316 children. As of the end of 2001, air bags had saved approximately 8,000 lives.

But despite the progress, motor vehicle crashes continue to be one of America's biggest preventable killers. We say *preventable* for the same reason we adopt the word *crash* instead of the word *accident*. Motor vehicle safety experts and law enforcement officials say that nearly all crashes can be avoided, and that using the word *accident* implies that the crash was no one's fault. They say a crash is almost always someone's fault, that something might have been done to avoid it. While we acknowledge that sometimes accidents happen, we agree that calling them crashes instead of accidents will keep us all aware that there are things we can do to understand and reduce our risk.

THE HAZARD

Extensive nationwide data collection systems keep track of all crashes that kill, and many that injure or damage property. That information reveals some of the most important causes of motor vehicle crashes, information that can help put this major risk in perspective.

Number of Vehicles Involved

One thing the numbers reveal is that most crashes—roughly three quarters—involve multiple vehicles. These crashes are more likely to cause injury than death. Crashes between two or more vehicles were responsible for 80 percent of the nonfatal injuries in 2000, but of all the crashes that killed someone, multiple vehicle crashes were responsible for only about half. When it comes to the risk of dying in a vehicle crash, another significant category of hazard is a single vehicle hitting a tree or a pole or a bridge abutment or some other fixed object. One in four of the deaths in motor vehicle crashes in 2000—approximately 11,200—came from a single vehicle hitting a fixed object. The other major categories for fatalities were pedestrian deaths, at 12 percent, and noncollisions (rollovers, jackknifed trucks, for example), which caused about 11 percent of the deaths that year.

Type of Vehicle Involved

In the year 2000, passenger cars accounted for 60 percent of all vehicle registrations, 63 percent of all crashes, but only half of fatal crashes. Light trucks (vans, SUVs, and pickup trucks), 35 percent of vehicle registrations, accounted for 32 percent of all crashes and 35 percent of fatal crashes. Medium and heavy trucks were the vehicles most likely to kill. They were only 2 percent of vehicle registrations, but accounted for 3 percent of all crashes and 8 percent of all motor vehicle crash deaths. Motorcycles and scooters were dangerous too. They were 2 percent of the registered vehicle fleet in America in 2000, and accounted for only 0.5 percent of all accidents. But motorcycles and scooters accounted for 6 percent of motor vehicle deaths.

Alcohol

In 2000, alcohol use at any level, not just legal intoxication, was involved in 40 percent of the crashes that killed people. Alcohol-related crashes killed 16,653 Americans, 1 every 32 minutes, and injured another 310,000 people, 1 person every 2 minutes. Fatal crashes at night were three times more likely to involve alcohol, and fatal crashes on the weekend were 70 percent more likely to involve alcohol. Drinking played a huge role in pedestrian deaths too. But it was more often the pedestrian who was impaired. In alcohol-related crashes that killed pedestrians in 2000, 31 percent of the pedestrians were intoxicated but only 13 percent of the drivers were.

Speed

You have probably heard the slogan "Speed kills." For motor vehicle crashes, it's true. Excessive speed was a factor in about one third of all fatal crashes, and killed 12,350 people in the year 2000. And the deadly consequences of speeding don't require a highway: 85 percent of speeding-related fatalities occurred on roads that were not interstate highways.

Speeding is often coupled with another motor vehicle risk that some people take, not wearing safety belts. Drivers in fatal crashes who were speeding were only half as likely to be wearing their safety belts as drivers who were obeying the speed limit.

And speed and alcohol also appear to go together. If drivers are drunk, they're more likely to be speeding. And if they're speeding, it's more likely they're drunk. Drivers in fatal crashes who were speeding were roughly twice as likely to be drunk as drivers in fatal crashes who were not speeding. This combination gets much worse at night. After dark, three quarters of speeding drivers in fatal crashes were intoxicated.

Driver Age

Another major risk factor for motor vehicle crashes is driver age. The youngest drivers, with the least experience, and the oldest drivers, who may have problems with vision, reaction time, and general awareness, are much more likely to be involved in a fatal crash than drivers in the middle age groups.

Drivers between 15 and 20 years of age represent about 7 percent of all drivers, but they're involved in 15 percent of fatal crashes and 18 percent of all police-reported crashes. The rate of involvement in fatalities for drivers aged 16 to 19 is three times higher than the national average for all ages.

The other age group disproportionately involved in fatal crashes are those 75 and older. The fatality rate for these drivers is four times greater than the national average. Older drivers may be involved in more fatal crashes not because they crash more, but because they die more often than other age groups when they're involved in crashes because of their frail health.

Gender

The sex of the driver is another clear risk factor. And both genders share some blame. Males are much more likely to be involved in fatal crashes than females. Based on miles driven, the rate of involvement in fatal crashes in the year 2000 was 27 for men and 16 for women, according to the National Safety Council. But women are more responsible than men for causing nonfatal crashes. In 2000, women were involved in 100 crashes per 10 million miles driven. Men were involved in 90.

THE RANGE OF EXPOSURES

Several factors that don't directly cause crashes still contribute to your risk of being in a motor vehicle crash, or being injured or killed when you are.

Proper Restraint

You are at significantly greater risk of being hurt or killed in a motor vehicle crash if you are not properly restrained. Safety belts reduce the risk of injury or death for front seat occupants in passenger cars by 45 to 50 percent. They're even more protective for people in light trucks (vans, SUVs, pickups), reducing the risk of injury or death by 60 to 65 percent.

Safety belts saved roughly 11,200 lives in 1999 among passenger car occupants over four years old. But while safety belt use is increasing, it's still only 73 percent, despite the fact that 49 states and the District of Columbia require safety belt use. In 16 states and the District of Columbia, primary enforcement laws allow police to stop someone for not wearing his safety belt, without any other reason for stopping him. States with these laws have a 78 percent safety belt use rate. In 33 states with secondary enforcement laws, police have to have another reason for stopping the motorist before they can enforce the safety belt laws. These states have a 67 percent usage rate. According to the National Highway Traffic Safety Administration, 9,553 Americans could have been saved in 2000 had all passenger vehicle occupants over four worn their safety belts.

Safety belts reduce one of the biggest risks of death in the event of a crash, being thrown from the vehicle. Roughly one quarter of the people killed in passenger vehicle crashes in 2000 were ejected from the vehicle. You're at much greater risk of being ejected from a vehicle if you're not properly restrained. Among crashes in which people were killed, just 1 percent of occupants who were wearing safety belts were ejected from their vehicle, but 22 percent of the unrestrained occupants were thrown from a car, van, SUV, or truck.

The protective value of restraint goes for kids too. Small children, who are not adequately restrained by normal safety belts, need to be placed in child safety seats, which need to be properly strapped to the car seat. These devices reduce the chance of fatal injury from motor vehicle crashes by 71 percent for infants less than one year old, and by 54 percent for toddlers between one and five, in passenger cars. In light trucks, safety seats reduce the risk of fatal injury by 58 percent for infants and 59 percent for toddlers.

It's also a risk factor for children 12 and under just to sit in the front. Kids 12 and under who aren't fully restrained by shoulder and lap belts run a 27 percent higher risk of being killed in a crash if they're sitting in the front than if they're sitting in the back. Some of this risk comes from front seat air bags. (We examine the risks and benefits of air bags more completely in Chapter 2.)

Vehicle Type and Size

Many people assume that riding in a bigger, heavier vehicle automatically makes them safer than riding in one that's smaller and lighter. But it's more complicated than that. In collisions between two vehicles, the larger vehicle and its occupants will sustain less damage because it's bigger and heavier. But some heavier vehicles in the light truck class, particularly some SUVs, have a combination of high center of gravity and width from tire to tire that makes them more prone to rollovers. There were approximately 4,600 deaths in 2000 from noncollision crashes, most of which were rollovers, and SUVs had the highest rollover involvement rate of any vehicle type in fatal crashes: 36 percent of the fatal accidents involving SUVs were rollovers, compared to 24 percent for pickups, 19 percent for vans, and 15 percent for passenger cars.

The way your vehicle is built, specifically the frame, also contributes to how much risk you're exposed to in a crash. Cars, vans, and smaller pickup trucks are built with "unibody construction." That essentially means that there is no separate underlying frame on which the body of the vehicle rests. The frame and body are all one piece. Larger pickup trucks, and many SUVs, on the other hand, are built with "ladder-frame construction." That means the body of these vehicles rests on and is at-

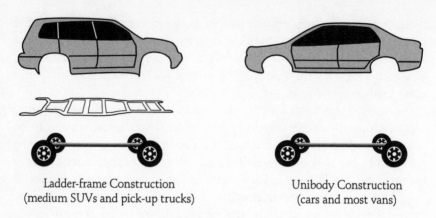

Ladder-frame Construction
(medium SUVs and pick-up trucks)

Unibody Construction
(cars and most vans)

Motor vehicle frames

tached to a frame made of steel beams, shaped like a ladder, under the vehicle.

Ladder-frame vehicles are stiffer than unibody construction vehicles. That helps the occupants of the ladder-frame vehicle if it collides with a unibody construction vehicle because more of the inertia will be absorbed by the more flexible unibody vehicle. But when a ladder-frame vehicle hits a fixed object like a tree or a pole or an abutment that is stiffer than the vehicle (one fourth of all fatal crashes are of this type), because the frame has less give, more of the impact is absorbed by the vehicle, increasing the likelihood of injury for its occupants. Also, riding in a bigger, stiffer vehicle doesn't help much if you're hit from the side, especially if your vehicle is hit by another ladder-frame vehicle.

Pedestrians

Just because you're not in a motor vehicle doesn't mean you're not exposed to the risk of being hurt or killed by one. In the year 2000, more than 4,700 pedestrians were killed in the United States, one roughly every two hours. Approximately 78,000 pedestrians were hurt, one every seven minutes. About half of these deaths and injuries occur as people are crossing the street. The other major risk factor is walking along a road, which accounts for 1 pedestrian death in 10. And if you have to walk along a road someday, you should know that the risk of walking with traffic, when you can't see what's coming, is twice as high as walking against it.

Two thirds of pedestrian fatalities take place at night. Two thirds of pedestrians killed in 2000 were male. And the death rate for pedestrians

75 and older was more than twice the average for all age groups combined.

Bicyclists

The first motor vehicle crash ever recorded, in 1896, involved a car and a bike. In the year 2000, 690 bicycle riders were killed in motor vehicle crashes and another 51,000 were injured. And 9 out of 10 of the bike riders killed, and 8 of 10 who were injured, were males. A quarter of the bike riders killed were between 5 and 14 years old. The federal government's data indicate that 98 percent of the bike riders who were killed were not wearing helmets. But the Bike Helmet Safety Institute says those numbers aren't reliable because police reports on fatal crashes don't keep good details on helmet use.

Motorcyclists

Motorcycles are a symbol of freedom and individuality. They are also the riskiest class of vehicle to ride, with the highest rate of fatality and injury of any vehicle category. Motorcycles make up less than 2 percent of registered vehicles, and less than 0.5 percent of vehicle miles traveled, but in 2000 they accounted for 7 percent of motor vehicle deaths. Per vehicle mile, motorcyclists are approximately 18 times as likely to die in a crash as passenger car occupants.

Motorcyclists in fatal crashes are more likely to be intoxicated than drivers of other motor vehicles. In 2000, 27 percent of motorcycle drivers involved in fatal crashes were intoxicated, compared with 20 percent of light truck drivers and 19 percent of passenger car drivers. Drunk motorcycle drivers were 50 percent less likely to be wearing a helmet than sober ones. Motorcyclists are more likely to be speeding too. In 2000, about 4 in 10 motorcyclists involved in fatal crashes were speeding. That's twice as high as for drivers of passenger cars or light trucks.

REDUCING YOUR RISK

You can reduce your risk of being in a motor vehicle crash, and reduce the risk of injury if a crash occurs, in many ways.

Impairment

Avoid driving when you're not as alert as you need to be. Many of us realize we shouldn't drive while impaired by alcohol, but there are other things that can limit our level of alertness too. Over-the-counter or prescription drugs that warn about drowsiness can increase your risk when you drive. So can illegal drugs, fatigue, or the buildup of carbon monoxide if your motor vehicle exhaust system isn't in good repair. Avoid driving when your ability to be physically and mentally alert is impaired.

Restraint and Seating Patterns

Wear your safety belts, including both the lap belt and shoulder belt. Infants and young children should always ride in child safety seats. The federal government suggests that kids should be in booster seats until they are eight years old, or 4 feet 9 inches tall. And make sure that child safety seats are strapped to the vehicle properly. Many police departments and car dealers are beginning to offer help with installation.

Don't forget, kids 12 and under should always ride in the back. If a child must ride in the front, and there are multiple children in the car, the largest child should get the front seat. And the seat should be pushed back as far as it can go.

Driving Behaviors

Travel close to the speed limit. These numbers aren't just picked out of a hat. They're actually figured out by highway traffic safety engineers to maximize safety for you and pedestrians around you, based on factors like the width and curviness of the road, and the likelihood of entering traffic or pedestrians. Remember, this precaution includes not driving too far *below* the speed limit too, especially on highways. And try to maintain an even speed, so drivers around you aren't surprised if you speed up or slow down for no apparent reason.

Drive defensively. Be aware of everything around you, including what's going on next to you and behind you. Check your mirrors regularly. Defensive driving means that you should expect the unexpected, anticipating the possibility that somebody could pull out of that hidden side street, or the guy in front of you might jam on his brakes, or a child might jump out from behind a parked car. Be alert to the possibility that unexpected events can pop up at any time.

Another part of defensive driving is keeping a safe travel distance from vehicles in front of you. Driving safety experts suggest the two-second rule. Pick an object on the side of the road up ahead, like a pole or tree or building, and when the vehicle in front of you passes it, it should take you at least two seconds before you get to that object. (The old "one-one thousand, two-one thousand" method for counting off the seconds works fine.) If you're going 25 miles per hour, that'll put you 74 feet behind the vehicle in front. At 35 miles per hour, you'll be

104 feet back. At 45 miles per hour, you'll be 132 feet back, and at 55 miles per hour, you'll be 162 feet back. Those distances may seem excessive, but they're pretty close to your stopping distance if you slam on the brakes. Anything closer and you could hit the vehicle ahead if it stops short. Add a couple of seconds for conditions such as darkness, rain, or fog, which reduce your reaction time and stretch out your stopping distance.

Another big part of driving defensively is not driving aggressively. Treating driving like a competition surely increases your risk of accident, and it rarely gets you where you're going much faster. Using turn signals is not giving away secret information to the enemy! Tailgating the vehicle in front of you is dangerous behavior, for you as well as the other motorist, and rarely gets the other vehicle to go faster or move over. Letting motorists into traffic in front of you and obeying yield rules at traffic circles and intersections reduces your risk, lowers your stress, and does not threaten your manhood or womanhood.

Avoid provoking others who are driving aggressively. Along with the obvious, such as avoiding hand gestures, avoid traveling in the far left lane on highways when you're not passing someone. This high-speed lane is designated for passing. (State police say one of the largest sources of road rage on highways comes from motorists in the high-speed lane who do not move into lower-speed lanes when someone wants to pass.)

Avoidance of Distraction

We all do many things when we're driving, besides driving. We eat, fiddle with the audio system, fuss with makeup, talk to friends or family members who are riding with us, use the cell phone. (See Chapter 7, "Cellular Telephones and Driving.") Some people read or take notes while they drive! These activities can distract us visually by drawing our eyes away from the road. They distract us biomechanically, by making us take one or both hands off the steering wheel. Or they distract us cognitively, by making us pay mental attention to something other than traffic and road conditions. All these distractions are dangerous and should be avoided or minimized.

Awareness of Risk Factors

Some risk factors are obvious, like a road surface slippery with snow or ice. But we should also be aware of less obvious risk factors and change our driving accordingly. For example, as we noted earlier it's more dangerous to drive at night, and more dangerous on the weekends, so when you're driving at these times, you should be more defensive and alert. Other risk factors, like driving near teenagers or the elderly, should also make you more cautious.

Graduated Licensing for Young Drivers

Law enforcement and public safety officials in some jurisdictions have adopted graduated licensing programs as a way to reduce the risk posed by inexperienced young drivers, to themselves and others. These programs allow young drivers to gain experience in steps before earning an unrestricted driver's license. The first step is a learner's permit, in which the young driver cannot drive without a licensed driver alongside. In most cases the licensed driver has to be 21 or older. Some learner's permit programs don't allow the learner to drive at night. This is followed by a provisional license that allows the young driver to drive alone, but under other restrictions. Provisional licensees are usually not allowed to have passengers 18 and under, and many programs limit the number of passengers in the vehicle. A full and unrestricted license is awarded only after one or two years of infraction-free driving. Zero alcohol tolerance and mandatory seat belt use are also common provisions of these programs. In the first year after such a program was introduced in Florida, the involvement rate in crashes for those 18 and under went down 9 percent.

Vehicle Safety

Once a month, you should check your tires, battery, operating fluids, wiper blades, and your lights. You should also periodically inspect your exhaust system. The older your vehicle, the more important these quick and easy inspections become.

There is also, of course, the matter of choosing your vehicle for its safety characteristics in the first place. The selection process can be complicated, but the federal government has a vehicle safety ranking system to help: www.nhtsa.dot.gov/ncap.

You can also check vehicle safety ratings at the Insurance Institute for Highway Safety website: www.highwaysafety.org/vehicle_ratings/ratings.htm.

FOR MORE INFORMATION

Insurance Institute for Highway Safety
www.highwaysafety.org

1005 North Glebe Road, Suite 800
Arlington, VA 22201
(703) 247-1500

National Highway Traffic Safety Administration
www.nhtsa.dot.gov
400 7th Street SW
Washington, DC 20590
(888) 327-4236

This chapter was reviewed by Jim Simons, Director of the Office of Regulatory Analysis and Evaluation, Plans, and Policy, at the National Highway Traffic Safety Administration; and by Alan Williams, Chief Scientist at the Insurance Institute for Highway Safety.

17. SCHOOL BUSES

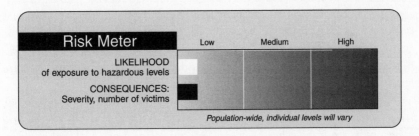

Risk Meter	Low	Medium	High
LIKELIHOOD of exposure to hazardous levels			
CONSEQUENCES: Severity, number of victims			

Population-wide, individual levels will vary

IT'S REASSURING to know that, given their young and important cargo, school buses are one of the safest modes of transportation on the road today. Each year in the United States, more than 23 million children travel more than 4.3 billion miles, with an average of just 27 fatalities, 7 of which are occupants of the bus and 20 of which are pedestrians entering or leaving the vehicle. Per vehicle mile traveled, school buses are eight times safer than passenger cars. Another way to put the risk to kids in school buses in context: an estimated 600 children die each year on the way to or from school in cars or other vehicles.

But wouldn't school buses be even safer if they also had safety belts for their young passengers? The question seems straightforward enough: If safety belts save thousands of lives and are required in cars and vans, why don't school buses have them? What seems like common sense is actually more complex, because the effects of a crash in a large, strongly built school bus with specially designed padded seats are quite different from those in a passenger car, van, or SUV.

THE HAZARD

School buses are designed to protect occupants through closely spaced, high-backed padded seats with energy-absorbing backs, a design called "compartmentalization." School buses are also built with a stronger roof structure and a more hardened passenger compartment than regular buses to protect occupants in case of collision or rollover. Some studies indicate that safety belts might indeed save lives or prevent injuries in certain kinds of crashes, in which the occupants may be thrown sideways or ejected from the bus. But if not properly fastened or adjusted to a child's size, belts in some cases could actually increase injuries. And, remember, many school buses carry students ranging in age from 5 to 18 to elementary, middle, and high school every day. So equipping the buses with the right size belts, and having kids of such different heights and weights adjust the belts properly, would be a big challenge.

The seats in school buses may present an extra risk of head injuries if a child is wearing only a seat belt. In the event of a crash, the head and neck would be thrown forward. Without any belt, the child's whole body is thrown forward and more of the impact is absorbed by the legs, hips, and torso as the knees hit the padded seat back. These injuries are less likely to be fatal than injuries to the head and neck.

The National Highway Traffic Safety Administration concluded that there wasn't enough evidence to justify a federal mandate requiring safety belts on large school buses. (Smaller buses, under 10,000 pounds, are required to have them.) The National Transportation Safety Board came to the same conclusion. So did the National Academy of Sciences (NAS). The Canadian government decided not to require them after a 1985 study concluded that safety belts on school buses might actually have an adverse effect on safety, increasing some head and neck injuries. In a National Education Association poll, school bus drivers were strongly opposed to a safety belt requirement.

But a detailed study at UCLA which used test dummies of different sizes and different bus seating configurations strongly recommended a combination of 28-inch-high safety backs, armrests, and safety belts. Installation of safety belts in school buses has been recommended at various times by the American Medical Association, the American Academy of Pediatrics, the American Academy of Orthopedic Surgeons, and the national Parent Teacher Association.

THE RANGE OF CONSEQUENCES

The kids most likely to be killed in school bus crashes are between ages five and eight. School bus–related deaths are roughly twice as likely to occur in the afternoon as in the morning. As we stated at the beginning of the chapter, about three times as many children are killed when they are pedestrians—an average of 20 a year—than when they are occupants in the bus. Two thirds of these pedestrian deaths are caused by the bus itself, and one third are caused by other motor vehicles, some of which are illegally passing the bus. While school buses provide excellent safety for children when they're inside the vehicle, the large size of the vehicle makes it hard for bus drivers to see kids—particularly younger, shorter children—if they are in the "danger zones" around the bus (about 6 to 10 feet from the sides and front, and anywhere in back). As a result, the NAS recommends that money spent on safety belts could save more lives if it were spent instead on improving safety in school bus loading zones.

THE RANGE OF EXPOSURES

About half of all U.S. schoolchildren ride to school on about 450,000 buses, 90 percent of which are the large buses (over 10,000 pounds) not currently required to have safety belts. New York State has required belts in new school buses since 1987, and New Jersey since 1992. Anecdotal reports by school bus system operators suggest that most kids don't wear them. Florida, Louisiana, and California passed laws in 1999 requiring safety belts on school buses, but those laws have not yet taken effect. At least 23 other states have legislation under consideration.

REDUCING YOUR CHILD'S RISK

The most important way to reduce the risk to children in and around school buses is to teach them how to stay safe around all vehicles, including school buses. They should be told to step far from the bus before crossing the road, to be sure the driver can see them, and never to cross behind the bus. They should also know that if they drop something near the bus, they should alert the driver before bending down to retrieve it.

Teaching children to use safety belts whenever they are available, and to use them correctly, may be one of the most important things parents can do to protect their kids. Some advocates of safety belts in school buses say that one of the strongest arguments in favor of installing them is the message it will send to kids, whose parents have trained them to use safety belts whenever they are in any other kind of vehicle. A U.S. Department of Transportation study found that children who had safety belts in their school buses were more likely to use safety belts in other vehicles. And it has been shown that safety belt use in passenger vehicles reduces fatalities for front seat occupants by 45 percent, and the risk of moderate to severe injury by 50 percent. So if safety belts in school buses encourage even a small improvement in the number of children who get into the habit of using them in all vehicles, this behavior modeling might indirectly save more lives than the safety belts on the school buses themselves. This argument, however, is counterbalanced by studies that show that seat belts on school buses might cause more injuries, particularly head and neck injuries that have a greater likelihood of fatality.

FOR MORE INFORMATION

National Highway Traffic Safety Administration
www.nhtsa.dot.gov/people/injury/buses
400 7th Street SW
Washington, DC 20590
(800) 327-4236

National Association for Pupil Transportation
www.napt.org
1840 Western Avenue
Albany, NY 12203-0647
(800) 989-NAPT or (800) 989-6278

National Coalition for School Bus Safety
www.ncsbs.org

P.O. Box 1616
Torrington, CT 06790-1616
(860) 567-2030

This chapter was reviewed by Richard Blomberg, President of Dunlap and Associates, Inc., who has worked extensively on the pedestrian safety issues associated with pupil transportation and is safety consultant to several school districts; and by Dr. Doug Robertson, Director of the Highway Safety Research Center at the University of North Carolina at Chapel Hill. He has more than 35 years experience in transportation safety and currently chairs the National Research Council's Committee on School Transportation Safety.

18. TOBACCO

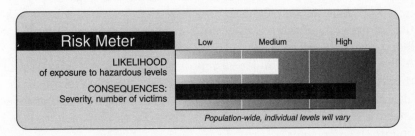

IF SOMEONE suggested that by continuing to participate in a behavior that you started when you were a kid, that gives you only very modest benefit, and that carries a one in two risk of killing you, you probably wouldn't continue that behavior. But roughly one in five Americans take that gamble with their life. They use tobacco. Half of the 47 million Americans smoking today, who continue to smoke, will die because of it.

There is no hazard in this book that does more damage to public health than tobacco. In exchange for the benefits people derive from tobacco, including relaxation from stress, mild mood stimulation, alertness, appetite suppression, and weight loss, tobacco users are subjecting themselves to a lethal product that is the greatest single preventable cause of death in America. Tobacco use kills more than 400,000 Americans each year from several cancers, heart disease, and respiratory diseases. One in five deaths in America is tobacco-related. Each year, tobacco use kills more Americans than the combination of motor vehicle crashes, murder, suicide, alcohol, drug abuse, AIDS, and fire.

This chapter is about the hazard of tobacco, not just the hazard of smoking. The majority of tobacco consumers smoke cigarettes, followed by cigar smokers, and then users of smokeless tobacco.

THE HAZARD

Arguably the most notorious ingredient in tobacco is nicotine. It occurs naturally in tobacco leaves at levels of 2 to 7 percent. Nicotine is biphasic, which means it has both stimulant and depressant effects on the body. In relatively small doses, and in the immediate time after exposure, it is a stimulant. In larger doses, and several minutes or more after exposure, it becomes a physical depressant. At first nicotine raises levels of adrenaline in the blood, which temporarily accelerates the heart rate as much as 10 to 20 beats per minute and raises blood pressure as much as 5 to 10 points. It causes the blood vessels to constrict and narrow. It raises blood glucose levels and increases insulin production. It changes blood chemistry in a way that enhances the aggregation of platelets in the bloodstream, contributing to the formation of blood clots.

But after an initial stimulation period, nicotine causes a relaxation of the respiratory muscles. It suppresses appetite for certain foods, particularly simple carbohydrates—sweet food. It relaxes the valve that separates the stomach from the esophagus, allowing digestive acid to move up into the esophagus.

Along with its stimulant and depressant effects, nicotine can increase sweating, nausea, and diarrhea. Cognitively, in regular smokers who adjust to the presence of nicotine and then go without it for a while, reintroduction of nicotine stimulates memory and alertness and enhances skills that require reaction time and mental concentration. It tends to alleviate stress and boredom.

Nicotine is powerfully addictive. Nicotine binds to certain brain cells through receptors similar to those that bind with heroin and cocaine. These receptors are involved in the release of the neurotransmitter dopamine, which is associated with pleasurable sensation. Nicotine has addictive potential similar to (some say greater than) cocaine,

morphine, heroin, and alcohol, and even minimal amounts of tobacco exposure can initiate the process of this addiction. The brain
chemistry of nicotine and other physically addictive drugs is such that a
tolerance develops to any given level of exposure. As a result, progressively higher levels are necessary to produce the same pleasurable effects. Once this tolerance has developed, it can take months to disappear from underlying brain biochemistry after nicotine exposure ends,
which is why withdrawal from nicotine addiction is so difficult for
many people.

~

Nicotine is not, however, the only harmful substance in tobacco smoke.
Among the 4,000 individual compounds that have been identified are
known or probable carcinogens such as benzo(a)pyrene, arsenic, cadmium, polyaromatic hydrocarbons, vinyl chloride, formaldehyde, and
nitrosamines. Toxic or irritant materials in tobacco smoke include hydrogen cyanide, carbon monoxide, nitrogen oxide, acetone, methane,
phenols, carbolic acid, and ammonia. Some of these compounds are in
the gaseous part of the smoke. Some are in the particulate phase of tobacco smoke, the tiny particles that make smoke visible.

Cumulatively these particles form what is known as "tar," a nonsci-
entific term for the dark, thick, sticky substance that goes deep into the
tiniest parts of the lungs, the little air sacs called alveoli, and adheres
to their cells. This phenomenon not only irritates the tissue directly,
but allows the carcinogenic or toxic chemicals in the tar to stay there
longer, increasing the likelihood of harm.

THE RANGE OF CONSEQUENCES

The list of harm done by tobacco is extensive. Much of this damage is
temporary and abates when tobacco use ends. And the earlier a smoker
quits, the better. But some of the damage from tobacco exposure is
long-lasting, and in a few cases it's permanent. So while all the risks
from tobacco use go down when the exposure stops, the risk for some
diseases remains elevated for ex-smokers, compared with those who
never smoked, years after tobacco use has ended. In some cases, particularly the cancers associated with tobacco use and the effects on infants
whose mothers smoked while they were pregnant, the elevated risk remains with ex-smokers the rest of their lives.

We discuss the consequences of tobacco exposure by category.

Cancer

It is estimated that smoking accounts for roughly 30 percent of all deaths due to cancer in the United States. It is the leading cause of lung cancer, which is the leading cause of cancer death in the country, estimated to kill roughly 155,000 people in 2002. Cigarette smoking is believed to cause cancers not only of the lung but of the larynx, oral cavity, and esophagus. It is believed to be involved in the development of cancers of the pancreas, bladder, kidney, stomach, blood and lymph system (leukemia), and cervix.

SMOKING AND CANCER MORTALITY				
Cancer		Relative Risk Compared with Nonsmokers		Percent of Cases Attributed to Smoking
		CURRENT SMOKER	FORMER SMOKER	
Lung	Male	22.4	9.4	90
	Female	11.9	4.7	79
Larynx	Male	10.5	5.2	81
	Female	17.8	11.9	87
Oral Cavity	Male	27.5	8.8	92
	Female	5.6	2.9	61
Esophagus	Male	7.6	5.8	78
	Female	10.3	3.2	75
Pancreas	Male	2.1	1.1	29
	Female	2.3	1.8	34
Bladder	Male	2.9	1.9	47
	Female	2.6	1.9	37
Kidney	Male	3.0	2.0	48
	Female	1.4	1.2	12
Stomach	Male	1.5	?	17
	Female	1.5	?	25
Leukemia	Male	2.0	?	20
	Female	2.0	?	20
Cervix	Female	2.1	1.9	31

Source: American Cancer Society, Newcomb, PA. Carbone PP, The health consequences of smoking: cancer. In: Fiore MC, ed. Cigarette Smoking: A Clinical Guide to Assessment and Treatment. Philadelphia, PA: WB Saunders Co; 1992: 305–331, Medical Clinics of North America. Updated, 2000.

Heart Disease

But cancer is by no means the only way that tobacco exposure causes harm. The U.S. surgeon general says smoking is "the most important of the known modifiable risk factors for coronary heart disease in the United States." (See Chapter 41, "Heart Disease.")

As many as 30 percent of all cases of coronary heart disease each year are attributed to the effects of smoking. The risk of heart attacks for smokers is more than twice that of nonsmokers, and smokers have between two and four times greater risk for sudden death—massive heart attack.

Tobacco smoke damages the cardiovascular system in several ways: It stimulates blood clotting and atherosclerotic deposits. It causes scarring of the lining of the blood vessels. Nicotine increases the heart rate and blood pressure and constricts the capillaries and arteries. Carbon monoxide in the smoke effectively reduces the level of oxygen in the blood. (See Chapter 23, "Carbon Monoxide.")

Quitting smoking has dramatic benefits for heart disease risk. After one year away from smoking, the risk of heart disease is cut in half compared with people who continue to smoke. Five years after quitting, former smokers have about the same risk of heart disease as people who never smoked. Male smokers between ages 35 and 39 who stop smoking add an average of five years to their lives. Women in this age group who quit add three years to their lives. People who quit later in life, at ages 65 to 69, increase their life expectancy by a year.

Respiratory Disease

Chronic obstructive pulmonary disease includes emphysema and bronchitis. It is the fourth leading diagnosis among U.S. residents who died in 1999, killing more than 124,000 people. Tobacco smoking causes an estimated 80 to 90 percent of cases of chronic bronchitis and emphysema. The risk of dying from these diseases is approximately ten times higher for smokers than nonsmokers.

Emphysema is a disease in which the lung loses its elasticity. Damage to the small air sacs within the lung, the alveoli, causes those spaces to become enlarged. They then trap stale air and can't effectively exchange it with fresh air. This causes difficulty in breathing and can cause an insufficient supply of oxygen to the blood. Emphysema killed 17,555 Americans in 1998. Nearly 2 million Americans are living with this disease.

Bronchitis is an inflammation of the lining of the bronchial tubes (which connect the trachea—the windpipe—with the lungs). It restricts air flow in and out of the lungs and produces a heavy phlegm or mucus and coughing. Anyone who has experienced mild difficulty breathing or coughing with mucus has probably experienced acute bronchitis. Chronic bronchitis occurs when something, usually tobacco smoking, causes these conditions to persist. Chronic bronchitis kills approximately 3,000 Americans a year and an estimated 9 million cases are reported annually.

Effects on Pregnancy

Some of the substances inhaled during smoking reach the developing fetus through the placenta. Nicotine and carbon monoxide deprive the fetus of the oxygen and nutrients necessary for normal healthy development. Children born to mothers who smoked sometimes suffer what is known as "fetal tobacco syndrome": deficits in growth, intellectual and emotional development, and behavior. Miscarriage is approximately two to three times higher in women who smoke, though it is not known if this is a coincidental association or if the smoking is part of the cause. Smoking during pregnancy accounts for 20 to 30 percent of babies born below a healthy weight, an estimated 14 percent of infants born prematurely, and 10 percent of stillbirths and infant mortality.

Effects on Children

The effects of smoking on children don't stop once they're born. When new mothers smoke they raise their baby's risk of sudden infant death syndrome. The chances of developing asthma go up too, doubling among children exposed to people who smoke more than 10 cigarettes a day. Cigarette smoke is an irritant that also triggers asthma attacks in kids who already have the disease, and it leads to more colds and ear infections. Ear infections are the most common reason why children need prescription medication each year. Surgical implantation of devices to help drain the ears of those with chronic ear infections is one of the most common operations young children undergo. Many studies have found that children growing up in a home with smokers have as much as two to three times the chance of developing these chronic ear infections as kids not exposed to smoking. An estimated 150,000 to 300,000 lower respiratory tract infections in children less than 18 months of age are caused each year by exposure to tobacco smoke.

Other Consequences

Smoking has still other effects. Because it raises heart rate and blood pressure and damages and constricts arteries and contributes to atherosclerosis, tobacco smoking doubles the risk of stroke. It contributes to the loss of supporting bone under the teeth and increases the risk of developing periodontal disease or increasing its severity. Other oral problems associated with tobacco use include slower healing of mouth wounds and a reduced sense of taste and smell.

THE RANGE OF EXPOSURES

Roughly 47 million Americans use tobacco (mostly smoking cigarettes). Prevalence is highest in those 18 to 44 years old, 5 percent higher than the average of all age groups. As many as 90 percent of people who smoke start before age 18. Approximately 35 percent of high school students and 9 percent of middle school students report that they smoke.

Exposure to tobacco is declining. The percentage of Americans who smoke has dropped from 42 percent in 1965 to roughly 23 percent now. In 1965, 52 percent of men and 34 percent of women were smokers. As of the year 2000, 26 percent of adult American males and 22 percent of adult American women smoked cigarettes. A number of factors may have contributed to these general declines. Since 1964, when the first U.S. surgeon general's report on smoking came out, the government has required hazard labels on tobacco products, banned advertising in broadcast media, banned smoking on commercial airlines, and federal and state governments have raised taxes on cigarettes dramatically. A legal settlement against major tobacco companies further limited their advertising and funded major antismoking public health campaigns in many states. Perhaps the most successful efforts at reducing tobacco use have resulted from increases in state and federal taxes on cigarettes. For instance, 30 years ago, a pack of cigarettes cost about 50 cents (about $1.25 adjusted for inflation.) Today, that same pack averages $3.25, and can range in price up to $7 depending on the brand and the store where it is sold. Studies find that for every 10 percent increase in the price of a pack of cigarettes there is a 3 to 5 percent decrease in cigarette consumption by adults and a 7 percent decrease in use by teens.

Another exposure to tobacco comes from secondhand or environmental tobacco smoke (ETS). Most studies suggest that ETS can cause all the harmful effects of primary smoke. Roughly half the smoke from a cigarette escapes into the environment rather than being inhaled by the smoker, although of course the concentration of pollutants in secondhand smoke isn't as high as for smokers because it's diluted by air. A 1991 study found traces of nicotine in 90 percent of people tested in a nationally representative sample. ETS is estimated to cause 3,000 lung cancer deaths a year, and thousands of heart disease deaths, though the science on these deaths is controversial. ETS is also the principal form of tobacco exposure for children.

In 1992, when the EPA classified ETS as a carcinogen, it triggered a massive shift in smoking exposure in the United States. Suddenly property owners and businesses had a legal liability for the health effects on nonsmoking occupants from the smoke of others. The EPA classification of ETS as a carcinogen prompted most government agencies and many private businesses to restrict or eliminate smoking from their premises.

REDUCING YOUR RISK

Tobacco users can reduce the risk of tobacco-related illnesses by quitting: stop smoking and stop chewing. But that's far easier said, or intended, than done. Three quarters of America's current smokers say they want to quit; 70 percent have tried at least once and about half of all smokers try to quit in any given year. But even those who have succeeded—in 1999, 45.7 million adults identified themselves as former smokers, 25.8 million men and 19.9 million women—illustrate how hard it is to break the addiction. Of these former users, 70 percent say they had made one or two previously unsuccessful tries, 20 percent say they had tried three to five times before finally quitting, and 9 percent say it took them six or more attempts before succeeding.

While you're using tobacco, your body adjusts to both the stimulant and depressant effects of nicotine. The more you use it, the stronger and more permanent these adjustments become. Then, when you stop using tobacco, these physiological adaptations for nicotine remain for up to several months. Without the nicotine, your body can't function the same way it did when nicotine was present, producing the symptoms of physical addiction withdrawal, including irritability, anxiety, and depression.

Many approaches to this difficult challenge have been found to help. In general, people find it easier to quit if they participate in smoking cessation programs than if they try to go it alone. Such programs have a

SOME BENEFITS OF SMOKING CESSATION	
After 1 day	Less risk of sudden cardiac arrest (massive heart attack)
After 2 days	Sense of smell and taste begin to recover
After 2 weeks–3 months	Improved circulation
After 1–9 months	Less coughing, respiratory irritation/infection, fatigue
After 1 year	Risk of coronary heart disease reduced by 50 percent
After 5 years	Lung cancer/mouth cancer death rates decrease roughly 50 percent

Source: MEDLINEplus Medical Encyclopedia

20 to 40 percent success rate, whereas only about 3 percent of those who try to quit without help succeed. Programs offer peer support, education about behavior modification, and preparation for dealing with potential relapse situations, which all raise the likelihood of success.

Several products are available to help people stop smoking. Nicotine replacement products, including patches, gums, sprays, and inhalers, supply the nicotine but eliminate other harmful contaminants in tobacco smoke. These products are meant to wean the smoker off nicotine gradually, over varying periods of time. Studies have shown that nicotine replacement products are only moderately successful. One study showed that after a year, only 10 percent of people who used these products to quit were still not smoking. And of course, these products continue to expose users to the harmful effects of nicotine. Nicotine replacement products are more successful when used in conjunction with a smoking cessation program and with support from people close to the person trying to quit.

FOR MORE INFORMATION

Centers for Disease Control and Prevention
Office on Smoking and Health
National Center for Disease Prevention and Health Promotion
www.cdc.gov/tobacco
Atlanta, GA 30303
(770) 448-5705

National Cancer Institute
NCI Public Inquiries Office
www.nci.nih.gov (use the word *smoking* in the site's search
    engine)
Building 31, Room 10A31
31 Center Drive, MSC 2580
Bethesda, MD 20892-2580
(800) 4-CANCER or (800) 422-6237

American Lung Association
www.lungusa.org
1740 Broadway
New York, NY 10019
(800) 586-4872 or (212) 315-8700

American Heart Association
National Center
www.americanheart.org
7272 Greenville Avenue
Dallas, TX 75231
(800) 242-1793 or (214) 373-6300

American Cancer Society
www.cancer.org
1599 Clifton Road NE
Atlanta, GA 30329-4251
(800) ACS-2345 or (800) 227-2345

This chapter was reviewed by David Burns, M.D., Professor of Medicine at the UCSD School of Medicine, who has studied and published extensively on tobacco issues and is a member of the Society for Research on Nicotine and Tobacco; and by Dr. Michael Thun, Vice President of Epidemiology and Surveillance Research for the American Cancer Society.

· I I ·

THE ENVIRONMENT

19. AIR POLLUTION (INDOOR)

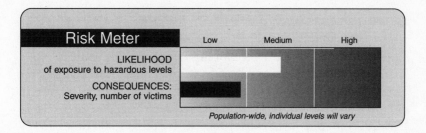

WE SPEND 90 percent of our time indoors. Whether at home, at work, at school, in stores, medical facilities, recreational facilities, or in some sort of vehicle for transportation, we're inside most of the time. So the quality of the air we breathe indoors is in many ways more relevant to our health than the air we breathe outdoors. Our exposure to air pollution—the concentration of pollutants per unit of air, multiplied by the time we're exposed to that air—may be 10 to 50 times higher indoors than outside. Many public health experts say indoor air pollution is one of the biggest environmental risks we face. Yet it is one of the least studied, most poorly regulated areas of public health.

THE HAZARD

Indoor air pollution is not just a hazard because we're inside a lot. For a number of reasons, concentrations of pollutants can be higher indoors than out. Compared with the outdoor environment, the indoor environment has a limited volume of air, so low levels of pollutants can lead to a higher concentrations of pollution per breath. The indoor environment also has less exchange with fresh air than the air outdoors. In addition, the range of indoor environments and the wide variety of products we use indoors subject us to a much broader universe of pollutants. And finally, indoor environments often have elevated levels of humidity and dampness, conditions that exacerbate some pollutants, like biological organisms, which grow better in the presence of moisture.

~

Indoor air pollutants fall into several general classes: gases (including environmental tobacco smoke), microbiologicals, pesticides, and particles.

Potentially Harmful Gases

- Environmental tobacco smoke (ETS). Also known as secondhand smoke, ETS includes smoke from cigarettes, cigars, pipes, and exhaled tobacco smoke. ETS contains dozens of compounds harmful to humans or animals. Several are known or suspected carcinogens. ETS contributes to respiratory infections in infants and children, and exacerbates asthma. However, the exposure to the harmful components of tobacco smoke is 100 to 1,000 times less from ETS than from smoking directly.
- Radon (see Chapter 35). A naturally occurring odorless, tasteless, invisible radioactive gas emitted from the earth which can seep into houses through foundations. Several epidemiological studies on uranium miners exposed to high levels of radon (which is a radioactive decay byproduct of uranium) show that it causes lung cancer. Radon and tobacco smoke interact synergistically, causing more lung cancer than these hazards would if you just added one on top of the other.
- Carbon monoxide (see Chapter 23). A product of inefficient combustion of carbon-based fuels. Indoor sources include heating or cooking devices like gas stoves and ovens, water heaters, and fireplaces. Carbon monoxide is also generated when motor vehicles operate in garages, even when doors are open. It interferes with the blood's ability to carry oxygen, and at high levels it is fatal.
- Nitrogen dioxide. This gas is a lung and respiratory tract irritant produced by combustion from heating and cooking devices such as stoves, water heaters, and fireplaces.
- Organic gases. Many household products emit vapors both when they're stored and when they're used. Paints, varnishes, cleaning solutions and disinfectants, cosmetics, and stored fuels emit vapors of chemicals that can irritate the eyes, respiratory tract, and, at high levels, temporarily impair memory and balance. Some, like spray paints, paint removers, and stored fuels that contain benzene, a confirmed human carcinogen, may have effects with long-term exposure.

Microbiologicals

These include bacteria, mold, mildew, viruses, animal dander, dust mites, and pollen from plants (and not just house plants; outdoor flora can contribute to the air that gets into our homes and buildings too). Moisture and humidity help many of these biological agents proliferate indoors. Biologicals can grow inside the components of central ventilating systems, in carpets, walls, upholstery, and any location damaged by water. Biologicals cause infections and disease, allergic reactions, exacerbate asthma, and irritate the eyes and respiratory tract.

Pesticides

These include insecticides, herbicides, miticides (used for lice), and disinfectants approved for household and indoor use. As with all indoor air pollutants, the levels of these individual pollutants are usually far below anything that could be harmful. But in the indoor environment, they contribute one more source to the mix of compounds in the air we breathe. One study estimates that 80 percent of our exposure to pesticides occurs indoors.

Particles

There are three types of indoor particle pollution of the most concern.

- Asbestos (see Chapter 21). A family of mineral fibers that are strong, thin, resistant to heat or chemical breakdown, and which don't conduct electricity very well. Asbestos is frequently found in insulation, soundproofing, fireproofing, and in floor and ceiling tiles and plastics used as decoration. More common in schools, office buildings, and other workplace settings than in homes, inhalation of asbestos causes lung damage that can lead to a dangerous form of lung cancer called mesothelioma or to asbestosis, a permanent scarring of the lungs.
- Lead (see Chapter 29). Lead was used in paint because it helped cover surfaces effectively and wore well. Though lead in paint was phased out as of 1978, any building constructed before then may well still contain lead paint. In the normal use of the building (opening windows or doors, bumping furniture against walls) or during building renovation, lead paint is turned to dust. These fine particles are inhalable and contribute to indoor air pollution. Lead

impairs central nervous system function, and even very low levels of lead in young children may interfere with normal cognitive development.

- Respirable particles (see Chapter 20). Recent scientific studies have identified a serious health risk from airborne particles, microscopic bits of solid or liquid matter 7 to 700 times smaller than the width of a human hair. Indoors, these particles are generated principally by combustion from stoves, water heaters, boilers, or fireplaces, and from household activities like cleaning which stir up even the finest dust. Particles generated by motor vehicle exhaust and in the emissions of power plants can also enter indoor environments. Particles are associated with respiratory irritation, more frequent and more severe asthma attacks, and heart problems.

For the most part, indoor air pollution comes from indoor sources. But pollutants in outdoor air (see Chapter 20) can get indoors through windows, doors, and ventilating systems, and add to the mix of contaminants in indoor air.

THE RANGE OF CONSEQUENCES

The consequences of exposure to indoor air pollution are complex because of the wide range of pollutants, the diversity of indoor environments, and the varying susceptibilities of specific groups of people. While many of these consequences have not been studied enough to quantify them, even the partial estimates we do have suggest that indoor air pollution is one of the most serious environmental risks to public health.

- ETS, almost exclusively an indoor air quality risk, is estimated by some research to cause 3,000 lung cancer deaths per year, as well as 150,000 to 300,000 lower respiratory tract infections in children. (Some dispute these findings.) It aggravates asthma in 200,000 to 1 million children a year. The American Heart Association estimates that ETS causes 35,000 to 40,000 heart disease deaths a year. The National Cancer Institute estimates that postnatal ETS exposure has been implicated in 1,900 to 2,700 cases of sudden infant death syndrome.
- Radon causes an estimated 3,000 to 32,000 lung cancer deaths a year, with the most likely range being 15,400 and 21,800. Nearly 9 in

10 of these estimated deaths are in smokers because of the synergy between radon and tobacco smoke.
- Blood lead levels are decreasing in the United States, but the federal government estimated that in 1997, 900,000 American children between the ages of 1 and 5 had blood lead levels above the limit that policy guidelines deem safe.
- Carbon monoxide poisoning kills an estimated 700 to 1,000 Americans each year, and sends about 10,000 to 40,000 people to hospital emergency rooms.
- Asthma affects nearly 15 million Americans, and 10.5 million of these people are under 45. The disease killed approximately 4,700 U.S. residents in 1999, and 1.8 million Americans require emergency room treatment for severe asthma symptoms. Asthma is the leading cause of school absenteeism. A majority of indoor air pollutants aggravate asthma, and repeated exposure to many of these pollutants is believed to cause not just asthma symptoms but the underlying disease itself.

The following table describes common indoor air pollutants and their effects. Remember that the symptoms and illnesses listed under "Potential Health Effects" are often caused by a combination of pollutants rather than just one. Also remember that this area of public health is not well understood. Finally, remember that individual sensitivities

INDOOR AIR POLLUTANTS AND THEIR EFFECTS	
Pollutant	Potential Health Effects
GASES	
Environmental "secondhand" tobacco smoke	*Immediate:* Respiratory irritation, respiratory infection in children, decreased lung function, eye irritation *Delayed:* Lung cancer, heart disease, chronic respiratory problems
Radon	*Immediate:* None *Delayed:* Lung cancer (much higher rate in smokers)
Carbon monoxide	*Immediate:* Low levels of nausea, fatigue, headache, chest pain in people with heart disease; high levels of impaired vision, dizziness, confusion, seizures, brain damage, death *Delayed:* None

Nitrogen dioxide	*Immediate:* Respiratory irritation, impaired lung function, increased respiratory infections in children
	Delayed: May contribute to onset of asthma
Organic gases	*Immediate:* Eye, nose, and respiratory irritation; headaches; nausea
	Delayed: At higher levels, damage to liver, kidneys, central nervous system; possibly cancer

MICROBIOLOGICALS

Bacteria, viruses	*Immediate:* Influenza and other airborne infectious diseases, infections, digestive problems, humidifier fever (see the bulleted list that follows)
	Delayed: None
Molds, mildew, fungi	*Immediate:* Eye, nose, and respiratory irritation; skin rash, allergic reactions; hypersensitivity pneumonitis (see the bulleted list that follows)
	Delayed: None
Animal dander, dust mites, cockroach residue, pollen from indoor plants	*Immediate:* Allergic reactions, asthma attacks
	Delayed: May contribute to onset of asthma

PESTICIDES

All pesticides	*Immediate:* Eye, nose, and respiratory irritation
	Delayed: At high levels, damage to central nervous system and kidneys, small increased risk of cancer

PARTICLES

Asbestos	*Immediate:* None
	Delayed: Asbestosis, lung cancer
Particles	*Immediate:* Eye, nose, and respiratory irritation; asthma attacks; increased respiratory infections; chest pain; heart arrythmias; heart attack
	Delayed: None
Lead	*Immediate:* None (except acute poisoning)
	Delayed: Impaired mental functioning, especially in children, impaired hearing and motor control, decreased growth rate, behavior problems, impaired vitamin D metabolism; at higher levels, kidney damage, anemia, severe brain damage, coma, death

and widely varying pollution concentrations determine whether an individual suffers some, all, or none of these effects, and how severe the effects may be.

In addition to these general health impacts, several specific diseases are linked to indoor air pollution. They include:

- Legionnaires' disease, so named because its first outbreak in 1976 killed 29 American Legion members at a convention. Legionnaires' disease is a form of pneumonia caused by inhaling *Legionella* bacteria, which can be present in contaminated water from building cooling systems, whirlpools and spas, humidifiers, food market vegetable misters, and other indoor moisture sources. It is estimated that Legionnaires' disease affects 10,000 to 15,000 U.S. residents per year, killing between 500 and 2,250, largely those who are middle-aged or older, those who smoke or drink heavily, and those with weakened immune systems.
- Hypersensitivity pneumonitis. A unique lung disease often confused with typical pneumonia. It has the same general symptoms as bacterial and viral pneumonia but has different indicators when doctors examine the lungs. Hypersensitivity pneumonitis is associated with bacteria and mold contamination in air conditioning and humidification systems.
- Humidifier fever. This illness has the same symptoms as hypersensitivity pneumonitis—fever, ache, and fatigue, which abate after a couple of days. But humidifier fever affects more people and may be caused by toxic byproducts of bacteria.

In addition to these specific diseases, two widely publicized health syndromes are also associated with indoor air pollution: sick building syndrome and multiple chemical sensitivity. These illnesses are known as syndromes rather than specific diseases because they exhibit a wide range of sometimes vague symptoms and are triggered by a wide range of causes, which have not been clearly identified.

Sick Building Syndrome (SBS)

Symptoms of sick building syndrome include fatigue; difficulty concentrating; nausea; headache; eye, nose, and throat irritation; dry or itchy skin; and sensitivity to odors. There is sometimes no single set of symptoms, and no single identifiable source for those symptoms. Though SBS sufferers experience symptoms in direct relation to time spent in the building, the syndrome often affects people who occupy one part of

the building more than others. SBS is far more common in office buildings and schools than in homes or other indoor environments. Symptoms usually abate after the person is out of the building for several
hours or more. Some SBS episodes are eventually explained by exposure
to certain materials, ventilation conditions, or a combination of factors.

Multiple Chemical Sensitivity (MCS)

MCS is a controversial medical issue. Some people who suffer a wide
range of illnesses affecting a number of bodily systems report that their
illness is due to prolonged exposure to chemicals or other pollutants in
indoor air. The syndrome is also sometimes called environmental illness, chemical AIDS, and twentieth-century disease.

But established medical associations such as the American Academy
of Allergy, Asthma and Immunology say this condition does not exist.
No clinical proof of this syndrome has been established and no widely
accepted definition even exists. One that is often cited was written by
the American Academy of Environmental Medicine, a group of health
care providers known as clinical ecologists, including allergists and physicians from other specialties, who treat people with this condition.
They call MCS "an ecologic illness that is a polysymptomatic, multi-
system, chronic disorder characterized by adverse reactions to environmental excitants, as they are modified by individual susceptibility in
terms of specific adaptations. The excitants are present in air, water,
drugs, and our habitats."

Theories suggest that sufferers, who are far more commonly women
than men, may be experiencing dysfunction of the immune system or
neurological abnormalities brought on after either severe chemical exposure (like a chemical spill) or after chronic low-dose exposures, including indoor air pollution. MCS patients say their symptoms are triggered by exposure to low levels of everyday chemicals such as those
found in cosmetics, soaps, and newspaper inks. They report a range of
symptoms that commonly includes headaches, rashes, asthma, depression, muscle and joint aches, fatigue, memory loss, and confusion. But
nearly all the research on this issue indicates that these symptoms are
too broad and the exposures that apparently cause them far too vague
to support calling MCS an actual medical condition.

THE RANGE OF EXPOSURES

"Indoors" means more than being at home or work. We're also indoors
when we shop, when we travel by car or bus or plane, when we go to

the movies or the gym or the ice rink or the doctor or the hospital, and so on. Each of these types of indoor environment presents unique exposures. In some settings the issue is the sources of pollution—based on the materials used inside the building or vehicle, or the construction materials used to manufacture it. Sometimes the most important determinant of exposure is the general nature of the building or vehicle — how it's used, where it's located, how it's ventilated, and how it's maintained. Many indoor environments have air quality exposure issues because of the susceptible populations that live in or visit those settings. We present this section in terms of the general types of indoor settings.

Homes

Residential environments usually aren't ventilated as well as offices, factories, or public buildings, especially in colder weather when homes are sealed more tightly to avoid air leaks in order to save energy and maintain comfortable temperatures. Humidity and moisture are a particular concern in residential buildings, generated by sources like kitchens, laundry rooms, bathrooms, and household humidifiers. Houses with below-ground basements are a uniquely residential source of indoor air moisture. Leaky roofs, walls, and windows also contribute to the presence of biologicals in residential indoor air.

Radon exposure is greater in homes than other types of buildings. With the growing number of restrictions on smoking in public or workplace settings, ETS exposure is also more common in both houses and apartments. Organic gases are emitted by some products most heavily used at home, such as pesticides, deodorants, cosmetics, household furniture, draperies, and stored fuels. Organic gases can also come from household cleaners and disinfectants, paints and varnishes, adhesives, and carpeting. An Environmental Protection Agency (EPA) study found levels of these gases in some homes as much as 10 times higher than outdoor levels, even when those homes were near a significant outdoor source of such gases, like petrochemical plants.

Carpeting on foundation floors, where moisture from the ground dampens both the foundation and thus the carpeting, is principally a residential problem. Pet dander is too. Cockroach allergens are generally highest in residential buildings. Lead is principally a residential indoor air pollutant, caused when lead paint in buildings painted before 1978 is released by abrasion of painted surfaces. Something as simple as raising or lowering a window, rubbing against a wall, or household renovation work can disperse fine lead particles.

Other respirable particles are generated indoors by cooking and heat-

ing appliances, in addition to smoking. Particle pollution from outdoor sources can also become a problem inside homes near busy streets
or highways. So can pollutant gases that come from motor vehicle combustion like sulfur dioxide, nitrogen oxide, and carbon monoxide. These
outdoor sources are abated if the home is air-conditioned.

Schools

The General Accounting Office estimates that half the schools in the
United States have some sort of indoor air problem. Many schools
suffer from poor ventilation, often due to efforts to save energy by
minimizing the influx of outdoor air. Individual ventilation units designed to serve just one classroom are often simply turned off. Inadequate cleaning of filters, poor maintenance of ventilation equipment, or
failure to repair leaks in roofs, walls, and windows are also common
problems for schools.

The general usage patterns of schools also contribute to unique indoor air pollution issues. Schools are crowded buildings, permitted to
have higher rates of occupancy than almost any other class of building.
Since the young populations in schools regularly introduce biological
pollutants—bacteria, viruses, and pet dander—crowding makes exposure worse. Also, children breathe more air per unit of body weight
than adults, and they more readily absorb contaminants into their still
growing bodies. While these factors don't make the air pollution in
schools worse, it means that lower concentrations of contaminants
may have more impact on children's health.

Schools have a unique range of potential pollutants, which include:
markers, paint, adhesives, and other art materials; science lab materials;
vocational area materials; industrial cleaners and disinfectants; floor
wax; and emissions from copying and printing machines. Asbestos is
still found in schools because it was used as a fire retardant on boilers,
pipes, in wall and ceiling tiles during construction, and in floor tile. Asbestos in schools was so common that in the mid-1980s the federal
government required that schools be tested for asbestos. Money was
made available for abatement. These programs significantly reduced
the problem nationwide. Radon can also be an air pollutant in schools,
which frequently have no basements: radon seeping into the building
from underground immediately encounters occupied space.

Office Buildings

The HVAC (heating, ventilation, and air conditioning) systems of
many office buildings simply do not bring in enough fresh air. Some-

times this failure is the result of efforts to save energy and costs. Sometimes it's because of poor maintenance of the ventilation system. HVAC systems, with their extensive networks of ducts and pipes, are often not cleaned. This includes failure to disinfect water supplies for cooling. Inadequate maintenance also means that sources of moisture in HVAC systems which can breed biologicals are not identified and repaired. Office buildings are frequently situated in environments where HVAC intakes recruit outdoor air contaminated from nearby sources of pollution such as areas with heavy traffic or dumpsters. Poorly designed office building HVAC systems sometimes locate their intake ducts too close to outflow vents, circulating some contaminants back into the building.

Office buildings sometimes contain spaces dedicated to special uses, like underground garages, restaurants, cleaners, or printing shops. Potentially harmful emissions from these sources are sometimes carried in the HVAC systems of the overall building. With higher permissible occupancy rates than homes, offices are often converted buildings originally erected for some other purpose. Ventilation systems are sometimes not adequately adjusted for the new, higher occupancy. Furnishings in office buildings, like movable partitions, often interfere with designed airflows, reducing ventilation. Air ducts are sometimes simply blocked by filing cabinets, partitions, or other equipment or furniture.

In terms of unique sources of pollutants, asbestos is sometimes found in office buildings because it was used as a fire retardant on boilers, pipes, and structural components. Industrial carpeting used in offices emits organic gases. These gases also come from printing and copying operations, and industrial cleaners and floor waxes. Biologicals come from inadequately cleaned and maintained HVAC systems; pipe, roof, or window leaks; and from water-damaged walls, ceilings, or carpets.

Industrial Facilities

There are a wide variety of indoor industrial settings, from factories to smelters to warehouses. In general, levels of air pollution in indoor industrial environments are regulated by the Occupation Safety and Health Administration. OSHA standards are usually more permissive than those of agencies that monitor other environments, because the most vulnerable populations to air pollution—the young, the elderly, or those with compromised immune systems—are usually not the kinds of people who spend significant time in workplace settings.

Common indoor air pollutants in industrial settings include emissions from loading and hauling equipment like forklifts or from power equipment like generators, chemical emissions from manufacturing processes or raw materials, or dust or textile fibers.

Hospitals

Hospitals have a number of unique challenges, as many people who spend time in them have infectious diseases. Their illnesses make them particularly susceptible to indoor air pollution, but it also means that the germs they carry are a source of that pollution too. In addition to the germs that people bring in, hospitals have unique sources of pollution. They use much more chemically aggressive cleaners and detergents, and use them much more frequently, than most other types of buildings. Disinfectants to clean and sterilize equipment include glutaraldehyde, formaldehyde, and volatile organics, which can be harmful at high levels. Topical antibiotics also contribute to indoor air quality issues in hospitals. The dust from latex gloves used by medical providers has been shown to cause allergies. Hospitals also use a large number of individually wrapped disposable items, and the cellulose fibers from these packages contributes to particle pollution in the air.

Hospitals also have unique challenges to ventilation systems, which have to be much more advanced than in most other buildings. Hospital ventilation systems have to segregate airflows from various parts of the building. They have to filter the air much more aggressively. And in many locations like operating rooms and intensive care units they have to maintain air pressure in such a way that the air flows into the room and up into the ventilation system, not out of the room and into the general building. This heavy reliance on aggressive mechanical ventilation systems means that any failure can have serious consequences for indoor air quality.

Transportation

Millions of people spend several hours each week in their cars, vans, or trucks. They are indoors, but the air they're breathing contains many outdoor air pollutants brought in by the vehicle's circulation system. One study found that in Boston, levels of two outdoor air pollutants known or suspected to be carcinogens were higher for people driving to work in their cars than they were for people walking or riding their bikes. Any leaks in your vehicle's exhaust system or problems with your emissions control equipment can expose you to elevated levels of carbon monoxide, nitrogen oxide, particles, and organic gases that are supposed to be going out the tailpipe. That "new car smell" is actually organic gas from plastic components inside the car. Biologicals are emitted from vehicle cooling and heating systems. And don't forget that the volume of air inside a motor vehicle is pretty small, so concentrations of pollutants, including carbon dioxide, can build up quickly.

Studies have found that levels of many typical urban air pollutants are lower in mass transit vehicles than they are in cars, possibly because of larger volumes of air in these vehicles and because their doors are opening all the time, increasing ventilation.

Have you ever gotten a cold or some other infectious illness after an airplane trip? The close proximity of airplane passengers for extended periods is just one indoor air quality issue on planes. Concentrations of some chemicals and particles can rise dramatically while the plane is on the ground. A mix of sources exists on planes, where tests have found chemicals associated with cleaners, cosmetics, air fresheners, dry cleaning fluid, and jet fuel. Carbon dioxide levels on commercial planes can reach levels that indicate inadequate fresh air ventilation.

Recreation

Small local ice rinks have a particular indoor air problem. Ice cleaning machines, most commonly known by the trade name Zamboni, drive around hourly, as many as 14 to 15 times each day, emitting carbon monoxide and nitrogen dioxide, often at high levels if the machinery isn't equipped with pollution control equipment. These gases build up because many rinks do not ventilate, often because outside air is either too warm, requiring more cost to keep the ice surface refrigerated, or too cold, requiring rink managers to heat the seating and observation areas. Concentrations are greatest near the ice, where skaters are exerting themselves and breathing heavily because the cold from the ice tends to make the air in the rink sink, and the boards around the rink contain the pollutants, kind of like water in a bathtub. Both conditions make it harder for ventilation systems to remove contaminants.

"Hockey headache" and "skater's cough" are common complaints, describing the effects of carbon monoxide and nitrogen dioxide exposure. Many rinks are solving the problem by putting pollution control equipment on their ice cleaning machinery, by purchasing pollution-free electric-powered machinery, and by ventilating their rinks more often.

This problem does not affect large professional ice hockey arenas because the volume of air inside is greater, because such arenas are ventilated, and because the ice cleaning machinery is properly equipped and operates only three or four times a day.

Another type of indoor recreational activity may not seem as if it's indoors at all. But when you are camping inside a tent, you are in an enclosed indoor environment where pollutants can build up, especially if the tent is not vented. Propane heaters and stoves can quickly produce dangerous levels of carbon monoxide. This gear should not be used in a tent at all.

REDUCING YOUR RISK

With all of the various sources of pollution, and all of the different kinds of indoor environments, we could devote an entire book to detailing the specific steps you can take to reduce the risk from poor indoor air quality. But regardless of the specifics, some general steps are often helpful. (These steps are the same ones that you should request for indoor environments you don't control.)

First, eliminate the sources of contamination.

- If you choose to smoke, don't do it indoors.
- Make sure your gas heating and cooking appliances are running efficiently. (A yellow-tipped flame needs adjustment. A blue-tipped flame is optimal.)
- Install a ventilation hood over your stove, if possible.
- Follow the steps for radon remediation listed in Chapter 35, asbestos abatement in Chapter 21, and lead abatement in Chapter 29.
- Clean and disinfect home humidifiers and central ventilation systems regularly. Repair leaks to central heating and cooling systems.
- Repair and seal any leaks in your roof, foundation, walls, or windows to control biologicals that thrive in the presence of moisture.
- Remove carpeting on concrete or stone foundation floors, or any carpet that's gotten wet from flooding.
- Empty drip pans under refrigerators.
- Ventilate your attic to prevent moisture buildup.
- Vacuum carpets regularly.

- If installing new carpeting, roll it out outdoors first and let it air out for several hours to reduce the organic gases it will emit once installed.
- Keep food areas clean to avoid cockroaches and the allergens they spread.
- Buy limited quantities of cleaners, disinfectants, and pesticides, and use them according to directions and under good ventilation conditions. Properly dispose of unused portions you're unlikely to need.
- Don't idle your vehicle inside the garage.

Second, ventilate. Even if the outdoor air is polluted, concentrations are probably not as high as they are inside. An EPA study found that concentrations of a dozen common organic gases in homes were two to five times higher indoors than outdoors, whether the home was in the city, suburb, or country.

- Open windows or doors, even just briefly if it's cold out.
- Install attic fans.
- Run air conditioners with the vent open to increase exchange with outdoor air.
- Maximize ventilation when you're painting or doing something that generates pollutants, such as cooking, cleaning (stirring up dust), using pesticides, or renovating.
- Vent clothes dryers to the outside.

Finally, while air cleaners and filters are helpful, without these other steps they won't solve the problem. Neither will houseplants.

FOR MORE INFORMATION

Environmental Protection Agency
Indoor Air Quality Information Clearinghouse
www.epa.gov/iaq/search.html
P.O. Box 37133
Washington, DC 20013-7133
(703) 356-4020

This chapter was reviewed by Brenda E. Barry, Ph.D., Senior Associate and Toxicologist at Environmental Health and Engineering, Inc., in Newton, Massachu-

setts, one of the leading indoor environmental quality consulting firms in America; and by Jack Spengler, Professor of Environmental Health and Human Habitation and Director of the Environmental Science and Engineering Program, Harvard School of Public Health.

20. AIR POLLUTION (OUTDOOR)

THE DAYS of belching smokestacks in the United States are gone. Air quality is far better than it has been in decades. Controls on power plants, motor vehicles, and factories have dramatically improved the safety of the 3,400 gallons of air an average adult breathes each day. Still, nearly all that air contains compounds that are potentially harmful to human health. In addition to "natural" components like oxygen and hydrogen, the air we breathe contains sulfur dioxide (SO_2), nitrogen oxides (NOx), polyaromatic hydrocarbons (PAHs), carbon monoxide (CO), particulates (microscopic solid or liquid particles), volatile organic compounds (VOCs), and, depending on the weather, ozone, which most of us know as smog. And those are just the major pollutants in outdoor air. There are hundreds more.

The issue of air pollution is complex, so as an organizing device we group the major air pollutants into three categories: particles, ozone, and air toxics. We focus on pollutants that have a direct link to human health. We do not discuss the damage that many air pollutants do to the environment, such as acid rain or climate change. This is not to ignore the risk from these problems, both to the environment and ultimately to human health. Rather, it is to stick by the organizing principle of this book—to help you understand the *direct* risks to your health.

PARTICLES

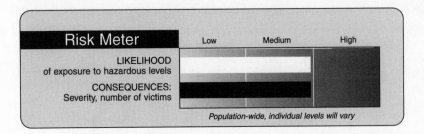

THE HAZARD

Particles are simply microscopic bits of solid or liquid matter in the air we breathe. They are very small, measured in microns. (A human hair is 70 microns across. There are a million microns in a meter—39.37 inches.) Particles are produced from industrial, residential, and motor vehicle combustion of coal, oil, natural gas, wood or other plant matter, and from anything that stirs up dust, like construction, vehicles traveling on unpaved roads, crushing and grinding operations, materials handling, even the wind. (Many natural processes also produce airborne particles.) Some particles are produced directly from these processes. Some, specifically chemicals called sulfates and nitrates, are known as "secondary" particles because they are the product of chemical reactions in the atmosphere between other "primary" air pollutant gases. The specific chemical composition of particles is not well understood, since there are so many processes and fuels that create them.

But it's not what they're made of as much as their size that seems to matter to our health, and it appears that the smaller they are, the more dangerous they are. Air pollution particles are identified by size. Less than 0.1 micron (<PM 0.1) particles are called "ultrafine." Those measuring between PM 0.1 and PM 2.5 are called "fine." Between PM 2.5 and PM 10 are "coarse." When we breathe in, the air carries the fine and ultrafine particles deep into the tiny air sacs in our lungs. The smaller the particles are, the deeper into our respiratory system they can go and the longer they can stay there before our body's natural cleaning systems capture and remove them.

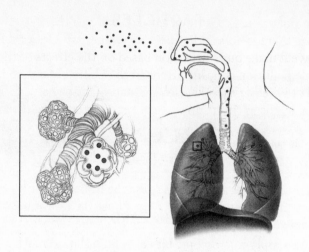

Particulates in the respiratory system

It is not known precisely how particles cause health damage, but there are basic theories. The longer these foreign objects stay in our lungs, the more chance they have to irritate and damage the delicate tissues of those tiny air sacs, which are known as alveoli. Ultrafine particles may also trigger a response from the immune system that changes blood chemistry and blood pressure, contributing to injury or death in people with cardiac, pulmonary, or other preexisting conditions.

THE RANGE OF CONSEQUENCES

A growing body of evidence strongly suggests that particles are closely associated with morbidity (illness) and mortality (death). Many studies find that the higher the concentration of particles in the air, the greater the number of illnesses and deaths. Studies have found that for every increase of 10 micrograms per cubic meter ($\mu g/m^3$) of particle pollution, the general death rate in an area goes up anywhere from 0.5 percent to 5 percent. (That wide range is explained by the nature of the way this problem is studied. Studies that look at the short-term relationship between yesterday's pollution and today's illnesses and deaths find the lower effect. Longer-term studies that follow populations for decades find the larger effect.)

To give you some perspective on the potential size of the problem, we'll assume, as some experts do, that no matter how low the level of particle pollution exposure, there are still health effects—in other words, that there is no safe level below which particles aren't hazardous. And we'll make our calculation based on the effects of the average fine particle pollution concentrations in the United States of 15 $\mu g/m^3$, regardless of whether the particles are man-made or natural. Taking all 2.3 million deaths in this country each year, particles could be responsible for between 17,250 and 172,500 annually. (Remember, this estimation is rough. Much about particle pollution remains uncertain.)

Particles may contribute to death in a number of ways:

- Research suggests that particles cause the heart to beat faster and that they interfere with its ability to adjust the rate at which it beats depending on your level of activity. Some scientists believe that more than half the deaths associated with exposure to particles are officially from heart disease, the number one killer in the United States.
- Particles inflame the lungs and breathing passageways, contributing to deaths among those suffering from chronic obstructive pulmonary disease (COPD), the fourth leading cause of death in the United States, and from pneumonia and influenza, the sixth leading cause of death.
- Particles are believed to cause systemic inflammation, based on the evidence of immune response in the bloodstream. Some studies suggest this phenomenon could contribute to diabetes.

Particles also contribute to nonfatal outcomes:

- Increased nonfatal heart attacks, angina, and other impacts of heart disease
- Hospitalizations and other impacts from COPD
- Increased respiratory infections
- Increased asthma attacks
- Decreased lung function (the amount of air we can take in per breath)

THE RANGE OF EXPOSURES

Because particles are the byproduct of everything we do in an industrialized society, as well as natural processes like wind, erosion, and forest and brush fires, they are everywhere. However, since some particles, especially the fine and ultrafine types, are produced by human activity, particle concentrations are higher in regions that are more developed and lower in regions that are less populated.

Exposure to the biggest and the smallest particles is usually highest near transportation corridors, in highly traveled urban areas, and near power plants and other combustion sources. The larger coarse particles are more concentrated in these places because they are bigger, heavier, and tend to fall out of the air more readily. Ultrafine particle concentrations are highest near emission sources for a completely different reason. Within a few hours of being emitted, ultrafine particles bump into one another and clump together, aggregating into bigger and bigger particles, growing into the larger PM 2.5—fine—category. Before they can move too far from where they were emitted, they've outgrown the ultrafine size.

Fine particles are carried anywhere on the wind and are the most common form of air pollution in the United States. Most of the country west of the Mississippi, more sparsely populated, has PM 2.5 levels between 9 and 17 $\mu g/m^3$. The more densely populated areas of southern California and regions east of the Mississippi—excluding New England, Florida, and eastern North Carolina—have PM 2.5 levels between 17 and 22 $\mu g/m^3$. Central and eastern Ohio, eastern Kentucky, northwestern Indiana, and central Alabama, regions with a high number of older power plants that have less restrictive emissions requirements, have fine particle pollution levels between 23 and 30 $\mu g/m^3$.

In 1997, based on mounting evidence of the association between particle pollution and risk to human health, the Environmental Protection Agency (EPA) proposed new rules to reduce particle exposure. Several industry groups, led by the American Trucking Association, which operates diesel-powered vehicles, sued to overturn these regulations, claiming their costs would outweigh their benefits. The Supreme Court ruled that the Clean Air Act doesn't allow balancing of costs and benefits and allowed the EPA to move ahead with the regulations. While the agency finalizes the particle rules, the federal government is also enacting programs to require cleaner burning fuel for diesel engines and tighter emissions standards for new diesel engines. Together these steps are expected to have a significant impact on reducing particle pollution since diesel exhaust is a major source of emissions. (See Chapter 25, "Diesel Emissions.")

SMOG

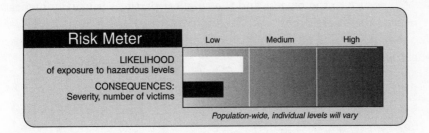

Risk Meter	Low	Medium	High
LIKELIHOOD of exposure to hazardous levels			
CONSEQUENCES: Severity, number of victims			

Population-wide, individual levels will vary

THE HAZARD

Smog is a term that people sometimes loosely (and imprecisely) use to describe any kind of air pollution. The word was coined in the early 1900s to describe the combination of smoke and fog over British and Scottish cities and towns which killed more than 1,000 people in Glasgow and Edinburgh in 1909. Other famous smog events include "The Great London Smog" in December 1952, which killed more than 4,000, and the "Donora Smog" that hit Donora, Pennsylvania, near Pittsburgh, in October 1948, killing 20 and sending approximately 7,000 to the hospital.

That's not the type of smog we worry about in America today. The kind that hit London and Donora was a thick mixture of SO_2 and particles that built up from home heating with coal and from industrial plant emissions under unique weather conditions that included little or no wind and heavy humidity and fog. The air got so thick from contamination that some of the victims literally asphyxiated.

The kind of smog with which we are most familiar is photochemical smog. It is the product of a chemical reaction in the air between certain air pollutants, in the presence of sunlight, warmth, and weather conditions that cause the emissions to concentrate. When NOx and VOCs mix and "cook" in still, sunny conditions, they produce ozone, a molecule with three oxygen atoms (O_3). Ozone is the principal ingredient in photochemical smog and the one that does the most damage to human health. (Ozone in the air we breathe can be harmful. But in the higher levels of the atmosphere, the same O_3 molecule is vital to protect our health, filtering out potentially harmful forms of ultraviolet radiation from the sun. We deal with the depletion of this upper-level ozone layer in Chapter 32, "Ozone Depletion.")

Nitrogen Oxides + Volatile Organic Compounds + _ Ozone
UV Radiation + Temperatures Above 65° F + Little Wind = (Smog)

The formula for smog

The problem with the O_3 molecule is that it is highly reactive. It binds to and damages or destroys other molecules. This is called oxidation, the process that causes rust or makes colors fade in our clothes and paints. When oxidation happens to molecules in our body, it can cause health damage.

THE RANGE OF CONSEQUENCES

Ground-level ozone—smog—is perhaps the most extensively studied air pollutant. The effects of ozone exposure include:

• Irritation of the respiratory system. Symptoms include coughing, scratchiness or tightness in your throat or chest, aggravation of preexisting respiratory illness, and greater susceptibility to respiratory infection. Between 10 and 20 percent of the respiratory-related hospital admissions in the Northeast each summer are related to ozone exposure.
• Reduced lung function: you can't take in as much air because of inflammation in your lungs. Moderately exercising adults (light jogging, climbing stairs, pushing a wheelbarrow) can experience 15

to 20 percent reductions in lung function if exposed to low levels of ozone for at least 6 or 7 hours.

- Increased frequency of asthma attacks and increased severity of those attacks. Although most health effects from ozone are short-term, some studies have linked long-term ozone exposure with development of asthma (the underlying disease itself, not just the symptomatic attacks).
- Scientists have also found that long-term exposure in lab animals can affect the development of the lungs and lead to permanent changes in lung function. Also, some studies are beginning to find a possible association between elevated ozone levels and increased deaths from respiratory or cardiovascular diseases. (As this book was going to press, a paper by several Harvard School of Public Health researchers asserted that the evidence that ozone may cause death needs to be taken more seriously.)

Several population subgroups are most at risk from ozone:

- Children. They breathe more air per pound of body weight than adults, have higher rates of asthma than adults, and young children have respiratory and immune systems that are still developing and are therefore more susceptible to irritation from ozone.
- Exercising adults. People outdoors who do strenuous work, or who participate in vigorous recreational activity, increase their intake of ozone and thus their risk from exposure.
- People with respiratory problems. In addition to people with asthma, those with chronic respiratory disease or other lung problems are more vulnerable to the effects of ozone.
- People with unusual susceptibility. Scientists don't know why, but some otherwise healthy people seem unusually sensitive to ozone, experiencing effects sooner, and at lower levels, than the general population.

THE RANGE OF EXPOSURES

The basic ingredients of ozone are common (at varying levels) in most of the air in the United States. Fortunately, the conditions that create ozone are sporadic, temporary, and limited to certain areas where the weather and geography allow the chemical reactions that form ozone to take place. The components of photochemical smog, NOx and VOCs, come from a variety of sources. NOx are emitted from motor ve-

hicles and power plants. VOCs come from combustion sources like vehicles and power plants, as well as industrial facilities, consumer products, and even from the fumes of evaporating gasoline. (That's why
vapor recovery devices are required on the nozzles of gasoline pumps, to
capture the VOCs emitted as we "fill 'er up.") Some types of VOCs are
emitted naturally, including from trees.

After these ozone precursors are emitted, if the weather keeps emissions from rising more than a couple of thousand feet (these conditions
are called temperature inversions), and if wind patterns and hills or
mountains cause the chemicals to accumulate (as in Los Angeles, Denver, or Mexico City), and all these factors come together in the presence
of sufficient sunlight and warm temperatures over as little as three or
four hours, the soup of chemicals mixes and essentially cooks into
ozone. These requirements explain why those smog-alert days you may
hear about from time to time are a summertime phenomenon.

Concentrations of ozone precursors are greatest in urban areas near
roadways, power plants, and industrial facilities, but smog is by no
means a strictly urban phenomenon. Precursor chemicals are carried
great distances on the wind and lead to the formation of ozone in
regions that have no emissions of their own. And ozone itself, once
formed, can travel on the wind, contributing to levels of photochemical
smog in areas hundreds of miles away from where it first arose. One frequently cited example is the occasionally high ozone concentrations in
the relatively undeveloped region of Acadia National Park in Maine, a
day or two after the ozone levels rise in metropolitan areas to the south.
Because of wind patterns and the complex conditions needed for ozone
formation (including the passage of time), ozone levels are often higher
on the top of the highest mountain in this national park than they are
in downtown Boston.

～

As we said at the outset, air quality in the United States is better than
it's been in decades, and ozone levels indicate that the improvement is
continuing. Between 1980 and 1999, ozone levels nationally decreased
20 percent as measured on a one-hour exposure average, and 12 percent
on the newer (1997) rating system of an eight-hour exposure average.
The greatest declines were in the west, down 37 percent on the one-
hour exposure scale, and the Northeast, where one-hour exposure levels
went down 30 percent. There was only a 6 percent decrease in the
south, and a 10 percent decrease in the Mid-Atlantic region. The biggest

reason for the reduction has been lower levels of VOC emissions due to a range of federal controls. Between 1980 and 1999 VOC emissions fell 33 percent. NOx emissions rose 1 percent.

The EPA, with help from state environmental agencies, now tracks ozone levels in many areas on maps that are posted online. Many of these maps even predict ozone levels several hours in advance, based on weather forecasts. The maps are color-coded and list five levels of health effects, ranging from "Good" to "Very Unhealthy." Many local television, radio, and newspaper weather forecasts also offer the EPA's Air Quality Index (AQI), which rates local air pollution conditions, including ozone. The scale runs from 0 to 500, with 100 representing air quality that meets national standards. Higher levels mean progressively more polluted air. Levels between 150 and 200 are considered unhealthful. From 201 to 300, they're considered very unhealthful. Levels over 300 rarely occur in the United States.

AIR TOXICS

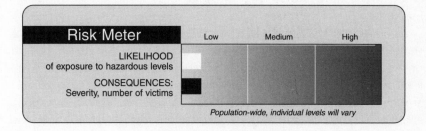

Risk Meter	Low	Medium	High
LIKELIHOOD of exposure to hazardous levels			
CONSEQUENCES: Severity, number of victims			

Population-wide, individual levels will vary

THE HAZARD

The battle to clean the air began with the passage of the Clean Air Act in 1970, a time when smoke billowed from tailpipes and smokestacks. The targets were obvious and the solutions relatively simple. But as we have learned more about air pollution, new classes of invisible pollutants have been discovered, such as air toxics. This group of air pollutants includes some of the more common chemicals in the modern environmental risk lexicon: benzene, dioxin, lead, mercury, polychlorinated biphenyls (PCBs), perchloroethylene, and vinyl chloride. Air toxics are a designated class of air pollutants known or suspected to cause cancer or other acute and chronic health effects, including damage to the im-

mune system, reduced fertility, birth defects, and respiratory and neurological problems.

Most air toxics are the result of human activity. Stationary sources
include factories, refineries, power plants, and numerous small businesses like dry cleaners and auto body shops. Mobile sources include
most motor vehicles, construction equipment, and industrial and farming machinery such as forklifts and harvesters. Air toxics are also emitted by some construction and household materials like glues and solvents. Nature also emits air toxics from forest fires and volcanoes.

It is certain that every American is exposed to some degree of air
toxic pollution, though it is likely that these exposures rarely occur at
hazardous levels.

THE RANGE OF CONSEQUENCES

Air toxics are principally a concern because they can contribute to an
increase in the risk of cancer. In fact, the specified goal of the government's air toxics reduction program is to reduce the potential cancer
risk from air exposure by 75 percent. But of the 188 air toxic chemicals,
only one is a known carcinogen. Most are only suspected based on animal tests.

As of the end of 2001, the EPA was still preparing a risk assessment of
the consequences of air toxics exposure in the United States. Exposure
previously hadn't been studied, and there is still uncertainty about the
level of hazard these chemicals present, especially at the very low levels
at which we're probably exposed. Based on what we know now about
emissions and the hazard of these chemicals in tests on lab animals, it is
believed that the risk of cancer or other effects from air toxics is very
low.

THE RANGE OF EXPOSURES

In December 1984, 40 tons of methyl isocyanate gas leaked from a pesticide factory in Bhopal, India, killing as many as 4,000 and injuring
hundreds of thousands more. In the wake of what many still call the
worst industrial accident in history, Congress and the EPA created the
Toxics Release Inventory (TRI), which tracks the amount of certain
chemicals released into the air, the water, and onto the land. Based
on TRI information and other input, the EPA is able to estimate
the sources of air toxic emissions. Data indicate that mobile on-road

sources contribute 30 percent of the air toxic pollution in the United States. It is important to remember that all of these pollutants are not equally toxic—a pound of dioxin is not the same risk as a pound of perchlorethylene—so the TRI numbers are not guides to risk. Small businesses (which emit less than 10 tons of any one air toxic, or 25 tons of combined air toxics, per year) contribute about one quarter. Major sources, with emissions higher than those thresholds, contribute another quarter. Nonroad mobile sources contribute about 20 percent.

If you look at just the 33 toxic air pollutants of most concern, the numbers change. These priority air toxic chemicals make up about 25 percent of all air toxics by volume. Sources for these 33 pollutants are: small businesses (40 percent), mobile on-road sources (29 percent), mobile nonroad sources (22 percent), and major sources like factories, refineries, and power plants (9 percent).

Of the 4.6 million tons of total air toxics emissions in the United States in 1996, the year the EPA used as a baseline in its first study of air toxics, roughly two thirds were emitted in urban areas and one third in rural areas. And as with many types of air pollution, the emission of air toxics is decreasing. In the period between 1990 and 1993, approximately 6 million tons per year were emitted. In 1996, emissions were an estimated 4.6 million.

Again, just because something is released into the air doesn't mean it's a hazard to human health. Without further information on exposure and on the acute and chronic toxicity of many of these chemicals, the actual risk will remain unknown. As of the beginning of 2002, the EPA was working on developing such information for the 33 air toxics of priority concern.

CONCENTRATIONS OF SELECTED AIR TOXICS IN URBAN AREAS ($\mu g/m^3$)		
Compound	1994	1999
Benzene	3.25	2
1-, 3-Butadiene	7	5.5
Perchloroethylene (PCE)	1.35	0.95
Styrene	1.1	5.75
Toluene	8.25	4.5

Source: EPA

REDUCING YOUR RISK FROM ALL TYPES
OF AIR POLLUTION

Basically anything you do to reduce air pollution makes the risk of air toxics, smog, or particles that much lower. While the EPA and other federal and state agencies have numerous programs to reduce air pollution, there are a number of things you can do as an individual. No one thing you do will make that much difference. But if everyone does just a few things, it could add up to cleaner air and greater public health for you and your neighbors.

For instance, most of us drive.

- Keep your vehicle tuned up. That reduces emissions. Tune-ups are more important for older vehicles (more than four to five years old) in which emissions control systems are starting to malfunction, or which have emission control equipment that isn't as sophisticated as that found on newer vehicles.
- Avoid operating at high speeds, including dramatic accelerations. Driving at the speed limit not only reduces emissions from your vehicle but optimizes fuel economy, saving you money.
- Keep your tires well inflated. This reduces friction and improves fuel economy.
- If you're stopped and not going anywhere, turn off your vehicle while you're sitting in it. You can run music systems off the battery for a while. You can even keep things warm in cold weather by occasionally running the vehicle for a couple of minutes, rather than keeping it running all the time.
- You can even reduce air pollution just by being careful when you refuel. Don't top off the tank, because you'll probably spill a small amount of fuel, which evaporates and contributes to air pollution. That may not seem like much, but states that have required fueling stations to fit their fuel pump nozzles with vapor recovery hoods have experienced reductions in some pollutants of up to 5 percent.

Of course, driving less is a major way to reduce air pollution. Carpooling, mass transit, bicycles, even just maximizing your efficiency when you take your vehicle out to run errands, all help.

You can actually reduce outdoor air pollution by choices you make indoors at home. Most have to do with using less energy, which requires less fuel to be burned at power plants. Most of these steps also save money on your fuel bills.

- One option is to use energy-efficient light bulbs, like compact fluorescents, and energy-efficient appliances.
- Turn off VCRs, computers, and stereos when you're not using them.
- Cooking small meals in the microwave uses less energy than using the stovetop or oven.
- Turn your thermostat down a couple of degrees in the winter and up a couple in the summer. Insulate your home, and your hot water pipes and water heater.
- Make sure any household chemicals, paints, and solvents are in well-sealed containers to keep fumes from escaping.

Consumption contributes to air pollution. Anything we buy, after all, is made of raw materials that had to be mined or farmed or industrially produced. Products have to be processed or assembled. They have to be transported to our local stores. All those steps require the burning of fuels, and most require the use of industrial chemicals and intermediary products, which all contribute to air pollution. So recycling creates less air pollution since it's a less energy-intensive way to produce raw materials and products. When you can, buy appliances and other devices that are more energy-efficient. They sometimes cost more up front but pay for themselves in energy savings over time.

Those of us who live in areas that experience smog can alter our behavior when weather conditions favor the development of ground-level ozone. Defer gardening chores with gasoline-powered equipment until later in the day. (As with refueling your vehicle, try not to spill fuel from your power equipment.) Use an electric starter for outdoor grills, instead of lighter fluid. Drive a little less on days when there are government smog warnings, and put off refueling, if possible, during smog-alert conditions.

FOR MORE INFORMATION

National Safety Council
www.nsc.org/ehc/mobile/airpollu.htm.
1121 Spring Lake Drive
Itasca, IL 60143-3201
Air Pollution Hotline: (800) 557-2366

Environmental Protection Agency
www.epa.gov/airnow (current air pollution conditions where
   you live)

www.epa.gov/airnow/webcam.html (real-time webcam)
www.epa.gov/oar/agtrnd00/pmatter.html (particle-pollution
  maps)
www.epa.gov/airnow/ozone.html (ozone-level maps)
www.epa.gov/autoemissions (vehicle ratings for fuel economy)

This chapter was reviewed by Jonathan Levy, Assistant Professor of Environmental Health at the Harvard School of Public Health; by Jack Spengler, Professor of Environmental Health and Human Habitation at the Harvard School of Public Health; and by Dan Greenbaum, President of the Health Effects Institute and former Commissioner of the Massachusetts Department of Environmental Protection. Andrij Holian, Professor, Department of Pharmaceutical Sciences, and Director of the Center for Environmental Health Sciences, University of Montana, reviewed the section on Air Toxics. Maria Costantini, Toxicologist and Principal Scientist at the Health Effects Institute, offered valuable input on the effects of particles.

21. ASBESTOS

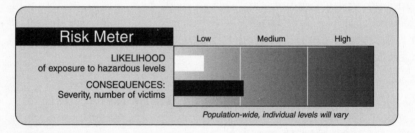

IT IS LITTLE WONDER that asbestos is so widely used. It's a very handy mineral. There are six different mineral forms of asbestos, but all of them are long fibers that are strong, thin, highly resistant to heat and chemicals, and poor conductors of electricity. That makes them very useful in a wide range of products, including heat-resistant fabrics and insulation, friction products like motor vehicle brake linings and clutch pads, and building materials like floor tiles, exterior siding, and cement pipes. Asbestos has been used in everything from schools to offices to ships to nuclear reactors to cars and trucks. But the properties that make it so useful to industry are precisely why asbestos fibers pose a hazard to human health.

THE HAZARD

First, asbestos fibers are exceedingly thin, 1,200 times thinner than a human hair. That makes them a good material for a variety of uses because they don't take up much space. But it also means that when they are inhaled, some of these tiny fibers can get past the body's natural filtering systems and burrow deep down in the lungs. Second, asbestos is highly resistant to chemicals. It doesn't break down. That makes it a sturdy material for industrial use, but it means that once inside the body, a small percentage of the tiny mineral fibers embed in the tissue and are not broken down by the body's chemistry. Though the precise mechanism by which they cause illness is still unknown, over time the presence of these fibers irritates the cells in the area where they are embedded, causing changes that lead to disease. The three major illnesses caused by asbestos are illnesses of the lungs.

Anything that contained asbestos that was being made and sold as of June 1989 can still be manufactured and sold. But the discovery that asbestos can cause serious lung damage and death has prompted a significant reduction in the use of the material. At least two dozen companies involved in asbestos manufacture have had to declare bankruptcy in the face of declining sales, and because lawsuits claimed that some of these companies were knowingly manufacturing and selling products with health risks without warning workers or consumers.

Though asbestos was first used in the early 1900s to insulate steam engines, and then in World War II for shipbuilding, it wasn't widely used in consumer products until the 1950s. Since then, one of its main applications has been in schools and other public buildings, as insulation, soundproofing, fireproofing, and in tiles and plastics as decoration. The Environmental Protection Agency (EPA) estimates that there is asbestos-containing material in most of the nation's 107,000 public schools and in 733,000 public and commercial buildings. But most of that asbestos is embedded in the material it is part of, and that's important because asbestos is only a health hazard if inhaled. Remember, an otherwise hazardous material is not a risk if you're not exposed to it at levels that can do you harm. The EPA reports that the prevailing levels of airborne asbestos in these buildings, and the health risk from them, are very low.

THE RANGE OF CONSEQUENCES

The most important thing to remember about the consequences of asbestos is that the fibers are a health threat only when they are released into the air. Asbestos may occur in many materials, but unless those materials are disturbed, damaged, or deteriorate with age, it does not pose a health risk.

When fibers do break free and are inhaled, asbestos is known to cause *lung cancer, mesothelioma*—a rare and nearly always fatal cancer of the lining of the lung, chest, and abdomen—and *asbestosis,* a serious, chronic, often progressively disabling and sometimes fatal respiratory disease. Asbestosis is the scarring of lung tissue, which leads to shortness of breath, increased risk of respiratory infection, and increased risk of cardiac failure. It is not curable. In 1999, for the entire U.S. population 15 and older, there were 1,265 deaths from asbestosis and 2,484 deaths from mesothelioma. In 1996, the last year for which there are such records, asbestos was the cause of approximately 6,000 lung cancer deaths in the United States.

All three of these diseases occur almost exclusively in people who are exposed to asbestos on the job. People who work in mines, who apply asbestos as insulation on ships or in buildings, who work on brakes and clutch pads, who work on building renovation or demolition, or people who work with asbestos in a manufacturing process, make up the overwhelming majority of those suffering from asbestos-related illness. Asbestos-associated disease in nonoccupationally exposed individuals is uncommon.

And the overwhelming majority of people who get asbestos-related illnesses are smokers. Asbestos not only adds to the risk of lung cancer in smokers. It somehow interacts with the effects of smoking and multiplies the risk. The EPA finds that people who smoke and who are exposed to asbestos have a 50 to 90 times greater chance of developing lung cancer than those who don't smoke and are not exposed to asbestos.

The latency period between exposure and the onset of lung cancer is generally 15 years or more, with lag periods of 30 to 35 years not uncommon. The lag period for mesothelioma and asbestosis is generally greater, often between 40 and 45 years. In other words, the number of cases we're seeing now reflects exposures long ago. As the use of asbestos declines and remediation programs reduce exposure levels, it's probable that the number of people suffering the consequences of asbestos exposure will go down.

THE RANGE OF EXPOSURES

Since asbestos was used for decades in as many as 5,000 products, and since it occurs naturally in rock and soil, it is ubiquitous in our environment. In rural areas, tests have found levels of 0.00000003 to 0.000003 fibers per cubic centimeter (f/cc) of outdoor air. Urban levels have been measured between 0.000003 and 0.0003 f/cc of air. Levels near asbestos mines can reach 0.002 f/cc or higher. Indoors, levels measured in schools, homes, and other buildings range from 0.000007 to 0.006 f/cc.

There is no safe level of exposure to asbestos. The EPA states that any exposure might be harmful. But to put the exposure numbers in perspective, according to the Agency for Toxic Substances and Disease Registry, the lowest level at which people have been known to develop health problems from asbestos is 1 f/cc over an exposure period of at least 20 years. The Occupational Safety and Health Administration (OSHA) has established a workplace limit of 0.1 f/cc per 8 hours of exposure (for a 40-hour workweek). Normal exposures, even in occupational settings, are many times lower than that.

~

A principal source of exposure to asbestos is building renovation or demolition. Such work can damage asbestos-containing material and produce high levels of airborne fibers in the immediate area of the work for a relatively short period of time.

Asbestos-containing material can be found in the following places in homes, schools, or other buildings:

- Insulation on pipes or heating devices
- Loose-blown insulation in wall spaces (especially in buildings built between 1930 and 1950)
- Fire-retardant material on structural elements of the building
- Exterior shingles
- Vinyl floor tile
- Older electrical sockets and fuse boxes
- Built-in ovens and dishwashers pre-1975

- Older portable devices that generate heat, like clothes dryers, electric blankets, and toasters. Most newer versions of these devices contain no asbestos.

In addition, brake pads and clutch pads in motor vehicles may contain asbestos. Some vehicle manufacturers still use asbestos in brake and clutch pads.

Asbestos was also used in cement water pipes. But most asbestos that is ingested rather than inhaled is excreted by the body. Studies have not confirmed any link between ingested asbestos and disease.

REDUCING YOUR RISK

OSHA estimates that 1.3 million employees in construction and general industry in the United States face significant asbestos exposure on the job. Basic safety procedures for these workers require that, in areas where exposure exceeds the federal standard of 0.1 f/cc of air, respirators should be used, protective clothing should be worn, and eating, smoking, and drinking at the workplace should be avoided. Even more restrictive rules apply when asbestos removal is under way, including negative pressure inside the workspace to make sure that air flows into, not out of, the area.

Public schools are required by federal law to investigate their buildings for asbestos and hire trained mitigation professionals if removal or encapsulation of the asbestos-containing material is appropriate. Federal grants are available to assist with this work.

If you suspect there is airborne asbestos in your home, school, or workplace, government health agencies recommend that you contract an analytical laboratory to collect and analyze samples. The cost runs between $20 and $40 per test. Remember that the asbestos fibers that are dangerous are too small to be seen by the naked eye. If you see dust particles, you are not seeing the asbestos fibers you need to worry about. But if you see dust in areas where asbestos is likely to be, there might be invisible asbestos fibers in that dust, so don't disturb it and have it checked.

Also remember that since asbestos is only a health risk if it becomes airborne, removal is recommended only in cases of significant exposure such as building renovation or demolition. Other forms of remediation involve encapsulating the material within a protective outer shell that can be sprayed or painted on. Even a coat of latex paint can help contain asbestos fibers. Often the material that contains the asbestos is solid enough that fibers are not escaping from it, in which case the recommended response is simply to monitor the condition of the material over time.

Government agencies recommend against doing asbestos mitigation work yourself. They suggest that you hire a certified asbestos treatment professional who will have the right vacuums, featuring high efficiency particulate air (HEPA) filters, and other equipment to do the work safely.

Finally, if you think you have been exposed to asbestos in high enough levels to threaten your health, the National Cancer Institute recommends:

- Stop smoking.
- A chest X ray may be appropriate, read by someone experienced in diagnosing asbestos exposure.
- Get prompt medical attention for any respiratory illness.
- Be alert to symptoms that may include shortness of breath, pain in the chest or abdomen, difficulty swallowing. (Remember that the lag between exposure and symptoms is often decades.)
- Use all protective equipment and safety procedures for further work around asbestos-containing material.

FOR MORE INFORMATION

Occupational Safety and Health Administration
www.osha-slc.gov/SLTC/asbestos
OSHA has regional offices in most states.
In Washington, DC: (202) 693-1999

Environmental Protection Agency
www.epa.gov/opptintr/asbestos/help.htm (about asbestos)
www.epa.gov/opptintr/asbestos/contacts.htm (list of local
    contacts)

This chapter was reviewed by David Christiani, Professor of Occupational Medicine and Epidemiology at the Harvard School of Public Health; and by Gregory Wagner, Director of the Division of Respiratory Disease Studies at the National Institute for Occupational Safety and Health, a division of the Centers for Disease Control and Prevention.

22. BIOLOGICAL WEAPONS

Risk Meter	Low	Medium	High
LIKELIHOOD of exposure to hazardous levels	?	?	?
CONSEQUENCES: Severity, number of victims	?	?	?

Population-wide, individual levels will vary

THE ANTHRAX-LACED MAIL that killed five Americans in the fall of 2001 made the threat of biological weapons frighteningly real in the United States for the first time. But battlefield use of living microscopic organisms as weapons goes back at least to medieval days, when some armies used catapults to hurl the bodies of plague victims into enemy lines. In the French and Indian War in the mid-1700s, British soldiers sent blankets that had been used by smallpox victims to American Indians, with murderous effects. The idea of pathogens as weapons is hardly new.

THE HAZARD

In theory, biological weapons could be even more devastating than chemical or nuclear weapons. That's because some of them can spread far beyond the initial point of release through ongoing and multiplying human-to-human transmission. They are also incredibly powerful for their weight and cost. One government analysis showed nuclear weapons cost $1,500 per victim killed, while anthrax deaths might cost a penny each.

Fortunately, carrying out an attack with biological agents which kills large numbers of people is difficult. Distributing these pathogens in a way that exposes large numbers of people is not simple. You don't just brew up some deadly germs in a lab and go somewhere and shake them out of a jar. For most biological weapons to reach more than just a few people, they have to be dispersed in the air. To accomplish that, the agent has to be dried, then ground up or "milled" into tiny particles that can remain airborne for days, and in some cases further treated to control clumping. These steps take time, money, special equipment, and expertise. They also require sophisticated protective clothing, filters, and containment equipment if the people who want to use them as weapons don't want to become their own first victims.

The Japanese terrorist group Aum Shinrikyo, before its Tokyo subway attack with the nerve gas sarin, attempted several attacks with botulinum toxin, anthrax, and other agents but couldn't manage to cause a single death. And the 2001 mailborne anthrax attacks in the United States demonstrated how difficult it is to use even potent "weaponized" agents to kill more than a small number of people.

However, the 2001 attacks showed that there is still much we don't fully understand about bioweapons. Scientists did not expect anyone to be exposed from a letter unless he held it to his nose. They now know this is not true. They also learned that it takes fewer spores of anthrax to cause disease than previously believed. And the risk of an airborne anthrax attack is clear in light of evidence from an unintentional human experiment in the Soviet Union in 1979. A Soviet biological weapons plant in Sverdlosk accidentally released "aerosolized" anthrax. Although the breach lasted only hours, and involved less than a gram (one twenty-eighth of an ounce), the wind carried a narrow plume of the pathogen across one section of the city. At least 66 people died.

～

The U.S. Centers for Disease Control and Prevention (CDC) has named six pathogenic microbes that pose the greatest risk as bioweapons: anthrax, smallpox, plague, tularemia, botulism, and viral hemorrhagic fever (such as Ebola or Marburg). These agents are the easiest to disseminate and transmit, and they're the most deadly. A second group of agents includes Q fever, brucellosis, glanders, and toxins such as ricin or staphylococcal enterotoxin. They also pose a severe risk, but aren't as lethal, are tougher to spread, and require less stockpiling of antibiotics or vaccines. A third group of potential biological weapons includes such agents as hantavirus and Nipah virus. They are not necessarily less dangerous, but they are far less well understood. In their present form they pose less of a threat, but the risk with these pathogens is that someday they could be developed to be more deadly and more easily spread.

We examine the first group of pathogens in detail, along with a brief overview of some other possible bioweapons.

Anthrax

In its natural form, anthrax is widespread in the environment, infecting sheep and cattle, and occasionally people, throughout the world. (It was once known as "wool-sorter's disease.") Onset of illness usually occurs one to five days after exposure, but symptoms can take as long as six weeks to develop. They initially resemble those of a cold or flu—nasal congestion, joint pain, fatigue, and a dry cough. Once symptoms begin, respiratory difficulty can get worse quickly, with death often following within one to three days. Anthrax is not transmissible from person to person, so the victims are only those directly exposed in some way to a sufficient number of spores. But spores can lay dormant for quite some time and become "re-aerosolized," so an ongoing risk of exposure exists unless contamination is thoroughly cleaned up.

Anthrax can be contracted in three different ways, and the outcome varies dramatically depending on how it enters the body. You can breathe it in, eat it, or get it on your skin. Inhalation anthrax has by far the worst prognosis, with mortality of about 90 percent if not treated early, before significant symptoms begin. Mortality from ingestion is 25 to 60 percent, and 20 percent from cutaneous exposure. With prompt antibiotic treatment, deaths from all forms are rare.

To be an effective weapon, the particles must be made extremely fine, about 1 to 10 microns (the average human hair is about 70 microns thick), to remain airborne and travel deep into the lungs. Spores can remain viable in the environment for 40 years or more, and they're tough to eliminate. Elaborate and expensive cleanup procedures are needed for exposed buildings. It took the federal government more than four months to decontaminate the Hart Senate office building after just a few envelopes carrying anthrax spores were sent to just two locations in the building.

A vaccine exists for anthrax, but until the attacks in the fall of 2001, it was available only for the military and for some emergency medical personnel. After the attacks, the vaccine was offered to thousands of people who were potentially exposed to anthrax. In addition, the government began stockpiling enough vaccine for the entire U.S. population. The vaccine might aid treatment for people even after they've been exposed, if it's administered before the onset of symptoms. But the best treatment for anthrax after exposure is prompt use of antibiotics. Ciprofloxacin got a lot of notice in the press, but several other antibiotics also work well against anthrax infection.

Smallpox

Smallpox is the first and only disease to have been eradicated in the world, with no natural cases reported since one in Somalia in 1977. (There was one case in a laboratory worker in England in 1978, after which most remaining laboratory samples were destroyed.) The only remaining authorized stocks are closely guarded laboratory samples in the United States and Russia, but according to CIA reports, a half dozen or so other nations may also possess the virus in defiance of international treaties. It has been more than 50 years since the last case of smallpox was seen in the United States.

Smallpox vaccine is one of the most successful global public health interventions of modern times. Before it was conquered through diligent tracking and vaccination, smallpox is estimated to have killed more people than any other infectious disease—about 500 million in the twentieth century alone, perhaps twice as many casualties as in all that century's wars.

Smallpox is a viral disease that killed about 30 percent of those infected. Some strains were much worse, with mortality approaching 100 percent. Survivors bear disfiguring scars for life. The incubation period is 7 to 17 days, and the virus spreads readily from the victims (who in the early stages might not even realize they are infected) to anyone nearby, through tiny exhaled droplets.

Smallpox may be hard to diagnose when people first get it because it is never seen by physicians and might be easily mistaken for chicken pox. As the disease progresses, symptoms advance from high fever, headache, and fatigue to a distinctive rash that is initially concentrated on the face, arms, and legs but eventually covers the whole body. No effective treatment exists, but smallpox vaccine can reduce the severity of the disease or even prevent its symptoms altogether if given within four days of exposure. Routine vaccination was discontinued in 1972, but the U.S. government in late 2001 ordered production of enough vaccine to inoculate the entire population if necessary, including booster shots for those vaccinated more than 20 years ago.

Plague

This bacterial disease has killed an estimated 200 million people in three massive outbreaks in human history. Naturally occurring plague is rare today, with only 1,000 to 3,000 cases per year reported worldwide, but it is often cited as a potential biological weapon because it can spread easily and cause epidemics. The last one in the United States was in Los Angeles in 1924 and 1925, part of an outbreak affecting port cities in nations around the Pacific, which claimed 12 million victims. Currently in the United States, there are only 10 to 15 cases per year.

The natural form of plague is most often contracted from fleas that live on infected rats. But as a biological weapon it is most likely to be released in an aerosolized form that can be inhaled. (Japan tried a simpler method, dropping infected fleas on parts of China in World War II, causing plague outbreaks where none had previously been reported.) Symptoms develop in one to six days, beginning with chills, fever, headache, weakness, and cough. Without treatment, they progress to shock, delirium, and organ failure. Death can follow rapidly from respiratory failure.

Plague is much less transmissible than smallpox, but like smallpox it can be spread from person to person. Without prompt treatment with antibiotics, death can result in 50 to 60 percent of cases. There is no recognized vaccine against plague. Early treatment with antibiotics is usually effective.

Tularemia

This hardy bacterial disease infects small mammals (it's sometimes called rabbit fever) worldwide, so it is easy to obtain. Tularemia's potential to produce human epidemics was discovered in the 1930s and 1940s when large outbreaks occurred in Europe and the Soviet Union. It has long been considered a potential weapon, and was actively developed for that purpose by the United States, the Soviet Union, Japan, and other nations.

The former Soviet biological weapons scientist Ken Alibek says that tens of thousands of tularemia cases among German and Soviet troops in World War II may have been the result of intentional use by the Soviets against the Germans. If this indeed occurred it demonstrates another of the problems of using biological agents as weapons. In the scenario Alibek describes, the attacker became the attacked when the disease spread back to the Soviets who had used it in the first place.

Tularemia symptoms include chills, nausea, headache, and fever and usually begin within 2 to 10 days after exposure, which occurs through inhalation. The symptoms can persist for weeks to months, and pneumonia develops in many cases. Without treatment, the mortality rate is about 30 percent. But tularemia progresses more slowly than either plague or anthrax, so mortality is low if treatment begins promptly after exposure. A vaccine to prevent tularemia has been developed, and the disease is treatable with antibiotics.

Botulism

The bacteria that cause this disease are found in soil worldwide so the raw material for this potential bioweapon is easy to obtain. Botulism, which is usually caused by ingesting the bacteria rather than inhaling them, paralyzes the muscles, and ingestion of minuscule amounts can produce symptoms quickly, within a day, progressing rapidly to death from respiratory failure. By weight, it is the most potent killer known: one nanogram per kilogram of body weight can be lethal. For a 220-pound person, just 4 billionths of an ounce of the poisonous protein from the bacterium can be lethal. About 25 cases of foodborne botulism occur in the United States each year.

This microbe used to kill half the people who became infected. But over the last half century that rate has dropped to just 8 percent. Intensive medical care for months, including use of an antitoxin to fight the poisonous effects of the secretions of the bacteria, and the use of mechanical ventilation to compensate for respiratory failure during the acute phase, can keep most patients alive. The CDC has limited stockpiles of the antitoxin for botulism. Botulism is not transmissible from person to person.

Viral Hemorrhagic Fever

The best-known forms of this class of diseases are Ebola and Marburg viruses, but it also occurs in less severe forms, including yellow fever and dengue fever. They are characterized by muscle pain, headache, and abrupt onset of fever, followed by nausea, abdominal pain, chest pain, sore throat, and cough.

Ebola and Marburg viruses are quite similar. Initial symptoms include a rash on the torso that breaks out about five days after exposure. That develops into bleeding and hemorrhaging. In outbreaks in several African countries, these two types of viral hemorrhagic fever have proven to be fatal in 30 to 90 percent of cases, depending on the strain. The incubation period is usually 5 to 10 days, but can range from 2 to 19. There is no vaccine and no known effective treatment. These diseases are highly transmissible through the transfer of blood and body secretions, but airborne transmission has been shown only in monkeys. There is no evidence that these viruses have been worked on as potential weapons, but they're included on the CDC's list of agents of concern because of their potency and virulence.

THE RANGE OF EXPOSURES

During the Cold War, the United States and the Soviet Union developed large stockpiles of several different kinds of biological weapons. A NATO handbook on dealing with biological warfare lists 31 potential pathogens that could be used as weapons. A former Soviet bioweapons scientist says that the U.S.S.R. identified at least 70 potential biological weapons agents, 11 of which were developed as weapons by the Soviet Union in the 1970s and 1980s. At its peak, the former Soviet Union's bioweapons program had stockpiled hundreds of tons of anthrax, according to the former director of the program. Several of these agents were also developed into weapons by the United States.

A 1972 international agreement called the Convention on the Prohibition of the Development, Production, and Stockpiling of Bacteriological and Toxin Weapons and on Their Destruction banned the development and stockpiling of biological weapons. It is still in effect, but no verification rules have been agreed to. Even without formal international agreement on the details, the United States and former Soviet Union are eliminating these weapons. The United States has destroyed most of its stockpiles. The former Soviet Union, with financial support from the United States, is beginning to destroy its supply, but several agents remain. The deadline to eliminate them is 2007. But as the public learned in the wake of the terrorist attacks and anthrax-laced mail in 2001, as many as 13 nations have active bioweapons programs, and the successful use of some biological agents is within the capabilities of well-financed terrorist organizations.

Still, it would take a series of attacks by groups with far greater access to such weapons than has been suggested by events so far to expose a lot of people to bioweapons. As we've explained, it is difficult to find a way to expose a lot of people to these agents all at once. Airborne contamination would pose a serious risk because it can produce rapid and widespread distribution—but only if everything works right. For a massive attack, as modeled in a U.S. government study, a large quantity of specially engineered pathogens would have to be released from an aircraft directly over a large city—but without an explosion, which would heat-sterilize the agent. The wind conditions and humidity would have to be just right, with no precipitation to wash the agent out of the air before it could spread. Even sunlight could matter, since ultraviolet radiation from the sun causes some agents such as anthrax to break down. Despite these difficult but not insurmountable challenges, an airborne attack may be the most serious risk people face from bioweapons.

Contamination of water supplies would almost surely not work, because the enormous volume of water that passes through a municipal water system hourly would dilute any agent put into it. And the chlorination, ozonation, and filtration that keep our water safe from other pathogens would almost surely kill these too.

Food contamination is another way these agents could be spread. In late 2001, U.S. Secretary of Health and Human Services Tommy Thompson said, "I am more fearful about this than anything else." Food contamination has actually been tried before, with worrisome results. In 1984, 751 people in an Oregon community became sick after followers of the religious leader Bhagwan Shree Rajneesh contaminated salad bars at 10 local restaurants with salmonella in order to influence turnout at a local election. No one died, though several victims nearly did. And there was a purposeful contamination of pastries with shigella bacteria in a Dallas medical center in 1996 which left 12 people sick. Dozens of foodborne disease outbreaks each year further testify to the fact that food could be used as a delivery vehicle for a bioweapons attack.

REDUCING YOUR RISK

Many people responded to the anthrax attack in the fall of 2001 the way we always do when a new, widely publicized risk comes along and makes us feel vulnerable to a hazard we don't understand. Thousands sought to take control by doing what they thought might reduce the risk: taking or stockpiling antibiotics, buying gas masks and special air filters, or trying to get vaccinations against possible biological agents like anthrax or smallpox. But specialists say that simple, basic hygiene and commonsense precautions are the most effective countermeasures against bioweapons.

Gas masks are useless against biological weapons, most experts say. The masks would have to be worn full-time since there is no way to know when an attack might take place, and most disease agents are odorless and colorless. Widespread preventive vaccination against some biological weapons may be possible and may be recommended at some point, but individuals should balance the low but real risk of a negative reaction to the vaccine against the likelihood of exposure to a biological agent, which in most cases is extremely low.

If an outbreak of any disease begins, you can protect yourself simply by not shaking hands, by wearing gloves, or by washing your hands frequently. And a person known to have the disease can limit transmission by wearing a simple paper or cloth mask.

FOR MORE INFORMATION

Centers for Disease Control and Prevention
www.bt.cdc.gov
1600 Clifton Road
Atlanta, GA 30333
www.cdc.gov/hoax_rumors.htm (hoaxes relating to bioweapons)
Public Inquiry Hotline: (888) 246-2675

Center for the Study of Bioterrorism and Emerging Infections
St. Louis University School of Public Health
www.bioterrorism.slu.edu
3545 Lafayette, Suite 300
St. Louis, MO 63104
(314) 977-8257

_____

This chapter was reviewed by Robyn Pangi, Research Associate, Executive Session on Domestic Preparedness at Harvard University's Kennedy School of Government and coauthor of "Preparing for the Worst: Mitigating the Consequences of Chemical and Biological Terrorism," by Greg Koblentz, who studied these issues for years at the Executive Session on Domestic Preparedness at the Kennedy School of Government; and by Greg Evans, Director of the Center for the Study of Bioterrorism and Emerging Infections, St. Louis University School of Public Health.

23. CARBON MONOXIDE

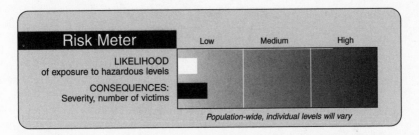

Risk Meter	Low	Medium	High
LIKELIHOOD of exposure to hazardous levels			
CONSEQUENCES: Severity, number of victims			

Population-wide, individual levels will vary

YOU HEAR about these tragic accidents from time to time on the news: a family overcome in their sleep, a motorist found dead in her car, a group of skaters taken from a local ice rink by ambulance, half-conscious. The cause is carbon monoxide, an odorless, colorless, tasteless gas that is around us all the time. Fortunately, levels are usually quite low. But if anything that burns gas or oil, coal or wood, malfunctions, or even if the combustion is fine but the fumes are not ventilated, higher levels can build up and carbon monoxide can become deadly.

THE HAZARD

Carbon monoxide, CO, is produced by the incomplete burning of carbon-based fuels, including gasoline, natural gas, kerosene, oil, propane, coal, or wood. Forest fires and volcanoes produce carbon monoxide too. Incomplete burning occurs when there is not enough oxygen to allow all the fuel to be consumed. That produces CO, a molecule with one carbon atom and one oxygen atom. Breathing CO at high levels can be dangerous because it reduces the ability of our blood to supply oxygen to the body: it outcompetes oxygen for space on the hemoglobin molecule in the blood. Hemoglobin is the substance that binds to oxygen molecules and carries them from the lungs out to our tissues to supply our cells. (Hemoglobin is also what makes blood red.) But the receptors on the hemoglobin molecule that bind to oxygen in the air we breathe will also grab the oxygen atom in the carbon monoxide molecule. Since CO is 200 times stronger at binding to these receptors than simple oxygen, when CO and regular oxygen are both present, the CO will win and occupy the hemoglobin receptor. So when CO is present, the

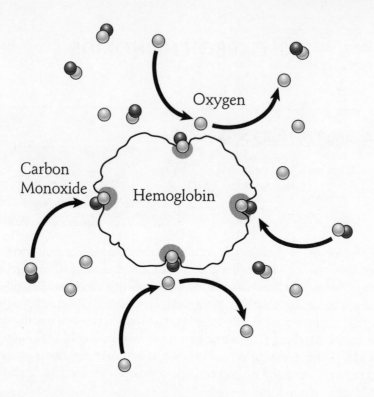

oxygen receptors on the hemoglobin are taken up by CO and the oxygen we inhale isn't picked up and carried to the rest of the body. The effect of carbon monoxide, then, is to deny oxygen to the body. It causes injury, or death, in the same way suffocation does.

THE RANGE OF CONSEQUENCES

According to the National Safety Council, an estimated 700 to 1,000 Americans die each year from accidental CO poisoning, and CO poisoning is responsible for about 10,000 to 40,000 emergency room visits per year. Among severely CO-poisoned victims who recover consciousness and survive, oxygen deprivation causes permanent neurological damage in approximately half of them. Even at the lowest levels of CO poisoning, permanent damage to organs particularly dependent on oxygen such as the heart and the brain may occur. Some studies have shown that low levels of CO poisoning can also damage the central nervous system, which can lead to partial or full paralysis. Other long-term effects include memory loss, personality changes, and neurological or psychiatric problems.

Fortunately, most CO poisoning is mild and produces far less dramatic effects. The most typical symptoms of low-level CO poisoning, including headaches, nausea, confusion, sleepiness, and dizziness, can be reversed if they're caught in time. Often, however, treatment is delayed in cases of low-level exposure, because patients and medical providers confuse these symptoms with flu or other common illnesses.

Damage from CO depends not only on how much you breathe, but also for how long. For instance, when people are said to be "overcome by smoke" in a fire, they are in fact overcome by high levels of CO, which, in a fire, can range from 10,000 to 50,000 parts per million (ppm) of air. At this concentration, with just a few breaths a person's blood CO level can reach almost 100 percent, leaving no room for oxygen in the blood and leading quickly to unconsciousness or death. On the other hand, cigarette smokers expose themselves to 1,000 to 2,000 ppm of carbon monoxide with every puff. But smokers don't lose consciousness because their exposure to this level of CO is usually only about 10 minutes per hour, and they inhale fresh air while smoking. Cigarette smokers typically have blood CO levels just below the concentration officially considered to be CO poisoning. These levels will reduce oxygen supply to the body but aren't enough to make smokers feel sick.

Another factor that helps determine how CO affects us is altitude. The higher you are, the more dangerous CO becomes. Partial pressure of oxygen at higher latitudes is less than the partial pressure of CO. (Partial pressure is the gas pressure that pushes the gas out of the lung tissues and into the bloodstream.) So at higher altitudes, not only does CO bind more successfully with hemoglobin molecules once it gets into the bloodstream, but it has a competitive advantage in getting there in the first place. For example, if you're at 4,000 feet above sea level and there are 50 ppm of CO in the air, your blood CO concentration will be 1 percent higher than it would be with the same CO exposure at sea level. Also, because there is less oxygen in the atmosphere the higher up you go, the physical effects of CO are more pronounced at higher elevations. At 14,000 feet, a 3 percent concentration of CO bound to hemoglobin feels like a 20 percent concentration at sea level.

An important factor that determines the health consequences for severely poisoned patients is how long it takes to receive treatment because while the CO levels in your blood stay high, you are slowly suffocating. Simpler cases of low-level exposure are treated just by giving the patient supplemental oxygen. But in about 5 percent of the cases treated in hospital emergency departments, levels are so high that the patient requires treatment with special equipment. The patient is placed in a hyperbaric chamber, the device used to treat scuba divers with "the bends." The patient is subjected to higher-than-normal atmospheric pressure and to pure oxygen, which helps force the CO off the hemoglobin, allowing it to resume its function of carrying oxygen from the lungs to the rest of the body. There are 400 to 450 hyperbaric treatment centers in North America. If you are suffering CO poisoning, seek immediate medical attention and let trained authorities decide on the proper course of treatment. Delaying treatment by trying to drive to the nearest hyperbaric treatment center could do more harm than good.

CO is measured in the body as a concentration of carboxyhemoglobin (COHb) in the blood. COHb is the molecule formed from the combination of CO and hemoglobin. High levels of COHb in the blood can result in loss of consciousness and death within minutes.

RANGE OF SYMPTOMS FROM CO POISONING

Blood COHb Level	Range of Symptoms
≤10 percent	May have no symptoms. May have headaches. In people with heart and respiratory disease, chest pains
20 percent	Dizziness, nausea, fainting, headache, fatigue
30 percent	Visual disturbances, confusion
40 percent	Confusion, loss of consciousness
50 percent	Seizures, coma
60 percent	Death

THE RANGE OF EXPOSURES

As we've stated, carbon monoxide is in the air we breathe nearly all the time. Poisoning occurs when emission sources produce large amounts and when CO accumulates in closed spaces, such as homes, offices, warehouses, barns, vehicles, or garages. The biggest exposure source is stationary motor vehicles. In fact, CO from motor vehicle exhaust is the most common cause of all accidental poisoning deaths in the United States. From 1979 to 1988, 11,547 deaths from CO were caused by motor vehicle exhaust. Eight in 10 of these cases involved vehicles that were parked. Most stationary motor vehicle–related CO deaths occurred even though the garage doors or windows where the vehicle was parked were open.

Indoor exposures to CO come from any appliance that uses gas, oil, wood, kerosene, or propane in devices including furnaces, water heaters, stoves, lamps, or work machinery. In some cases, high levels of CO are emitted if the appliance is not working properly. In other cases, lower levels of CO emitted from appliances that are working well can build up in spaces that don't have adequate ventilation.

CO exposure is a frequent problem in small indoor ice rinks, where ice-cleaning equipment, which runs hourly as many as 16 times a day, often isn't equipped with pollution control equipment, and where rink operators don't maintain adequate ventilation. Large arenas are not a CO exposure risk because the ice-cleaning equipment in these rinks operates only a few times a day and the exhaust from the equipment mixes with a much larger volume of air. These larger arenas also usually have much better ventilation.

Another source of indoor exposure is cigarette smoke, which is about 4 percent CO. Smokers can have a COHb level of up to 9 percent, four to five times higher than nonsmokers.

Carbon monoxide is an outdoor air pollutant regulated by the Environmental Protection Agency. Levels across the country are generally very low. Outdoor CO concentrations in urban areas are higher than in nonurban areas, mainly due to the higher concentrations of automobile exhaust, which contributes about 60 percent of all outdoor CO emissions nationwide.

REDUCING YOUR RISK

To reduce your risk of CO poisoning, you should prevent accumulation of CO in any indoor space and install equipment to detect levels that become elevated.

To prevent CO accumulation:

- Avoid operating fuel-powered engines indoors.
- Don't leave your car, lawn mower, gasoline engines or generators, or engine-powered tools and appliances idling indoors, even with garage doors and windows open.
- If you warm up a vehicle in cold weather, pull it out of the garage immediately after starting it up. CO can leak into homes through

cracks, vents, windows, and doors if the garage is attached to the home, even if the garage door is open.

- Gas-powered appliances such as furnaces, water heaters, stoves, or space heaters should be serviced by a qualified technician at least once a year. If the flame on your gas stove burns yellow, CO is being produced, and your burner needs a tune-up or a better supply of oxygen. A well-tuned flame burns blue.
- Have all fireplaces and chimneys inspected for cracks, leaks, or blockages once a year before cold-weather use. Ventilation ducts on furnaces and water heaters should be cleaned and operating properly.
- When camping, remember that recreational vehicles and tents are indoor spaces. They are a confined area in which CO can accumulate. Try to use only battery-powered heaters and lights in tents. If you use a device that burns fuel in a tent, make sure there is adequate ventilation and that the heater or stove is operating efficiently.

To detect CO, the Consumer Product Safety Commission recommends installing at least one carbon monoxide detector per household, preferably near the sleeping areas. Additional detectors on every level of a home and in every bedroom provide extra protection. CO detectors can be hard-wired to your home's electrical system, plugged into a standard electrical outlet, or battery-operated. The advantage of a battery-operated model is that it will continue to operate in the event of a power outage, which could also cause heating equipment in your home to malfunction and create a CO threat. Some hard-wired models come with battery backup. CO detectors generally sound an alarm when the concentration of CO corresponds to a 10 percent COHb level in the blood.

Should your CO alarm go off, check to see if anyone is experiencing symptoms. Remember, the symptoms of CO poisoning are headache, fatigue, dizziness, and nausea. If anyone feels symptoms, evacuate the building, leaving doors open as you go, and seek immediate medical care. If no one feels symptoms, open windows and doors, shut off all heating and cooking appliances, and call a qualified service technician to check your appliances. Be on the lookout for symptoms and follow the above instructions if any appear.

FOR MORE INFORMATION

National Safety Council
www.nsc.org/ehc/indoor/carb_mon.htm
1121 Spring Lake Drive
Itasca, IL 60143-3201
(630) 285-1121
Air Quality Hotline: (800) 557-2366

U.S. Environmental Protection Agency
Indoor Environments Division
www.epa.gov/iaq
1200 Pennsylvania Avenue NW
Mail Code 6609J
Washington, DC 20460
(202) 564-9370

Consumer Product Safety Commission
www.cpsc.gov
4330 East-West Highway
Bethesda, MD 20814-4408
(301) 504-0990

This chapter was reviewed by Ron Hall, Industrial Hygienist with the National Institute for Occupational Safety and Health, who has worked extensively on CO issues.

24. DDT

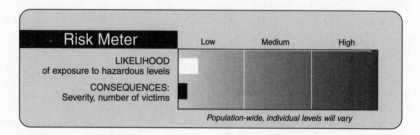

FOR MANY PEOPLE, DDT embodies everything worrisome about environmental risks from man-made modern industrial chemicals. But it also offers some valuable lessons about environmental risk, because the story behind DDT is not as simple as most people assume.

DDT was first synthesized in 1874, but it wasn't until just before World War II that its insecticidal properties were discovered. The "discovery of the high efficiency of DDT as a contact poison against several arthropods" won Paul Müller the 1948 Nobel Prize in Medicine and Physiology. DDT quickly became one of the most effective tools in the public health arsenal, saving millions of lives around the world as it was widely used beginning in WWII to control insectborne diseases like malaria and typhus.

Use peaked in 1962 when 37 million pounds were used around the world, both in agriculture to control insects and to control the mosquitoes that spread disease. DDT was so successful that a U.S. National Academy of Sciences panel wrote in 1970, "To only a few chemicals does man owe as great a debt as to DDT . . . In little more than two decades, DDT has prevented 500 million human deaths, due to malaria, that otherwise would have been inevitable." Yet just two years later, the United States government banned DDT because of concerns about its effects on wildlife, and possibly on humans.

THE HAZARD

Dicholorodiphenyltrichloroethane is a member of the organochlorine family of chemicals: "organo" because it contains carbon and "chlorine" because there are chlorine atoms in the molecule. Organochlorines have properties that cause concern when they are released in the environment. The chlorine in them makes them very stable, meaning they don't readily break down in the presence of sunlight, rain, or microorganisms. So DDT builds up in the environment.

DDT doesn't readily break down in the body either. Its half-life in the human body is approximately eight years, which means it takes eight years for half the DDT in the body to metabolize, then another eight years for half the remaining amount to be removed, and so on. In addition, DDT is lipophilic, which means it tends to build up in fat cells and be stored in fatty tissues. These characteristics create biomagnification in which DDT levels increase up the food chain. Insects eat small amounts, then small fish eat numerous insects and get a larger dose. Big fish feeding on lots of small fish receive an even larger dose, and predatory birds feeding on big fish get a larger dose yet. The amount of DDT that accumulates in an osprey or eagle is substantially greater than that found in the individual fish they eat.

DDT's apparent harm to birds brought this now notorious chemical to the public's attention as a threat, raising the possibility that what it was doing to birds it might be doing to humans. Publication of Rachel Carson's *Silent Spring* in 1962 cited the connection between DDT and thinner-than-normal eggshells of exposed birds. These defective shells tended to break in the nest. That was apparently contributing to the dramatic decline of some bird species. Extensive research has subsequently confirmed that DDT indeed has this effect, though more on certain bird species than others. Birds of prey seem most sensitive, while other species show little susceptibility. Birds of prey are likely to have greater exposure to DDT than other species because of the effect of biomagnification for species higher on the food chain.

In the wake of *Silent Spring,* which helped give rise to the modern environmental movement in the United States, the use of DDT was banned in the United States in 1972 in one of the first actions of the then new Environmental Protection Agency (EPA). But nearly 30 years later, because DDT is such a persistent substance, it can still be found in soils all over the country, though levels have dropped dramatically. Many scientists believe that reductions in the use of DDT have been responsible for the resurgence of eagles, peregrine falcons, and other predatory birds, particularly those that feed primarily on fish.

THE RANGE OF CONSEQUENCES

When insects come into contact with DDT, it kills them by interfering with their nervous system. But for humans, DDT is relatively nontoxic for acute exposure. In fact, it was once tested as a treatment for jaundice with no acute side effects. In very large amounts, far beyond what we are normally exposed to, DDT affects the nervous system. People who accidentally swallowed large amounts of DDT had tremors and seizures, both of which ended soon after the exposure stopped. People who in one early test took small daily doses of DDT by capsule for 18 months showed no effects. People who had higher-than-normal exposure to DDT in the workplace for a long time showed changes in the levels of their liver enzymes, effects that also ended once exposure stopped.

But most of the concerns about DDT are not for short-term, high-level exposures. The main concerns for human health, such as cancer and reproductive effects on newborns, are for chronic low-level exposure over an extended period. Some studies on rats exposed to very high doses of DDT found elevated rates of liver tumors and, in one study, lung tumors. In other studies on rats, also fed very high doses, no increase in cancer was observed. (Liver tumors are the most common tumors induced in rodents by chemicals, but liver cancer is very rare among people in the United States.) Based on findings of increased tumors in some of those rodent tests, the EPA calls DDT a "probable human carcinogen." These cancer effects have not shown up in studies of workers exposed to DDT, although those sample populations were not big enough to detect a small increase in risk if it did exist.

Some researchers also think DDT might contribute to breast cancer. Since DDT can, under some laboratory conditions, mimic the actions of the hormone estrogen in human cells, scientists theorize that it may be responsible for some cases of breast cancer. Results of studies investigating this idea are mixed, with some finding an effect, some not. For example, a study in New York City found higher average levels of a derivative of DDT—DDE—in the blood of women with breast cancer compared to those without disease. But more recent and larger studies have found no relationship between breast cancer and blood levels of DDT. In general, epidemiologic studies don't support the idea that DDT is a major factor in breast cancer.

More recently, wildlife researchers have found evidence that DDT acts like estrogen in animals in natural environments where DDT or its chemical cousins, DDE and DDD, have been detected at high levels. In one case, exposure to alligators is thought by some scientists to have caused demasculinization of males, abnormally high percentages of female offspring, and abnormally fragile eggshells. Preliminary research on humans suggests the possibility that DDT may impair immunity in babies exposed to high levels in the womb or from breast milk. Laboratory studies in rodents support the theory that very high doses of DDT can alter reproduction by acting like an estrogen. (For much more on the theory of environmental chemicals interfering with the human hormone system, see Chapter 26, "Environmental Hormones.)

THE RANGE OF EXPOSURES

Because DDT is persistent in the environment, and because it was used so widely for decades, both in the United States and around the world, scientists think that almost all of us carry some minute amount of this compound in our bodies. In this country, ongoing low levels of exposure are thought to come from eating fish grown in contaminated waters, or from meat or dairy products from cattle eating forage from contaminated ground. The tendency of DDT to bind to fat explains its presence in dairy foods. It is estimated that the average person in the United States is exposed to about 2 micrograms of DDT a day. (A microgram is one millionth of a gram. A gram is one twenty-eighth of an ounce.) This level is about 20 times lower than the level the EPA suggests is safe to protect against noncancer effects.

The U.S. Food and Drug Administration has set DDT levels for many foods, and government surveys show the amounts in what we eat are well below those standards. Levels of DDT in the average U.S. resident have been dropping since it was banned in 1972.

Nursing infants can be exposed to very low levels of DDT in their mother's milk. Nonetheless, it is uniformly recommended that babies be breastfed because the benefits of nursing far outweigh the risks of DDT.

REDUCING YOUR RISK

As we said, the story of DDT is not as simple as many people assume. It may be surprising, given DDT's reputation, but there is little evidence of human health effects at current levels of exposure. Exposures to DDT continue to decline in the United States. As we mentioned, ongoing exposure comes from fish, meat, or dairy products. But since levels consumed in food are in general very low, changing your diet will have practically no effect on what is already a very low risk.

FOR MORE INFORMATION

Agency for Toxic Substances and Disease Registry
Division of Toxicology
www.atsdr.cdc.gov/tfacts35.html
1600 Clifton Road NE, Mailstop E-29
Atlanta, GA 30333
(888) 422-8737
Fax: (404) 498-0057

Environmental Protection Agency
www.epa.gov/history/topics/ddt/02.htm
Ariel Rios Building
1200 Pennsylvania Avenue NW
Washington, DC 20460
(202) 260-2090

Science News July 1, 2000, article by Janet Raloff, "The Case for
    DDT: What Do You Do When a Dreaded Environmental
    Pollutant Saves Lives?"
www.malaria.org/raloff.html
1719 N Street NW
Washington, DC 20036
(202) 785-2255

---

*This chapter was reviewed by Francine Laden, Instructor at Harvard Medical
School and author of several studies about links between breast cancer and environmental factors, including DDT; and by Dr. Barbara D. Beck, Risk Assessor, Toxicologist, and Principal at the Gradient Corporation.*

25. DIESEL EMISSIONS

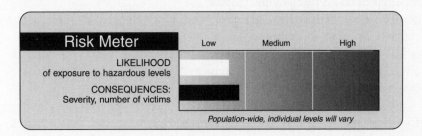

MOST PEOPLE who have driven behind a bus or truck have probably seen the smoke and smelled the smell of diesel exhaust. That pollution is more than dirty to look at. It's a threat to public health.

In 1892, Rudolf Diesel came up with the engine that bears his name. The idea was to use the unique properties of diesel fuel to develop an engine with high efficiency. Diesel fuel offered that opportunity because it's made of heavier molecules with more carbon atoms in longer chains than gasoline. Because the average gallon of diesel fuel contains 17 percent more energy than a gallon of gas, diesels power nearly all heavy vehicles: large trucks, buses, bulldozers, construction and agricultural machinery, ships, and trains. Approximately 80 percent of all the freight in the United States moves by truck, and virtually all large trucks use diesel engines.

THE HAZARD

It's not just the fuel, but also the way they burn it, that gives diesel engines their extra power and explains why they emit the pollutants they do. A gas engine sprays a uniform mist of gasoline and air into a chamber, compresses the mixture, and ignites it with a spark. The diesel engine fills the chamber with only air. The air is then compressed, which heats it. Then the diesel fuel is injected into the chamber, hits the heated air, and ignites. This system allows the cylinders in a diesel engine to compress the contents in the cylinder—the air—more than a car compresses its air-fuel mix. Compression in a diesel engine can be 75 to 100 percent greater than in a gas engine. More compression gives diesel engines more power.

Diesels have lower emissions of some pollutants than cars, including carbon dioxide, a key greenhouse gas. But because diesel fuel doesn't mix as uniformly with the air in the cylinders as gasoline does, diesel combustion produces more particulates than gasoline combustion. Diesel engines could be designed to produce fewer particles, and emissions controls could reduce how many of these particles actually make it out the exhaust pipe. But emissions from diesel engines have been less regulated for the last 30 years, while regulations have been tightened on gasoline engines. So diesels have become a disproportionate emitter of certain types of air pollution.

Three types of pollution emitted by diesel engines have raised the most public health concern: particles, nitrogen oxides, and toxic chemicals, which are associated with chronic health effects, including cancer.

Particles are simply microscopic bits of solid or liquid matter in the air we breathe. They are very small, measured in microns. A human hair is 70 microns across. There are a million microns in a meter. Particles are associated with a wide range of health problems and, as we will discuss in more detail, may contribute to the death of tens of thousands of Americans each year.

Nitrogen oxides (NOx) are byproducts of the burning of fossil fuels. They are emitted by power plants and all motor vehicles, but diesel combustion is a significant source of NOx. NOx reacts with other chemical pollutants, in the presence of sunshine, warmth, and stagnant air, to form ground-level ozone, also known as smog. Smog is associated with a host of respiratory problems, and may be associated with death from respiratory and cardiovascular disease.

Air toxics are compounds in diesel exhaust that include formaldehyde, acetaldehyde, acrolein, benzene, and polyclyclic aromatic hydrocarbons (PAHs), which are associated with elevated cancer risk and other chronic effects.

Diesel engines are a significant source of all three of these emissions. Trucks and buses account for two thirds of all particulate pollution from transportation sources in this country—131,000 tons per year—even though trucks account for only 6 percent of highway vehicle miles in the United States. Diesel engines produce at least 6 percent of *all* particulate emissions nationwide, but in some areas it can be as high as 36 percent. Diesel combustion accounts for one third of all NOx emissions in the United States: 10 percent is from trucks and buses; 23 percent is from off-road vehicles, including farm and construction equipment. Diesel engines that power farm and construction equipment, which have been largely unregulated until recently, produce even more emissions per vehicle than trucks and buses, which have been regulated more.

THE RANGE OF CONSEQUENCES

The consequences of diesel exhaust come from exposure to particle pollution, smog, and air toxics. (See "The Range of Consequences" in Chapter 20.) Remember as you read that diesel exhaust is just one source of these pollutants.

Numerous studies suggest that diesel exhaust increases cancer risks. Exposure to very high levels of diesel exhaust causes lung tumors in rats when they are exposed every day for their entire lives. Studies of people routinely exposed to occupational levels of diesel fumes—garage workers, miners, and those who work with trucks, railroads, and docks—suggest an association between diesel exhaust and cancer in humans too. The World Health Organization, the U.S. National Institute for Occupational Safety and Health, and the International Agency for Research on Cancer classify diesel exhaust as a probable or potential human carcinogen.

In addition, diesel emissions can produce short-term, acute effects, including eye, throat, and bronchial irritation, lightheadedness and nausea, and respiratory problems, which can often be chronic. No data exist to quantify the prevalence of these effects.

THE RANGE OF EXPOSURES

Exposure to diesel emissions depends on a number of factors. The particulate emissions from diesel engines occur in three different basic sizes, known as coarse, fine, and ultrafine. The larger coarse particles, though they are microscopic, don't carry as far on light winds. Concentrations of coarse particles are greater in areas with high traffic like cities or along heavily used highways or railways, and in areas where weather conditions tend to concentrate these pollutants, like Denver and Los Angeles. The smallest of the particles, the ultrafines, bump into one another and aggregate into larger particles after a few hours, so ultrafine concentrations are also higher near emission sources like roads. The medium-sized diesel particles, called fine, travel well on any kind of wind and are found across a broader area. The NOx and air toxic emissions from diesels also travel farther. But it is thought that most exposure to air toxics in diesel exhaust occurs when these chemicals adhere to particles that we breathe in. If this is so, air toxic exposure from diesels is greater where particle concentrations are higher.

As of the end of 2001, there was no legal standard for particulate emissions from diesels, but the Environmental Protection Agency (EPA) was considering a standard of 15 micrograms of diesel particulate matter per cubic meter of air (μg/DPM/m^3). It's hard to measure the current levels, but EPA studies show annual average exposures typically range from 0.5 to 1 μg/DPM/m^3. In some urban areas, average exposures were as high as 4 μg/DPM/m^3, and the EPA estimates that levels could occasionally exceed the proposed federal standards in urban places like busy intersections, rail yards, or bus stations.

In 1999 the EPA issued new regulations that would require all light-duty diesel vehicles to meet the same tight emission standards as gasoline vehicles by 2005. Then, in 2000, additional EPA regulations required that all heavy-duty diesel vehicles achieve a 90 percent reduction in particle emissions and a 90 percent reduction in NOx. The EPA estimates these new rules for heavy-duty vehicles would eventually prevent 8,300 premature deaths a year, 5,500 cases of chronic bronchitis, 17,600 cases of acute bronchitis in children, and more than 360,000 asthma attacks. In terms of eliminating air pollution, the agency says it would be the equivalent of taking 13 million of today's trucks off the road. The EPA also required fuel manufacturers to reduce the sulfur content in diesel fuel by 97 percent. These restrictions were to be phased in, taking final effect in 2009.

But the rules were embroiled in controversy as of the end of 2001. The Bush administration announced it would reexamine the timetables for implementation pending an EPA report on the progress of diesel engine manufacturers and oil companies in coming up with cleaner-burning equipment and lower-sulfur fuel. Lawsuits were filed against the rules by engine manufacturers and oil refiners.

REDUCING YOUR RISK

In most urban environments, diesel exhaust, which contains particulates, is ubiquitous and inescapable. But it is higher in localized areas near emissions sources like railroad yards, highways, bus stations, and construction sites. In some cities and states, legal action by neighbors of these areas has forced operators to turn off diesel engines instead of leaving them idling. Operators argue that they have to keep the train, bus, and truck engines running to keep the diesel fuel warm, because the thicker, heavier fuel makes it harder to start diesel engines in cold weather, but batteries and other devices to warm the fuel have solved this problem.

The emissions from diesel engines are likely to drop dramatically over the next several years. Many of these improvements will come from what is called clean diesel technology. In the face of proposed EPA regulations to reduce emissions, manufacturers are designing changes in diesel engines that would mix the fuel with the compressed air in the cylinders more uniformly, reducing the formation of particles in the engine. Emissions controls, including filters and catalytic converters, are being designed to reduce particle and NOx emissions from the exhaust pipe. These technologies would also take advantage of the reduction in the sulfur content of the fuel. When (if) the new diesel emissions reduction rules are phased in, for the first time diesel engines will have to meet the same basic emission criteria gasoline engines do.

FOR MORE INFORMATION

Office of Transportation and Air Quality
Environmental Protection Agency
www.epa.gov/otaq
Ariel Rios Building
1200 Pennsylvania Avenue NW
Washington, DC 20460
(734) 214-4333

DieselNet (sponsored by diesel manufacturers and emissions
    control companies)
www.dieselnet.com

National Safety Council
www.nsc.org/ehc/mobile/airpollu.htm
1121 Spring Lake Drive
Itasca, IL 60143-3201
Air Pollution Hotline: (800) 557-2366

This chapter was reviewed by Jonathan Levy, Assistant Professor of Environmental Health at the Harvard School of Public Health; by Daniel S. Greenbaum, President of the Health Effects Institute and former Commissioner of the Massachusetts Department of Environmental Protection; and by Maria Costantini, Toxicologist and Principal Scientist, Health Effects Institute.

26. ENVIRONMENTAL HORMONES

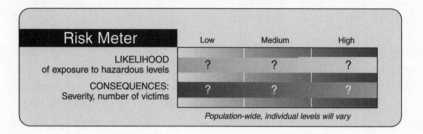

Risk Meter	Low	Medium	High
LIKELIHOOD of exposure to hazardous levels	?	?	?
CONSEQUENCES: Severity, number of victims	?	?	?

Population-wide, individual levels will vary

THE POST-WORLD WAR II industrial world provides us with a remarkable range of goods and services which greatly improve human health and quality of life. Everything from soda in cans to Velcro to plastics to computers to most modern pharmaceuticals, and so much more, are inventions of the past four or five decades. During these years, life expectancy has risen, agricultural productivity has increased, air and water pollution have declined, and scientific discovery has accelerated. But a byproduct of this progress has been exposure to new chemicals, and recently some scientists have suggested that these chemicals might interfere with the intricate hormone systems our bodies need for proper development and function. So far the evidence is sketchy, limited mostly to wildlife, and most experts who have reviewed the research say they're unconvinced. But supporters of what is known as the "endocrine disruption theory" interpret the clues from birds and fish and reptiles as a pattern of what might be happening to us.

THE HAZARD

Hormones are key biological messengers in any living organism. They are chemicals, secreted by certain glands, which are carried by the blood to special target cells. When they bind to those cells at special receptor sites, hormones prompt those cells into a certain kind of activity. Together the hormones, the glands that secrete them, and the receptor cells where they bind are known as the endocrine system, a vital part of how our bodies regulate much of what makes life possible. You may be familiar with hormones such as estrogen and testosterone or insulin or melatonin. And some familiar endocrine glands include the thyroid, the pituitary, the pancreas, and the ovaries.

Our various hormone levels go up or down constantly, depending on internal and external conditions. Incredibly small variations in these levels can have significant impacts on everything from heart rate to the immune system to emotions to the ability to sleep to reproductive health—almost everything connected with biological function. Hormones signal hour-to hour, month-to-month, and year-to-year biochemical changes in organisms throughout the plant and animal kingdoms. And with hormones, timing is critical. In general, the effect of a hormone is as much a matter of when it is produced and when it binds with receptor cells as it is a simple matter of how much hormone there is.

The endocrine disruption theory suggests that hormonelike compounds, which we eat, drink, or inhale—in other words, hormones that we take in from the external environment—can interfere with the delicate dynamic of our natural endocrine system in ways that threaten human health. These environmental hormones are suspected of raising or lowering hormone levels at times that aren't in synch with our natural rhythms and requirements. This can be particularly harmful during fetal development, when small amounts of specific hormones, present at very specific times during development, are vital to the proper development of the body, its internal organs, and its biochemical systems.

∼

During the early 1940s, Australian sheep farmers puzzled over occasional epidemics of stillborn lambs and unexpectedly high numbers of ewes that failed to conceive. Research ultimately traced the problem to the animals' forage. The clover they had been eating contained natural chemicals that mimicked the biological activity of estrogen, the primary female sex hormone. Since then at least a dozen of these phyto— or plant—estrogens have turned up in nearly 300 different plants, from soy, garlic, and wheat to chickpeas, ginseng, and apples. No one knows why plants produce them. However, when eaten in large quantities, these compounds have been found to affect reproduction in cattle, pigs, rodents, birds, and fish.

In the late 1970s, wildlife toxicologists began reporting other inexplicable phenomena, such as aberrations in the gonads of birds. Some researchers described finding "intersex" birds—males possessing ovarian tissue. Some female birds appeared to also have been overfeminized—

producing "superfemales." They possessed an excess of female gonadal tissue, and some even exhibited behavioral abnormalities such as a desire to nest with other females. In these cases it turned out the most likely cause was not phytoestrogens, but an accumulation of a man-made chemical, DDT, in the birds. It turned out that DDT and one of its breakdown products, DDE, can bind to receptors on endocrine target cells which normally bind with androgens, the male sex hormones.

Some time later, scientists studied a population of alligators in Florida exposed to a spill of chemicals that included a pesticide, dicofol, and its breakdown products. Some male alligator babies were feminized to the point that they developed what looked like ovaries. The size of their penises was far smaller than normal. The changes seemed to reflect a shortage of testosterone, the primary male sex hormone in these animals. Scientists speculated that something, possibly the chemical exposure, was blocking the androgen production in the alligators. Still other wildlife studies found similar abnormalities in fish in several places.

In 1991, a small group of researchers who had each detected similar results in separate studies met and agreed that the evidence suggested a pattern, and that environmental hormonelike compounds were probably the cause of the damage. They called for research to study effects in humans.

After nine years of research, in 2000, the leading experts in this field wrote a review for the National Academy of Sciences (NAS). The experts found that the evidence for the endocrine disruption theory is still sketchy for wildlife, and not proven at all for humans. They agreed that the clues from wildlife suggest the possibility that exposure to external hazards might affect the wildlife's hormone activity. And they agreed that in a small number of cases in which humans have been exposed to certain chemicals, the evidence suggests that problems, including low birthweight, decreased IQ, and premature birth may be associated with the hormonelike effects of those chemicals. The experts agreed that the theory should be taken seriously and research should be accelerated.

But they agreed on little else. Their main finding is that a great deal of doubt remains about the endocrine disruption theory. The experts had deep differences of opinion, a testament to the uncertainties that remain about the theory. Their executive summary reflects those disagreements: "Some of the differences reflect areas where additional research would help; others reflect differing judgments about the significance of the existing information." The experts also wrote: "(We have) . . . differing views of the value of different kinds of evidence obtained by experiments (and) observations . . ." And the experts said they weren't the only ones struggling this way. Their summary says, "The differences are not confined to this committee but are reflected in the scientific community at large."

THE RANGE OF CONSEQUENCES

Most of the theoretical human consequences of environmental hormones and endocrine disruption result from exposure of the developing fetus to hormonally active agents. Herein arises much of the scientific uncertainty and disagreement, because fetal development is controlled by incredibly small levels of hormones that are hard to understand and hard to study. Tiny variations in the level of hormones can have dramatic effects. And even if the levels are normal, fetal development depends on exquisite timing of those exposures. The right amount of hormone has to present at just the right time. And the time window for these exposures is very short, often only a few critical days or weeks as the fetal heart or brain or gonads or immune system are just starting to develop. The right amount of just the right hormone has to be present at just the right time for healthy fetal development.

These subtle conditions are challenging for toxicologists to recreate and test, and then retest to verify their findings. Some critical experimental findings showing these low-dose effects have been difficult to replicate and confirm. One thing that helps is that hormones are highly conserved across species, which means they're nearly identical in lab animals and humans, making animal test results highly relevant to humans. But as of this writing, while great uncertainty remains, the scientific evidence does not support the theory that low-dose exposure to these environmental hormones has significant effects on humans.

What follows are some of the potential effects of environmental hormones on humans. Most are the theoretical effects that children might suffer if they are exposed in utero during fetal development.

Reproductive Effects

Since many of the effects in wildlife have been abnormalities in reproductive organs or fertility of offspring, that's the kind of effect most research on humans has looked at so far. Perhaps most significant of all the postulated human effects, according to supporters of the endocrine disruption theory, is evidence of falling sperm counts among men in certain countries. But other research has questioned those findings, and challenges the basic idea that sperm counts are going down. The NAS report says: "[R]etrospective analyses of trends over the past half-century remain controversial. When the data from large regions are combined and analyzed, some data sets indicate a statistically significant trend consistent with declining sperm concentrations. However, aggregation of data over larger geographic regions might not be an appropriate spacial scale for this analysis, given the significant geographic heterogeneity."

Supporters of the theory also suggest that in utero exposure to environmental hormones might cause cryptorchidism, a condition in which the testes in young boys do not descend from the body cavity into the scrotum. This can lead to reduced fertility because the sperm-producing cells within the testes need the lower temperatures outside the body cavity to develop properly. The longer the testes stay inside the body cavity, the fewer of these sperm-producing cells will develop and the less fertile the man will be. Evidence suggests that rates of cryptorchidism are rising. So, apparently, is the prevalence of another condition dealing with the male genitalia called hypospadias, in which the urethra is not found at the tip of the penis but at the base or between the base and tip. But reviewing the evidence to date, the NAS experts found no convincing evidence that cryptorchidism or hypospadias is the result of environmental hormonally acting agents. However, they did note that lab tests of rats, mice, guinea pigs, and female rhesus monkeys find that in utero exposure to a number of hormonally acting compounds (including DDT, PCBs, dioxin, and others) can produce structural and functional abnormalities of the reproductive tract similar to hypospadias and cryptorchidism.

The one unfortunate and accidental endocrine disruption experiment on people involved the use of the prescription drug diethylstilbestrol (DES). DES acts like an estrogen and was given to an estimated 5 million pregnant women between 1940 and 1970 to decrease the risk of miscarriage. In the mid-1970s, the daughters of the women who took DES were found to have elevated rates of vaginal cancer. Subsequently, these women were found to have an increased risk of miscarrying themselves, at a late stage of pregnancy. Men born to mothers who took DES were found to have slightly elevated rates of malformed genitalia. Follow-up tests on mice found that DES influences a critical series of genes that control fetal development of parts of the reproductive system.

But the DES issue is a perfect example of the uncertainty over the whole theory of low-dose exposure to environmental hormones causing major problems in humans. DES is an extremely potent hormonally active agent, more potent than the natural female hormone estradiol on which it's based. Moreover, the prescribed DES doses for pregnant mothers were massive compared to those that would come from environmental sources. Still, despite the fact that women took these very powerful drugs at high doses during the critical early months of their pregnancies, their male children show very few effects (and no cryptorchidism or hypospadias). And their female children show only the effects of vaginal cancer and elevated rates of miscarriage. Practically none of the problems with sexual development occurring in wildlife have shown up in the children of DES mothers.

Neurologic Effects

Studies of pregnant women accidentally exposed to high levels of PCBs and pregnant women who ate foods contaminated with high levels of mixtures of PCBs, dioxin, and pesticides found suggestions that prenatal exposure to these hormonally active agents can cause subtle effects on newborns. These effects include reduced IQ, low birthweight, and premature birth. Mice and monkeys exposed to PCBs prenatally in controlled experiments also show reduced IQ and low birthweight. But many toxicologists point out that results from studies of humans are not always replicated when the tests are repeated and the animal results may reflect the fact that these compounds are toxic in other ways, and that these effects might not have anything to do with the fact that PCBs, dioxins, and pesticides might be hormone disrupters.

Immunologic Effects

Supporters of the endocrine disruption theory say that wildlife evidence suggests that environmental hormones damage the immune system. They cite evidence from the Arctic, where whales and seals are staples of the human diet and deliver concentrated doses of some environmentally persistent compounds like PCBs and DDT. Children there have elevated rates of some infections, and show less protective benefits from vaccines, which work by stimulating the immune system. The NAS report says: "In humans, data on the immunologic effects of hormonally active agents are inadequate to support any definitive conclusions." But the experts did agree that endocrine disruption, predominantly from PCBs, seems to cause immune suppression in birds in the Great Lakes and seals in the PCB-contaminated Baltic Sea.

Consequences in Adults

While most of the effects that have been postulated are to children due to exposure during fetal development, the endocrine disruption theory has also been suggested as a cause of breast cancer in adults because of the established relationship between exposure to estrogens and the elevated risk of breast cancer. But the NAS panel wrote: "An evaluation of the available studies conducted to date does not support an association between adult exposure to DDT, DDE, TCDD, and PCBs and cancer of the breast." Regarding other cancers, the experts found that not enough studies have been done to know whether endocrine disruption might be associated with the hormonally sensitive testicular, prostate, or endometrial cancers.

THE RANGE OF EXPOSURES

We are all regularly exposed to hormonally active agents in the food we eat, the water we drink, and the air we breathe, though these exposures are almost always at extraordinarily low levels. The list of common chemicals that may be hormonally active, according to the Tulane and Xavier Universities Center for Bioenvironmental Research, includes many widely used products and materials (normal exposure to these compounds is extraordinarily low):

- Pesticides such as DDT, endosulfan, dieldrin, methoxychlor, kepone, dicofol, toxaphene, and chlordane (most have been banned but residues remain in the soil)
- Herbicides such as alachlor, atrazine, and nitrofen
- Fungicides such as benomyl, mancozeb, and tributyl tin
- Ingredients in plastics like bisphenol A and phthalates
- Pharmaceuticals including estrogens like birth control pills, DES, and cimetidine
- Ordinary household products (breakdown products of detergents and some of the chemical additives that help them work, including nonylphenol and octylphenol)
- Industrial chemicals like PCBs, dioxin, and benzo(a)pyrene
- Heavy metals such as lead, mercury, and cadmium

Chemicals are not the only hormonally active agents to which we are exposed. At least in animals, the human papillomavirus (a common sexually transmitted disease agent), bright nighttime illumination, and certain types of magnetic fields have all been found to interfere with the function of natural hormones. And there is exposure to phytoestrogens from many fruits and vegetables, exposures usually much higher than from chemical or industrial sources.

In addition, we are exposed to low levels of man-made environmental hormones that are specifically manufactured for use as hormones. Humans excrete hormones, such as the estrogens contained in birth control pills and postmenopausal hormone replacement therapy. British scientists discovered male fish that demonstrated female biology (producing vitellogenin to nurture eggs); the cause was estrogen, apparently from birth control pills, in the sewage effluent of local streams.

Exposure to some of these agents continues long after we first inhale or ingest them. Many hormonally active chemicals are persistent in the environment and in our bodies, which means they don't break down quickly. Also, many of these compounds are lipophilic, which means they have an affinity to bind to fat cells. They are sequestered in parts of the body where we store fat, and the levels build up, a process called bioaccumulation. Over time, small amounts of these chemicals are released into the bloodstream as some fat cells are liberated for use as energy. This phenomenon is potentially problematic for endocrine disruption because, as we've explained, the timing of the body's hormone system is vital. It's not just how much of a hormone is present. It's also a matter of when it's present. If bioaccumulated environmental hormones are released in our bodies at the wrong times, particularly during critical periods during the first six months of fetal development, very low levels might interfere with proper fetal growth.

∿

Measuring the actual levels of the compounds we're exposed to is difficult, since dozens—if not hundreds—of chemicals and natural agents can be hormonally active. The NAS report includes a table listing potential daily human exposure to various estrogens, which is just one class of environmental hormones. The academy notes that these estimates are crude, and likely to be overestimates.

(Note the relatively high numbers for DES and phytoestrogens in the diet. Also note that while each of these compounds has been shown to be estrogenic, the potency varies dramatically. It may take 100,000 times more of an environmental estrogen to occupy an estrogen receptor on a hormone target cell.)

ESTROGENICALLY ACTIVE AGENTS IN THE ENVIRONMENT	
Source	Exposure Level ($\mu g/d$—millionths of a gram per day)
Oral contraceptives	20–50
Hormone replacement therapy	50–200
DES to prevent spontaneous abortion	5,000–150,000
Bisphenol A in food cans (used to line the cans)	6.3
Bisphenol A in beverage containers (used to line the cans)	<0.75
Bisphenol A in dental sealant (in the hour after application)	90–131
Nonylphenol (an ingredient in some soaps, detergents, and spermicides) in river water	0.01
PCBs in total diet	0.002
Phytoestrogens in 110 grams of wheat	200
Phytoestrogens in total diet	1,000,000

Source: National Academy of Sciences, 2000

Low levels of several hormonally active agents have been detected by several studies in the blood, urine, and tissue of people around the globe, even in nonindustrialized countries where these chemicals are never used. Because hormonally active agents are so common in the products and foods of our modern world, continued exposure to trace amounts is nearly certain for all of us. Determining the actual levels of exposure is one part of the government's accelerated research into the endocrine disruption theory. Under congressional orders, the Environmental Protection Agency is initiating programs to screen more than 60,000 chemicals for their ability to behave like hormonally active agents and the Centers for Disease Control and Prevention (CDC) is beginning to screen people for the presence of hormonally active agents in their blood.

REDUCING YOUR RISK

It is impossible to eliminate contact with many of the compounds we've discussed. They are ubiquitous in our modern world. But there is no firm evidence yet that exposure is a problem. However, there are some ways to reduce exposure to at least a few of these materials, should you choose.

The same compounds that bioaccumulate in humans also do so in fish. The higher up the food chain those fish are, the more they have accumulated larger doses of these substances from eating quantities of smaller fish. So fish that are top predators in their ecosystems—freshwater species like largemouth and smallmouth bass, perch, pickerel, lake trout, and walleye pike, and marine species including shark, swordfish, and tuna—are more likely to have higher levels of these bioaccumulative compounds in their flesh. However, no evidence exists that eating normal amounts of fish exposes anyone to harmful levels of hormonally active agents.

There are also hormonally active agents in some kinds of plastic. Some plastic wrap and plastic containers can leach trace amounts of phthalates or bisphenol A into food, especially when the food is hot or when you put these containers in your microwave oven. You can reduce exposure to these infinitesimal amounts by heating food in glass or ceramic containers, rather than plastic.

Since plants contain hormonally active agents, you may wonder what to do about exposure to phytoestrogens in fruits and vegetables. Soy, for instance, contains hormonally active agents. One study has shown that soy can alter the length of the menstrual cycle. Another small study shows higher rates of hypospadias in the children of women who are vegans. On the other hand, some epidemiological studies have found that soy phytoestrogens may reduce the risk of heart disease and prostate cancer. We mention these conflicting associations to reiterate the controversy still surrounding the whole endocrine disruption issue. But the question of eating fruits and vegetables and fish is an easy one. There is overwhelming evidence of widespread health benefits from eating these foods, and no evidence that effects from the hormonally active agents in these foods, even if confirmed, outweigh those benefits.

Another unresolved question deals with the use of cosmetics, some of which contain low levels of phthalates. These substances have a very short half-life in the body, less than a day. Yet the CDC has found phthalates in the urine of sample populations, with higher levels in women in their twenties and thirties. That means many of these people, particularly the women, are being freshly exposed to phthalates on a daily basis. Research is under way to determine if cosmetics may be the source of this exposure. But the evidence at this time doesn't suggest that such exposure would pose any risk.

FOR MORE INFORMATION

Center for Bioenvironmental Research at Tulane and Xavier
    Universities
e.hormone.tulane.edu (endocrine disruption)
1430 Tulane Avenue, SL-3
New Orleans, LA 70112
(504) 585-6910

Environmental Protection Agency
www.epa.gov/scipoly/oscpendo/index.htm
Ariel Rios Building
1200 Pennsylvania Avenue NW
Washington, DC 20460
(202) 260-2090

This chapter was reviewed by Lorenz Rhomberg, Adjunct Assistant Professor of Risk Analysis and Environmental Health, Department of Health Policy and Management at the Harvard School of Public Health, and Risk Assessor and Principal in the Gradient Corporation; and by Robert Kavlock, Director of the Reproductive Toxicology Division of the National Health and Environmental Effects Research Laboratory at the U.S. Environmental Protection Agency and a steering committee member of the International Programme on Chemical Safety, World Health Organization, Global State-of-the-Science Assessment on Endocrine-Disrupting Chemicals.

27. HAZARDOUS WASTE

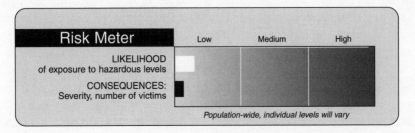

IN 1978, President Jimmy Carter declared a state of emergency in the city of Love Canal, New York, because chemical wastes, buried years before, were seeping into basements and waterways. Love Canal was partially built on an old landfill into which chemical companies had dumped tons of industrial waste. As the water table rose, chemicals in the landfill were discovered throughout the community. They seeped up into homes, schools, and other buildings. The episode established a new phrase in the American lexicon of risk: hazardous waste.

We have been throwing all sorts of things away for a long time. Sometimes these wastes are chemicals or other poisons. Not until the late 1970s and early 1980s did we finally start to understand some of the risks posed from these materials, and treat such wastes as a unique category. *Hazardous waste* is a catchall phrase that describes anything that can be harmful to human health or the environment if improperly handled. Thousands of waste compounds are considered hazardous.

Under federal legal definitions, waste is hazardous if it is corrosive, reactive (can explode), or ignitable (can burn readily). The waste also qualifies as hazardous if it contains chemicals that are acutely poisonous, or chemicals associated with human or ecological health problems. Hazardous waste can be liquid or solid. (Airborne hazardous materials are considered air pollution and regulated separately.) Hazardous waste can come from anywhere: industry, the home, mines, agriculture, small businesses, gas stations, schools, hospitals, your motor vehicle, and so on.

THE HAZARD

There are really two parts to the risk from hazardous waste. The first comes from sites where the waste has been improperly disposed of. Small and large businesses poured waste onto the ground or into streams, dumped it in pits, piled it up on their property with nothing to keep it from seeping into the ground, or paid "midnight dumpers" to take it away and dispose of it, no questions asked. Sometimes businesses wanted to get rid of materials that would have cost more to treat or store or dispose of more carefully. And in many of these cases, the people responsible for the dumping knew the material was dangerous and did it anyway. When this kind of dumping was revealed, it fueled public outrage and fear.

But many hazardous waste sites were created less by greed and more out of ignorance. Much of what we call *hazardous* waste today was just waste a few decades ago. Excess chemicals, used solvents, dirty oils, chemical-soaked solids, or remnants from leather tanning were often placed in large lagoons to evaporate, placed in barrels and buried, poured straight into rivers and harbors, or just piled up outside at industrial sites that were later abandoned. It seems thoughtless to us now, given what we have learned about the hazardous nature of some of these materials, but this was considered appropriate disposal at the time, before many of the dangerous properties of wastes like heavy metals, PCBs, and solvents had been realized. Those open lagoons and pits were actually considered a good way to keep the materials from spilling into local rivers and streams. They were an effort to protect the environment and dispose of hazardous materials safely.

A significant number of hazardous waste sites are facilities that were originally set up to solve the dumping problem: treatment or disposal sites, or businesses that took in the waste, separated it, and transferred it to other facilities for final disposition. Many of these companies went out of business, either because the technical challenges of dealing with dangerous materials were more than they could handle, or because costs rose so high after regulations required more careful handling of the materials and more paperwork to keep track of them. The wastes that were on site when the businesses closed remained, and now need to be cleaned up.

Regardless of the knowledge and motivation of those doing the dumping, the problem is that hazardous waste sites exposed us to these dangerous materials. Chemicals in storage lagoons sometimes leaked down into groundwater. Buried drums rusted and leaked, some-times polluting rivers and streams and the fish that live in them, some of which we eat. Leaking drums sometimes contaminated under-ground drinking water aquifers. Wastes piled above ground, or dumped in landfills, occasionally discharged their hazardous contents into the ground, also threatening groundwater. People using the land on which waste has been dumped came into direct contact with hazardous mate-rials. In the past, some of these sites were abandoned and became un-guarded areas in which children played.

We use the past tense for most of these exposures because they are all far less common today. At most of these sites, access is now restricted and cleanup is under way or finished. Monitoring and containment of underground water pollution means that contaminants rarely migrate in ways that can create human exposure.

Love Canal and dozens of other hazardous waste sites across the country were all discovered within a few years, and many of these discoveries drew widespread publicity. Together they fueled growing national outrage over this environmental threat to human health. In 1980, the Superfund law was created, named for the pool of money raised each year through a tax on chemical and petroleum products, among other sources. The fund was necessary since no responsible party existed to pay for the cleanup of many of the abandoned sites. Other sites were owned by people or businesses that didn't have the financial ability to cover the millions of dollars that cleanup would cost.

Officially known as the Comprehensive Environmental Response, Compensation, and Liability Act (CERCLA), the Superfund law em-powers the Environmental Protection Agency (EPA) to use the money to remediate risks from hazardous materials disposal sites, and then to sue the parties responsible for generating the waste—if they can be identified—in order to recover as much of the costs as possible.

Superfund was also designed to deter future dumping. It makes any-one liable for cleanup costs of any hazardous waste site where its mate-rial is found. Even if a business disposed of something legally, long ago, if that material ends up at a Superfund site, the company still has to pay. This is a powerful incentive for businesses to make sure their haz-ardous materials are safely and legally disposed of, and is a major reason why the kind of dumping that took place in the past has become rare today.

As of January 2002, there were 1,222 designated Superfund sites on the National Priority List (NPL major sites) and 72 more under consideration for official Superfund NPL designation. Over CERCLA's history, 257 sites have been deleted from the Superfund NPL list.

Superfund is only one of many government programs designed to solve the problem of hazardous waste sites. Under other programs, the EPA is investigating more than 11,000 other sites where some sort of hazardous waste has spilled. This list includes any kind of hazardous waste spill, even small ones, and while some might be earmarked for Superfund designation, most will be cleaned up or remediated in other ways. There are active state hazardous waste cleanup programs, federal and state efforts to clean up leaking underground fuel storage tanks, property transfer programs that require a piece of land or building to be cleaned of hazardous waste before it can be sold, and brownfield clean-ups in which government assistance helps clean sites in urban areas so they can be reused. Together with Superfund, these programs deal with thousands of contaminated locations.

Despite all this progress, people continue to rate abandoned hazardous waste sites as one of the most serious environmental threats to human health and demand aggressive government action. Experts in the field of hazardous waste remediation estimate that as much as $30 billion is spent yearly in the United States to clean up hazardous waste. Roughly $1.2 billion of that is government expense. The rest is paid for by private businesses.

~

But cleanup of hazardous waste sites is only one part of the issue. The other is the proper handling and disposal of the hazardous wastes still being generated today, byproducts of the modern world in which we live. They come from places we might suspect, like the chemical, electroplating, and petroleum industries. But they are also generated by less obvious sources such as dry cleaners, car repair shops, offices, print shops, photo processing centers, and hospitals and nursing homes, among many others.

Most hazardous waste, however, is generated by industry. The following table lists some of the industries and businesses that create hazardous wastes and the types of waste they produce. (Note the presence of "wastewater sludges." This type of waste is created when water is used to clean up other waste materials and is then filtered through treatment plants. This example illustrates that many hazardous wastes are actually the byproducts of compliance with other environmental laws.)

While CERCLA-Superfund is the law that empowers the EPA to

SOME MAJOR SOURCES OF HAZARDOUS WASTE	
Industry	Types of Waste Generated
Chemical	Acids and bases, spent solvents, reactive chemicals, spent catalytic chemicals, contaminated residues, wastewater treatment sludges
Petroleum/Refining	Acids and bases, spent catalytic chemicals, oily wastes, wastewater treatment sludges
Metal Manufacturing/Electroplating	Heavy metals, acids and bases, cyanide wastes, wastewater treatment sludges containing heavy metals
Mining	Metals, acids and bases, cyanide
Vehicle Maintenance	Paint wastes containing heavy metals, lead acid batteries, spent solvents, used oils
Printing	Heavy metals, acids and bases, waste ink, spent solvents
Dry Cleaning	Spent solvents, ignitable wastes

clean up hazardous waste sites, the Resource Conservation and Recovery Act (RCRA) is the legal authority designed to ensure proper handling of hazardous wastes still being produced. The EPA estimates that in 1999, approximately 20,000 large generators produced roughly 40 million tons of hazardous waste handled under RCRA regulations. Just 50 of those generators, mostly chemical plants and refineries, created about three quarters of that total.

More than half of this waste is water that is too acidic or basic or contaminated with hazardous chemicals to be treated. This wastewater is disposed of by high-pressure injection down into deep wells. These wells are far below any underground water aquifers, ranging in depth from 1,700 to 10,000 feet. To earn a permit, these facilities must prove that the waste won't affect the ground or surface water for at least 10,000 years. There are 163 of these wells at 51 facilities in 11 states. Most are in Texas or Louisiana. Deep-well injection disposed of 26 million tons of hazardous wastewater in 1999, most of which came from petroleum refineries, metal or chemical production plants, or pharmaceutical production.

The hazardous waste not disposed of in deep wells is either landfilled, burned, or recycled. About 1.4 million tons are buried in landfills that must have double liners to catch any leachate, and must have underground leachate collection and monitoring systems as an extra protection. As an additional precaution against leakage, no liquid hazardous wastes may be landfilled. Liquid wastes—a total of approximately 3 million tons per year—are incinerated (see Chapter 28.)

There are 70 hazardous waste landfills in the United States: 50 can bury only wastes that are produced on site, mostly at large chemical plants or refineries; the other 20 are commercial operations allowed to take waste from elsewhere.

Because it's so expensive to dispose of, and because hazardous waste contains so many commercially useful products, close to 3 million tons are recycled. About 1.5 million tons of solvents are cleaned so they can be reused. Sometimes they are recycled by mixing them with fuels. And more than 1.1 million tons of metals are captured from hazardous waste for reuse.

In all, there are 1,575 facilities that treat, store, or dispose of hazardous waste in the United States, but the top 50 handle about half of it.

THE RANGE OF CONSEQUENCES

Direct exposure to hazardous waste can be dangerous in many ways. As we mentioned, these wastes can be acutely and immediately poisonous, harmful to touch, to breathe, to drink or eat. Fortunately, the kind of direct exposure to hazardous waste that would cause such acute problems is rare. Many of the identified waste sites have been fenced off and posted, and hundreds of them are underground. Very few hazardous waste sites pose any kind of ongoing direct exposure, even to populations living nearby, much less to the general population. Rarely, however, dangerous materials from hazardous waste sites do spread directly to homes, businesses, schools, or other occupied buildings nearby.

More concern about hazardous waste sites focuses on effects from long-term low-level exposure to materials that leak off the site and contaminate water, fish, or soil. The mix of compounds found at hazardous waste sites includes chemicals suspected of causing cancer, respiratory diseases, developmental toxicity, and other health effects. The Agency for Toxic Substances and Disease Registry (ATSDR), a part of the Centers for Disease Control and Prevention created by CERCLA, keeps a priority list of hazardous chemicals based on their frequency of occurrence at Superfund sites, their toxicity, and the potential for human exposure to these substances. The top five compounds on the current

ATSDR list are arsenic, lead, mercury, vinyl chloride, and PCBs. Effects from exposure to these substances include cancer, neurological damage, developmental deficits in newborns, and skin problems.

Little specific information is available on how many people may suffer health consequences from hazardous waste, but several national risk assessments by the EPA indicate that the number is extraordinarily low. Even in unique circumstances of the highest risk—someone living very close to a waste site and drinking from well water—the chances of anyone suffering a health effect from hazardous waste is seldom estimated to be greater than 1 in 10,000.

Public health experts, including many at the EPA, believe that hazardous waste sites, both the old abandoned ones and the new controlled disposal facilities, pose a smaller risk to health than most other environmental threats. In 1987, less than 10 years after Love Canal, the EPA issued a report called "Unfinished Business," in which the agency asked its experts to rank various environmental and human health risks as part of an effort to evaluate priorities. The EPA officials ranked hazardous waste sites as one of the lowest-risk areas receiving EPA attention. A public opinion poll conducted as part of that review ranked hazardous waste sites near the top.

THE RANGE OF EXPOSURES

Except for people living very close to hazardous waste sites, or people using land where wastes have been dumped but haven't been discovered, direct exposures are unlikely. But people can also be exposed indirectly in a number of ways. As we've said, liquid waste can leak into the ground and get into drinking water, potentially contaminating public drinking wells. (Underground aquifers below hazardous waste sites are monitored, as are nearby wells, to reduce this exposure.)

Levels of some hazardous waste that contaminate streams, rivers, and lakes can be quite high in fish from those waters, because it biomagnifies. Small fish consume small quantities of chemicals like PCBs or pesticides, which are environmentally persistent, which means they don't break down and they last in the environment for years. Then bigger fish eat larger quantities of the smaller fish, and the dose magnifies up the food chain. Predator fish like bass, perch, pickerel, lake trout, and pike can have high doses of chemicals that came from improperly disposed hazardous waste. You may be aware that many states and the federal government have warnings, particularly for pregnant women, about eating certain fish. But most of the hazardous material that prompts these warnings is methylmercury, and the source of most of that contamination is not hazardous waste. It's airborne mercury that comes from the burning of coal or municipal waste and is washed out of the air and into water by rain or snow. (See Chapter 30, "Mercury.")

REDUCING YOUR RISK

If you live near a known or suspected hazardous waste site and have a well, experts suggest that you have your water tested. Also, if you live near a known or suspected site, very low levels of vapors from underground contamination can seep up into your building. If you live near such a site and detect chemical odors in your basement or first floor which are not from normal living activities, check with local health authorities about testing the air. Obviously you should avoid the disposal site property, and make sure your kids do too. If you suspect that a piece of land or an abandoned building might contain hazardous wastes, inform local or state public health authorities.

One way to become aware of hazardous wastes near you is through the EPA's Toxics Release Inventory (TRI). Created in the wake of the deadly release of chemicals from a factory in Bhopal, India, and a smaller release of chemicals from a plant in West Virginia, the TRI requires facilities that use particular chemicals above a set amount to report releases of those chemicals. This information is available to the public on the EPA's website. You can search the data by state, city, or Zip Code, so you can see what's being released near you. But the information gives you only half the risk picture because it lists only emissions, not actual exposure. To help, the advocacy organization Environmental Defense has a program called Scorecard, which combines TRI emission information with estimated exposure levels to come up with a more accurate suggestion of potential risk.

While the majority of hazardous waste is produced by large industries, individuals generate hazardous waste too, and there are things you can do to reduce your contribution. Many common household products, such as paints, oils, solvents, arts and crafts supplies, gasoline, pesticides, cleansers, lawn chemicals, rechargeable batteries, thermometers and thermostats, pool chemicals, antifreeze, fluorescent light bulbs, computers and computer monitors, games that include electronic parts, and appliances contain hazardous compounds. If these

materials are put into landfills, the potential exists that the hazardous materials might eventually leak and contaminate underground
aquifers. A growing number of communities are creating hazardous
waste collection sites, and many communities run special hazardous
waste collection days a few times a year, when you can bring these
items to a collection site for proper disposal.

FOR MORE INFORMATION

Agency for Toxic Substances and Disease Registry
Division of Toxicology
(Hazardous Waste Priority List)
www.atsdr.cdc.gov/cxcx3.html
1600 Clifton Road NE, Mailstop E-29
Atlanta, GA 30333
(888) 422-8737

Environmental Defense
www.scorecard.org
257 Park Avenue South
New York, NY 10010
(212) 505-2100

If you are interested in recycling your household hazardous wastes, call
your local public works department or recycling coordinator. If your
community doesn't have a program for recycling these wastes, you can
get advice on how to start one from your state environmental department, which often gives financial assistance to jump-start such programs.

---

*This chapter was reviewed by Barnes Johnson, Director of the Economics, Methods
and Risk Analysis Division, Office of Solid Waste and Emergency Response, Environmental Protection Agency; and by Ross Elliott, Environmental Affairs Specialist
at the EPA's Office of Solid Waste, who has worked in the field of hazardous waste
for many years.*

28. INCINERATORS

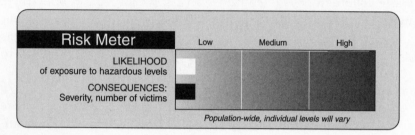

THINK ABOUT what you toss out each day: paper, plastic, food, metal cans, packaging, and so much more. The average U.S. citizen tosses out nearly five pounds of trash per day. As a nation, that comes to roughly half a trillion pounds of waste per year. Most of it goes to landfills, the cheapest form of disposal available. But through the 1980s, several environmental trends converged that gave rise to another form of waste disposal: incineration.

Landfills were found to leak, and in some places the leachate contaminated underground drinking water supplies. Environmental regulations were passed requiring expensive leachate collection and recovery systems. At the same time, in highly developed areas of the country, particularly the Northeast, land scarcity pushed up real estate values. These two pressures made landfilling a ton of waste so expensive that incinerators were built to provide a cheaper alternative.

Meanwhile, the energy crises of the late 1970s and early 1980s raised the price of electricity. That made the idea of incinerators even more attractive, because instead of just burning the waste to get rid of it, businesses could capture its heat, turn it into electricity, and sell it as fuel. And while all these trends were developing, we were trying to figure out how to deal with another new environmental problem—hazardous waste. Concern over burying these hazardous materials in landfills encouraged the development of incinerators to burn and destroy them.

So incinerators solved several environmental problems at the same time, but they created others, because the materials burned in these large facilities produce air emissions containing low but measurable amounts of several hazardous materials.

THE HAZARD

There are three main categories of waste incinerators.

Municipal Waste

In the United States, 109 large waste-to-energy incinerators burn about 14 percent of our solid waste. Some of these facilities have multiple units. There are 160 units in all that continuously burn from 200 to 750 tons of waste per day. These facilities convert the heat to electricity, enough to power 1.5 million American homes. There are roughly 75 other smaller municipal waste incinerators. Most are in regions where the water table is too high to allow for landfills. Florida, for example, has more of these facilities than any other state. Others are in places like Alaska, where cruise ships dock and by law have to send their waste to incinerators. Many of the smaller facilities operate only sporadically.

Municipal waste incinerators burn not only trash from homes, but from small businesses, and some nonmedical waste from hospitals: 35 to 40 percent of what they burn is paper, a little less than 20 percent is yard waste, nearly 10 percent is metal, about 8 percent is glass, 8 percent is plastic, and the rest is categorized as "other." The trash is piled on a slowly moving steel conveyor where it burns at temperatures of about 1800°F for half an hour. The waste has about the same energy content as wood and serves as its own fuel. The ash that's left is tested to make sure it doesn't contain any hazardous material that would require special handling and disposal, and is usually disposed of in a regular municipal landfill. Municipal incinerators reduce the volume of the waste they burn by roughly 90 percent.

The combustion of the mixed wastes in a municipal waste incinerator produces nine pollutants regulated by the EPA: dioxin, carbon monoxide, solid particles, gases including sulfur dioxide (SO_2), hydrogen chloride (HCl), nitrogen oxides (NOx), and metals, including mercury, lead, and cadmium. On the next page we list those pollutants and some of the burning wastes that produce them.

The emissions from these incinerators are low, but constant, and some of the pollutants they emit are persistent once they settle to the ground. That means they don't break down easily, so they build up over time. These include dioxin, lead, mercury, and cadmium. Those

INCINERATED WASTES AND POLLUTANTS	
Pollutant	Source
Dioxin	Incomplete combustion of any waste
Carbon monoxide (CO emissions from incinerators aren't very high, but CO levels are an indicator of combustion efficiency, which affects levels of other emissions)	Organics
Particles	Organics
Hydrogen chloride	Chlorine sources (salt, plastic, paper)
Sulfur dioxide	Sulfur sources (paper, rubber)
Nitrogen oxide	Nitrogen in the air, some waste materials (yard wastes)
Lead	Wood, paper, plastic (mostly color film), yard waste (plants take up lead from soil), metal
Mercury	Organics (yard waste, food, textiles), metal, batteries, thermostats, thermometers, fluorescent light bulbs, cathode ray tubes (from computer monitors)
Cadmium	Rechargeable batteries, plastic, organics

low chronic emissions have been drastically reduced under federal regulations that were phased in during the late 1990s and took effect in 2000. Total national dioxin emissions dropped from 1.8 pounds in 1990 to a little less than half an ounce in 2000. Lead emissions dropped from about 148 tons to approximately 34 tons. Emissions of particles dropped from roughly 16,900 tons to about 4,800 tons.

Incineration is a relatively simple matter of making sure the right amount of oxygen mixes with the waste—the fuel—to keep combustion at the proper temperature. Cleaning pollutants out of the combustion gases is a more complex process. Municipal waste incinerators use four basic technologies. First, urea is blown into the gases, reducing the nitrogen oxides to nitrogen and water. Second, the gases then enter devices called dry scrubbers, which blow in a slurry of lime and water. The lime, a chemical base, neutralizes the acidic sulfur dioxide and hydrogen chloride gases, and the water cools the gas, which condenses the metals from gas to solid. Third, activated carbon is injected into the gases, which captures mercury and dioxins. Finally, the gases pass through filters that catch the lime, metals, and particulates. The filters are taken to municipal waste landfills.

Hazardous Waste

There are 283 facilities in America that burn an estimated 3.6 million tons of hazardous waste a year, including solvents, pesticide wastes, and other industrial products that are corrosive, reactive (can explode), ignitable (can catch fire), or contain one or more chemicals listed as toxic by the Environmental Protection Agency (EPA). (See Chapter 27, "Hazardous Waste.")

Hazardous waste is burned in three different kinds of facilities. The most common are incinerators that burn it just to dispose of it: to destroy the hazardous substances and reduce their volume. In 1997, there were 149 of these facilities in the United States, running 189 incinerator units. Incinerators burn about 2 million tons of hazardous waste a year. They operate at temperatures of about 1,800°F, which is hot enough to destroy nearly everything in the waste. Liquid waste is destroyed quickly. Solid waste burns for about half an hour.

The next most common type of hazardous waste combustor is an on-site boiler or furnace at a large chemical, pharmaceutical, and petroleum plant. This is a waste-to-energy system, like a municipal incinerator, a big oven lined with steel walls embedded with water pipes that convert the heat to steam and then electricity. In 1997, 105 of these incinerators were in operation, running at temperatures averaging 1,800° to 2,500°F, burning about 700,000 tons of waste per year.

Many hazardous wastes, particularly certain liquids, make great fuel because they have twice the energy content of wood and the same energy content as bituminous coal. So certain kinds of liquid hazardous waste are preferred fuels for cement kilns and "lightweight aggregate" kilns that process minerals for construction, since these kilns need to maintain high temperatures to dry the limestone and clay they process. There are 24 of these facilities nationwide. They are huge rotary kilns, some longer than a football field, operating at temperatures of 2,400° to 3,000°F. Cement and aggregate kilns burn about 900,000 tons of hazardous waste a year.

All three types of hazardous waste combustors use several processes to control emissions. The first is simply the heat of the fire, which destroys most of the material. Then afterburners subject the gases to a second incineration, usually at a much lower temperature, below 400°F, still hot enough to destroy organic material but low enough to reduce significantly the formation of dioxins. (Afterburners are only used on incinerators.) Then the emission gases pass through electrostatic precipitators that use electrical charges to remove particles. Other filters, including activated charcoal, capture mercury and dioxin. Scrubbers similar to those used in municipal incinerators neutralize the acid gases. The dust from the filters and the lime from the scrubbers are collected, solidified, and disposed of in specially permitted hazardous waste landfills.

Emissions vary from unit to unit, but in general hazardous waste incinerators emit metals at less than 50 micrograms (millionths of a gram) per cubic meter, dioxin at less than 0.4 nanograms (billionths of a gram) per cubic meter TEQ (Toxic Equivalent Quantity, a measure of hazardousness), and particles at levels of 30 to 80 milligrams (thousandths of a gram) per cubic meter of exhaust.

Medical Waste

Medical incinerators burn infectious waste or materials used in treating patients or in research. Veterinary hospitals dispose of their infectious waste in incinerators too. Approximately 200 to 300 medical waste incinerators operate in the United States, but no one is sure of the precise number because it has been rapidly declining in recent years. There were as many as 6,000 medical waste incinerators in 1990, but many closed rather than comply with tighter regulations to reduce emissions. Furthermore, in the 1990s medical facilities became more cost-conscious and reduced their waste streams, so the need for medical incinerators was reduced. Many hospitals switched from using small on-site incinerators to a technology that destroys pathogens through the application of high-temperature, high-pressure disinfection. Waste treated in this way doesn't have to be disposed of in more expensive medical incinerators. So most of the remaining medical waste incinerators now serve an entire city or region rather than just one hospital. (As of the beginning of 2002, the EPA was designing regulations that would track these facilities more closely and tighten controls on their emissions.)

Medical waste incinerators vary a great deal in size and type of operation. Some, particularly the smaller ones located at hospitals, operate only a few hours a day, and not every day of the week, burning 100 to 200 pounds of waste per hour or less. Others, such as those operated by commercial waste disposal companies, may operate continuously five days a week and burn 1,000 to 2,000 pounds of waste or more per hour. Unlike the large municipal waste incinerators, which can be as big as a large multistory building, most medical waste incinerators are the size of a small truck or small building.

Most medical incinerators are not designed to convert the heat to energy the way municipal incinerators do. They burn the waste in two chambers. The first burns the waste for roughly half an hour at 1,200° to 1,400°F, which turns nearly all the waste into gas. The gas is then sent to a second chamber, which burns it at 1,600° to 1,800°F and completes the destruction of the biological materials in the waste.

It's impossible to know the levels of emissions from medical incinerators in the past because an accurate inventory of these facilities and their emissions was never kept. The EPA has developed estimates for what emissions likely were in the late 1980s and what they will be by the end of 2002.

ANNUAL MEDICAL WASTE INCINERATOR EMISSIONS (EPA ESTIMATES)

Pollutant	Late 1980s	End of 2002
Dioxin	6,500 grams TEQ	6 grams TEQ
Hydrogen chloride	45,000 tons	150 tons
Lead	85 tons	2 tons
Particles	12,000 tons	100 tons

Other Waste Combustors

Among the other kinds of facilities that burn solid waste, perhaps the most common isn't even officially considered an incinerator under government regulations. It is known as a commercial and industrial boiler and is permitted to burn any kind of nonhazardous waste generated on site in order to recover the heat from that material. There may be as many as 100,000 of these boilers, more than municipal, medical, and hazardous waste incinerators combined, but the EPA keeps no precise inventory of these facilities. Regulations to control emissions from boilers have been under development by the EPA since the mid-1990s.

Human and animal crematories are another type of combustion device. The EPA is considering regulations for crematories, along with several other types of combustion devices, including very small municipal waste incinerators and agricultural waste incinerators.

THE RANGE OF CONSEQUENCES

Incinerators produce a wide range of emissions. In addition to the products of combustion themselves, a variety of secondary chemicals are produced by chemical reactions in the flue as gases cool. Because the levels of emissions from incinerators are generally very low, the health consequences from incinerator emissions are believed to be quite small.

Particulate matter from incinerators, microscopic solid particles similar to those from diesel engines and other sources, are linked to aggravation of respiratory and cardiovascular disease and to increased incidence of premature death. Acid gases, including hydrogen chloride, sulfur dioxide, and nitrogen oxides, can cause gastritis, chronic bronchitis, dermatitis, and light sensitivity. Acute inhalation exposure can cause irritation and inflammation of eyes, nose, and throat. Exposure to dioxins and furans—which are typically produced in reactions in the gases in incinerator emissions—causes cancer in animals, and is believed to cause cancer in humans, though the link remains controversial. Dioxin and furan levels are also associated with changes in hormone and enzyme activity. Dioxins and furans can persist for decades, resisting breakdown or chemical reactions, and can bioaccumulate through the food chain.

And 35 metals have been documented in incinerator emissions. Most are toxic to humans at various concentrations. Several are known or possible carcinogens. As we specified, the EPA is most concerned about controlling emissions of lead, mercury, and cadmium. Exposure to lead causes central nervous system damage and anemia in children and can affect the nervous system and circulation in adults. Mercury can persist in the environment and accumulate in some organisms, and can cause developmental and nervous system disorders in humans, particularly in the children born to women exposed to very high levels during pregnancy if the mercury has been converted into methylmercury by aquatic organisms. (See Chapter 30, "Mercury.") Cadmium is both toxic and potentially carcinogenic.

The risk from incinerator emissions is extraordinarily low, but difficult to measure precisely. In one national risk assessment, the EPA calculated the potential health effects of emissions from normal operations of municipal incinerators. The results were given in a range, but the worst-case lifetime cancer risks for adults was 2 in 10 million, and for children it was 8 in 1 million.

THE RANGE OF EXPOSURES

According to both actual measurements and computer models, incinerator emissions disperse over an area of 10 kilometers or more, depending on how the facilities operate, the weather, and the type of pollutant. Municipal waste incinerators are most common in the Northeast and Florida. Hazardous waste incinerators are more common in Texas, Louisiana, and the Southeast. Medical incinerators exist largely in urban areas. Cumulatively, emissions from all incinerators are a small contribution to the overall outdoor air pollutant exposures in the United States. However, incinerator emissions may contribute a larger amount of pollutants to the air in the region around the facility if it is not operating according to regulations. This failure is more often the case with smaller waste combustion facilities.

REDUCING YOUR RISK

The per capita production of waste in this country rose from 2.7 pounds in 1960 to 4.62 pounds as of the end of 2001. But the amount of waste being burned has actually gone down. The environmental trends that made incineration so economically attractive in the 1980s have shifted, and since the mid-1990s no new, large incinerators have been built in the United States. The capacity of existing incinerators has actually gone down. Between those reductions and the tighter EPA air pollution standards, emissions from waste incineration are down dramatically. Still, you can reduce incinerator emissions further by reducing the amount of waste you toss out. Recycling is one way to do this, and recycling rates of glass, plastic, and paper are high and rising. But you can also recycle yard waste at home or at municipal recycling centers that turn the clippings, mowings, and brush into usable compost.

In addition to how much you throw out, be careful about what you throw out. Rechargeable batteries and electronic devices are made with nickel and cadmium, so keep them out of the trash and take them to collection sites to reduce cadmium emissions from municipal incinerators. Keep hazardous materials like pesticides, paints, and solvents out of the trash by taking them to municipal hazardous waste recycling centers if possible. More and more communities are organizing periodic hazardous waste collection days when you can drop off these materials.

You should also try to recycle electronics. Toys, games, appliances, computers, and stereo equipment all contain metals that, when burned, can contribute to hazardous emissions. Many hazardous waste collection programs also take these materials.

FOR MORE INFORMATION

Toxic Release Inventory—Community Right to Know Program
Environmental Protection Agency
www.epa.gov/tri
(202) 260-1488 or (202) 260-1531

Agency for Toxic Substances and Disease Registry
Centers for Disease Control and Prevention
www.atsdr.cdc.gov
1600 Clifton Road
Atlanta, GA 30333
(888) 422-8737

This chapter was reviewed by several government experts who have worked in this field for years and by Richard Anderson, who taught at the Boston University Department of Urban Affairs and Planning, worked in the incineration industry for 15 years, and is now a consultant. He has researched solid waste issues for 27 years, including hazardous, medical, industrial nonhazardous, and municipal solid waste.

29. LEAD

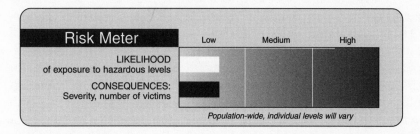

LEAD HAS BEEN USED for thousands of years. The ancient Egyptians used lead for sculpture, pottery, and jewelry. The Roman Empire used lead in shipbuilding, cooking pots, and to line the viaducts that carried its water. More recently, lead has been used in solders in electrical circuitry because it melts so easily, in paint because it covers surfaces effectively and wears well, in fuel to increase efficiency, in shielding to protect against radiation because it is so dense. It is still commonly used in batteries. Lead is an incredibly useful metal. But it is also a poison.

THE HAZARD

Lead has no natural function in the human body. It belongs to a class of compounds called heavy metals. Heavy metals can directly damage soft tissue, including the liver, kidneys, and nervous system. In addition, lead is known to alter the way nerve cells signal one another, and thus interfere with the way brain cells form connections. Lead also impairs the body's ability to use vitamin D.

Because lead is chemically similar to calcium, the body stores most of it in our bones and our teeth. As atoms of lead migrate into the bloodstream, they interfere with the creation of healthy red blood cells. The damaged cells aren't as efficient at picking up oxygen from the lungs and delivering it to the rest of the body. So lead affects our health by limiting our blood's ability to supply our cells and tissues with oxygen.

THE RANGE OF CONSEQUENCES

Lead is a risk to anyone, but it is particularly hazardous to children below the age of six because their brains and nervous systems are still developing and the presence of lead interferes with this process. Also, because the bones in children are still developing and growing, many researchers believe that a significant portion of the lead taken in by children remains stored in the body.

The Centers for Disease Control and Prevention (CDC) set 10 micrograms per deciliter of blood (10 μg/dL) or higher as the level of concern. (That equals 10 millionths of a gram in about three and a half ounces of fluid.) This figure is not an absolute cutoff above which lead is dangerous and below which it is not. Rather it is a general reference level for risk management decision making, above which effects are more significant, below which they are more subtle. Scientists believe that even very low levels of lead in the body may do some damage, to children or adults. At 10 μg/dL in blood or higher, lead in a child's body can slow mental development and result in problems with learning, motor control, and behavioral control. It can also cause some hearing loss and speech impairment. Some studies show that for each additional microgram of lead found in children's blood, there is a corresponding drop in math and reading test scores. According to one study, children with blood lead levels of 15 μg/dL have IQ levels three to five points below those with levels of 5 μg/dL. Several reviews of all the studies on lead find that in general, there is a one- to three-point decline in IQ for each 10 μg/dL in blood lead levels.

At higher blood lead levels, a child is at risk for more serious cognitive impairment. At very high levels, lead can cause brain damage, kidney damage, loss of consciousness, coma, and even death. These levels of exposure are now rare.

(The impacts at moderate levels of exposure—effects on fetal growth, intellectual ability, and hearing—are not "clinical." That means they're usually too subtle to notice in any one individual. Scientists measure them by noticing small shifts in the average characteristics in large groups of people who are studied.)

~

In adults, lead exposure raises blood pressure and increases the risk of cardiovascular disease. At high levels it can also cause digestive problems, kidney damage, sleep problems, muscle and joint pain, and neurological problems leading to mood changes, problems with motor function, and decreased fertility in males. High levels of lead interfere with nervous system signal transmission. Decades ago, professional painters who used lead paint suffered "painter's droop," the loss of motor control in their hands and feet, from lead exposure.

BLOOD LEAD LEVELS AND EFFECTS IN CHILDREN	
Blood Lead Level (in micrograms/deciliter)	Lowest Observed Effects in Children
≥10	In fetuses, associated with reduced birthweight. In children and infants, decrease in IQ, decrease in hearing, decrease in growth, decreased ability to maintain steady posture, increased behavioral problems
20–40	Decrease in vitamin D metabolism (soft bones), decrease in IQ
41–50	Anemia
51–100	Colic, anemia, kidney damage, brain damage
100+	Coma, death

Lead is a particular risk for pregnant, lactating, or menopausal women. These women release high levels of calcium stored in their bones, which also causes lead stored in the bones to be released into the bloodstream. The lead not only affects the adult, but also poses a threat to the fetus or, through breast milk, to a nursing infant.

Emotional stress that changes body chemistry is also known to release lead from the bones, sending it into the bloodstream.

THE RANGE OF EXPOSURES

Nearly all environmental lead contamination is due to human activity. The most common sources of exposure to lead are from lead paint, and the lead in the soil from motor vehicle exhaust. Lead contamination from these two sources is so widespread that even people who do not live in industrial areas have trace amounts of lead in their bloodstream. Most cases of clinical lead poisoning seen today are due to ingestion of leaded paint.

Paint manufacturers steadily reduced their use of lead in paint following World War II, and the federal government finally banned lead in household interior and exterior paints in 1978. But any residence painted before 1978 may well still contain lead paint. And lead is still widely used in commercial paints for bridges, the lines and markings on streets and roads, and boats and motor vehicles. Leaded gasoline was phased out in the late 1970s, but the phaseout wasn't complete until the early 1990s and the lead in the exhaust from the fuel that settled into the ground is still a contaminant in soil.

Lead in drinking water pipes is thought to be a minimal source of exposure for most people, though old houses with lead pipes carrying water in from municipal service lines can be an important source of lead exposure. This happens more often if the water has chemical characteristics that make it "attack" the lead in the pipe and release it into the water. Lead solder used in the manufacture of food cans was an important source of lead exposure until manufacturers in the United States stopped using it in the 1980s. It is possible that some imported food cans may still contain lead solder.

The primary exposure for children comes from swallowing soil or dust that carries small particles of lead. Though children also inhale dust that carries lead, their primary route of exposure is ingestion because they have more hand-to-mouth contact than adults, including eating dirt or sand with lead in it or mouthing objects or their hands, which carry dust that contains lead. The federal government estimated that in 1997, 900,000 children between the ages of one and five had blood lead levels of 10 μg/dL or higher.

The primary source of high-dose exposure to lead for adults is occupational. People who work in industries that use lead, such as battery manufacturing; lead paint manufacturing or application, such as in auto body shops; brass or bronze foundries; or steel cutting or welding, are at higher risk. Another significant source of occupational exposure to lead is construction and renovation work. Demolition or renovation of older buildings that contain lead paint is believed to be a significant source of occupational exposure. Between 500,000 and 1.5 million American workers are exposed to lead in the workplace. Lead gets into most adults by inhalation. But adults also ingest lead, particularly at construction sites, where workers have hand-to-mouth contact over the course of a day. Smoking is another potential cause of hand-to-mouth contact, where lead dust is present.

Although lead is found in the environment almost anywhere in the country, it is not evenly distributed. The highest-risk areas in the United States are urban areas with old housing. According to the CDC, more than 80 percent of all U.S. homes built before 1978 contain lead-based paint. The National Safety Council reports that two thirds of the homes built before 1940, and one half of the homes built from 1940 to 1960, contain lead paint. Lead in the soil is concentrated in areas near roads and highways or in urban areas, which have generally higher levels of motor vehicle–generated pollution. Experts say lead is a risk not only in inner-city neighborhoods but in any building built before 1978.

Renovation of buildings with lead paint, or even just simple wear and tear on painted surfaces in such buildings, creates particles of lead that become airborne in dust and can be inhaled or ingested. In Massachusetts, for example, one third of the children with elevated blood lead levels were exposed through renovation of owner-occupied homes in urban, suburban, and rural areas. In most cases, the amount of lead released from minor renovation or wear and tear is low. Levels can vary, however, and they can be quite high depending on the amount of lead in the building, particularly during renovation, and depending on the degree of disruption to the surfaces bearing lead paint.

Remember our definition of risk, however. There has to be a hazard—the lead—but also an exposure to that hazard. Lead paint that remains undisturbed is a hazard, but not a risk, unless we inhale or ingest it.

Much progress has been made in reducing lead exposure in the United States. A national study conducted between 1976 and 1980 estimated the average blood lead level for the entire child population at $16\,\mu g/dL$. By today's standard of $10\,\mu g/dL$ to define lead poisoning, the average American child in the 1970s was lead-poisoned. Since the banning of lead paint for residential use, elimination of lead from gasoline in most motor vehicles, and the elimination during the 1980s of lead solder in food cans, the average blood lead level in the United States is down to less than $4\,\mu g/dL$.

REDUCING YOUR RISK

Because even low lead levels can have long-term effects on children, the CDC recommends that all children ages one and two be screened for lead with a simple and inexpensive blood test. To reduce exposure, find out if there is lead paint in your home. The National Safety Council says the best way to do so is to hire a lead inspector to test the paint. Lead inspectors use X-ray instruments to determine the lead content in paint, and they can give you immediate results. This testing is strongly recommended if you are about to do any renovation work on your home, which could produce increased lead exposure if you don't first remove the lead paint, or contain it. The Environmental Protection Agency has not approved and does not recommend do-it-yourself lead test kits, which, it says, don't produce reliable results.

If your home does have lead paint, don't try to remove it yourself. Improper removal will release more lead particles into the air. Hire a qualified contractor. If the lead paint in your home is in good condition and not peeling or cracking, you can cover the painted surfaces to ensure that none of the paint is accessible to the touch. This procedure is called encapsulation. But since even this work can release some small amounts of lead into the environment, encapsulation should be avoided unless the lead-painted surface is exposed to use that would cause lead from the paint to be released.

Since the primary source of exposure to lead is in dust or soil, experts suggest several steps to reduce exposure to children:

- Do not let your children eat sand, dirt, or paint chips.
- Encourage them to play in grassy areas when available, and make sure they wash their hands after playing outside before eating or going to sleep.
- If possible, cover play areas with wood chips, grass, or even just a fresh layer of soil that has been tested to make sure it doesn't contain lead particles.
- Keep children's play areas as clean and dust-free as possible. The Massachusetts Department of Public Health recommends using a vacuum cleaner with a HEPA (high efficiency particulate air) filter. HEPA filters will trap 99 percent of all lead particles lying around. A conventional vacuum cleaner isn't nearly as efficient. Areas in houses with high lead-dust potential are windowsills and corners.
- A diet rich in iron and calcium reduces the amount of lead the body will absorb. Also, giving your kids regular meals will help. Children may absorb 10 to 100 times more lead on an empty stomach.

FOR MORE INFORMATION

Environmental Protection Agency
Office of Pollution Prevention and Toxics
www.epa.gov/opptintr/lead/index.html
1200 Pennsylvania Avenue NW
Mail Code 7401M
Washington, DC 20460
Lead Hotline: (800) 424-LEAD or (800) 424-5323
Safe Drinking Water Hotline: (800) 426-4791

U.S. Department of Housing and Urban Development
www.leadlisting.org (qualified lead removers)
451 7th Street SW

Washington, DC 20410
(888) LEAD-LIST or (888) 532-3547

National Safety Council
Lead Information Center
www.nsc.org/library/facts/lead.htm
1121 Spring Lake Drive
Itasca, IL 60143-3201
(800) 424-5323

This chapter was reviewed by Dr. Henry Falk, Assistant Administrator of the Agency for Toxic Substances and Disease Registry and a leading national expert on the risks of lead; by Dr. David Bellinger, Associate Professor in Neurology, Harvard Medical School, and Epidemiologist at Boston Children's Hospital; and by Kim Dietrich, Professor of Environmental Health and Pediatrics at the University of Cincinnati College of Medicine.

30. MERCURY

Risk Meter	Low	Medium	High
LIKELIHOOD of exposure to hazardous levels	▪		
CONSEQUENCES: Severity, number of victims	▪		
	Population-wide, individual levels will vary		

YOU ARE PROBABLY FAMILIAR with mercury as that silvery liquid metal in thermometers. Or maybe you've seen it as that rising column of silver in the blood pressure device your nurse or doctor uses. Mercury is a fascinating substance, a metal that is liquid at room temperature and which expands and contracts in direct relation to temperature and pressure. But it is also a heavy metal, a type of substance that can poison the human nervous system.

Today mercury is a commonly used substance in thermometers and

other temperature-sensing equipment like thermostats and temperature detectors in ovens. It is common in pressure-sensing equipment
like blood pressure cuffs and barometers. Mercury is also roughly half of
the material in dental amalgam, the fillings we get at the dentist, in
which it helps bond the other metals of the alloy. Mercury is used in tilt
switches inside the hoods of motor vehicles and washing machines to
turn the light on when the lid is open, and off when it closes. Because it
is a good conductor of electricity, mercury was heavily used in batteries
and in the switches of fluorescent light bulbs. It was employed in pesticides and fumigants, as a preservative in textiles and vaccines, and as a
preservative and a colorant in latex paints. Some of these uses of mercury, like batteries, are being phased out because of what we've learned
about mercury's poisonous effects.

THE HAZARD

There are two basic types of mercury, inorganic and organic. The most
abundant organic type is known as methylmercury. It forms when inorganic mercury gets into water systems, which happens in a variety of
ways. Some mercury leaches into these waters naturally from soil and
rocks. Of the anthropogenic sources—those resulting from human activity—the leading source is mercury vapor produced by the burning of
coal, which contains small amounts of natural mercury (the Environmental Protection Agency [EPA] says this is one third of all anthropogenic mercury), followed by mercury vapors from municipal waste
incineration, then medical waste incineration, and then from hazardous waste incineration.

Bacteria in the water and sediment convert the inorganic mercury to
methylmercury. These bacteria, and algae, take up the methylmercury
as they feed. They in turn are consumed by larger organisms. Moving
up the food chain, as bigger organisms consume larger quantities of the
smaller ones, the dose of methylmercury bioaccumulates—it becomes
concentrated. At the top of the aquatic food chain, predator fish have
levels of methylmercury in their muscle tissue between 10,000 and
100,000 times higher than the water itself.

These levels are still very low in terms of their hazard to humans.
While consumption of fish is the form of methylmercury exposure that
research has confirmed is the clearest risk to human health, most people
who consume average amounts of commercially available fish are not
exposed to dangerous amounts of mercury.

THE RANGE OF CONSEQUENCES

Mercury is principally neurotoxic: it damages the cells of the brain and central nervous system. One of the most dramatic cases of methylmercury poisoning occurred in the 1950s in Minamata, Japan, where industrial disposal of mercury into local waters created high methylmercury levels in fish, a staple of the local diet. Children and adults exposed to these very high levels experienced impaired vision; tingling or numbness in their hands and feet; impaired speech, hearing, and walking. Some pregnant women in Minamata bore infants with serious cognitive impairment. (Fish consumption in Minamata was more than 20 times greater than the average fish consumption in the United States, and methylmercury levels in the fish being eaten there were significantly higher than in this country because of the local dumping of industrial waste.)

A similar mercury poisoning episode took place in Niigata, Japan, in 1965. And two episodes occurred in Iraq in the 1960s and 1970s, in which a fungicide containing methylmercury was sprayed on seeds that were inadvertently ground into flour. The symptoms in these other episodes were similar to those experienced in Minamata.

Chronic low-dose exposure to mercury is of particular concern to women in their childbearing years. Mercury accumulates in the body and can cross the placenta and affect the developing fetus. Since the half-life of mercury in the human body is roughly two months (our body metabolizes half of the mercury in two months, then it takes another two months for half of the remaining mercury to be metabolized, and so on), a woman who has built up a level of mercury in her body will still have 25 percent of that mercury inside her four months later, 12.5 percent of the original level six months later, and 6 percent eight months later. In other words, at least some of the mercury a pregnant woman has in her body at conception will still be there when she delivers her baby, even if she ingests no more mercury during the pregnancy.

We know from the acute poisoning experiences in Japan and Iraq that high-level mercury exposure in pregnant women can effect the neural development of fetuses. Children of women who were exposed to these levels of methylmercury during pregnancy have exhibited delayed onset of walking and talking, reduced cognitive test scores, and learning difficulties. Mercury can also pass from the mother to an infant during breast-feeding, but research is inconclusive on whether this transfer causes any harm.

THE RANGE OF EXPOSURES

Fish is the greatest potential source of methylmercury exposure for the public. But the levels most of us are exposed to are not believed to be harmful. The National Academy of Sciences and the EPA report that the typical American is not in danger of consuming harmful levels of methylmercury from commercially available species of fish, in amounts equivalent to a quarter of a cup of tuna fish per day, every day for their whole lives. But certain species of fish, those that are top predators in their ecosystems, have higher levels of methylmercury than others. Freshwater species include largemouth and smallmouth bass, perch, pickerel, lake trout, and walleyed pike. Marine species include shark, swordfish, and tuna.

At current ambient levels, airborne mercury is not thought to be a health risk. (Airborne mercury used to be a serious problem, particularly in the hat industry, which used mercury for making felt. Thus the term "mad hatter's disease" and the Mad Hatter in *Alice in Wonderland*.) At current levels of exposure, mercury in drinking water is not thought to be a hazard either.

Mercury in dental fillings has caused much controversy. This exposure is thought to come from tiny amounts of mercury that vaporize from the filling, which we then inhale. There was some risk to dental hygiene professionals, who were exposed to higher concentrations of these vapors while preparing the amalgam material. But that occupational exposure has been significantly reduced now that most dentists are taking delivery of amalgam already mixed at industrial facilities that have safety systems in place for handling the mercury. As far as risk to the tens of millions of people who have fillings, there is not much hard science. A few studies of adults occupationally exposed (to levels much higher than a dental patient with amalgam fillings) suggest either no or very low-level effects. Two research studies with children were under way at the end of 2001, but no results have been reported.

Another form of mercury exposure has raised concern. In 2001 the Institute of Medicine investigated claims that thimerosal, a mercury-based preservative used in some vaccines, caused illness including autism in some children who got certain vaccines. They found no direct connection between the thimerosal and any health outcome, and found that even the theoretical risk was low. But they supported the decision of the Food and Drug Administration (FDA) to remove thimerosal from vaccines. Under that decision, as old vaccine stocks are used up, new ones may not use thimerosal, which should be completely out of the vaccine supply within a few years.

REDUCING YOUR RISK

As of July 2000, 40 states and American Samoa had issued advisories warning residents about consuming fish caught in local waters. These advisories included all of the Great Lakes, more than 52,000 other lakes, and 238,000 miles of rivers. These advisories suggest that everyone either reduce consumption of fish from specified waters or completely avoid eating fish caught in those waters. They specifically note the risk for women of childbearing age and for people of certain cultures, including Native Americans, Asians, and Pacific Islanders, for whom fish is a more common component of the diet.

For consumers in general, the EPA advises that if you are pregnant or could become pregnant, are nursing a baby, or if you are feeding a young child, you should limit consumption of freshwater fish caught by family and friends to one meal per week. For adults, one meal is six ounces of cooked fish or eight ounces of uncooked fish; for a young child one meal is two ounces of cooked fish or three ounces of uncooked fish.

The FDA suggests that women who are pregnant or could become pregnant, nursing mothers, and young children should not eat shark, swordfish, king mackerel, or tilefish. The FDA guidelines indicate that women of childbearing age and pregnant women can safely eat an average of 12 ounces of fish purchased in stores and restaurants each week. So if in a given week they eat 12 ounces of cooked fish from a store or restaurant, then they shouldn't eat fish caught by their family or friends that week. This calculation is important to keep the total level of methylmercury from all fish at a low level in the body.

～

The EPA has announced plans to reduce mercury emissions from coal-fired power plants. In 2004, the agency will make final general air pollution rules, which will require the burning of cleaner fuels in order to reduce a wide range of pollutants. The EPA estimates the rules will achieve a 50 percent reduction in mercury emissions compared to 1990 levels. It has already enacted regulations to reduce mercury emissions from municipal waste incinerators by 90 percent from 1990 levels. The agency predicts that regulations for medical waste incinerators will reduce levels by 94 percent compared with 1990. It estimates that those regulations will cut emissions from hazardous waste incinerators in half.

~

But there is still a lot of mercury out there. A significant amount is in products you may have around the house. Knowing what they are can help you handle them properly and dispose of them safely whenever you decide to get rid of them. Effective non-mercury-containing alternatives are available for all these uses.

In addition, the federal government has banned mercury additives in paint and pesticides. Battery manufacturers are drastically reducing their use of mercury, and some are helping to run battery recycling programs to recapture mercury rather than allowing it into the waste stream. Thermostat and thermometer manufacturers are turning to alternatives. Mercury tilt switches in the auto and appliance industries are being replaced with mechanical devices. Industrial demand for mercury in the United States dropped 75 percent between 1988 and 1997.

MERCURY IN HOUSEHOLD PRODUCTS

Product	Where to Find the Mercury	Recommended Disposal
Thermostats	Tilt switches: As the mercury expands and contracts, it tilts a switch to trigger the device. You should be able to see a little vial with mercury in it.	The Honeywell Corporation will take thermostats of all makers. Get a special mailing container through any local heating contractor. Or call (800) 345-6770, ext. 733, for a free mailer package.
Thermostat probes	Gas-fired appliances with pilot lights, like ranges, ovens, clothes dryers, water heaters, and furnaces: The mercury is in a metal tube and bulb attached to the gas line inside the device, which turns off the gas if the pilot light goes out.	Contact your health department to find the nearest hazardous waste recycling program.
Barometers, pressure gauges	In the visible indicator	Contact your health department to find the nearest hazardous waste recycling program.

Product	Where to Find the Mercury	Recommended Disposal
Thermometers	In the visible indicator	Contact your local health department to find the nearest hazardous waste recycling program. If a thermometer breaks, ventilate the room and pick up the mercury with an eyedropper, two razor blades or knives, or heavy paper, like playing cards. Don't use a vacuum cleaner, which can turn the mercury into an aerosol. Put the mercury in a plastic bag, inside another one, both tightly sealed. Contact your health department to find the nearest hazardous waste recycling program.
Batteries	Inside the device	Contact your health department to find the nearest battery or hazardous waste recycling program.
Fluorescent bulbs	Inside the bulb itself	If the bulb breaks, clean up the powder the way you would clean up a broken thermometer. Contact your health department to find the nearest hazardous waste recycling program.

FOR MORE INFORMATION

Environmental Protection Agency
Office of Science and Technology
www.epa.gov/OST/fish (fishing advisories)
401 M Street SW, Maildrop 4305
Washington, DC 20460
(202) 260-1305

For information on mercury in general, or up-to-date information about fish advisories in your area, contact your local or state public health or environment department.

National Academy of Sciences
www.nap.edu/books/0309071402/html (report on mercury)
2001 Wisconsin Avenue NW
Washington, DC 20007
(888) 624-8373
(202) 334-3313

This chapter was reviewed by Dr. David Bellinger, Associate Professor of Neurology, Harvard Medical School, and an epidemiologist at Boston Children's Hospital; and by Roberta White, Professor of Neurology, Environmental Health, and Psychology at the Boston Veterans Administration Medical Center and Boston University School of Public Health.

31. NUCLEAR POWER

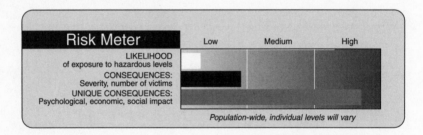

IT PROMISED a new age of energy, harnessing the potential of the atom to create electricity "too cheap to meter," from a source that wouldn't foul the skies and our lungs with air pollution. The first commercial nuclear power plant in the United States began operating in 1951. Thirty years later, the nuclear power industry was moribund, frozen in part by a series of accidents, but also by our fear of things associated with radiation.

THE HAZARD

No explanation of nuclear power can put this risk in perspective without a basic explanation of nuclear fission, the process by which atoms release energy when they split. (Fusion is the process of joining together parts of atoms. It releases much more energy. The sun works by fusion.)

The atoms of some elements are naturally unstable and tend to break apart and rearrange themselves into more stable forms. This phenomenon is called radioactive decay. As atoms decay, they emit subatomic particles. Those particles, and the energy they carry, are known as radiation. The uranium used in nuclear power is one of these unstable elements. One particular kind of uranium is particularly good for nuclear fission, because it emits subatomic particles—neutrons—that cause other uranium atoms to split when a neutron hits them. It is called uranium 235. If a U235 neutron flies off and hits the nucleus of another U235 atom, the nucleus in the second atom splits and sends some of its neutrons flying. If they go on to hit yet more U235 atoms, and there's enough U235 around, a self-sustaining reaction can start.

Each time an atom splits, the remaining atomic nucleus is slightly smaller than it used to be. That loss of mass gets converted into a huge amount of energy. A one-pound sphere of U235, about the size of a tennis ball, has the energy equivalent of a million gallons of gasoline, which would occupy a building 50 feet tall, wide, and long.

But the fuel in nuclear power plants can't explode the way atomic weapons do. In atomic weapons, designed to maximize the chain reaction in a very short period of time, the concentration of U235 is 90 percent or more. But in reactors, U235 makes up only 2 to 4 percent of the fuel. There isn't enough U235 in nuclear plant fuel to allow for enough fission to occur fast enough to produce an explosion.

Instead, the danger from nuclear power plants is from the radiation produced by nuclear fission. It's dangerous because it is ionizing radiation, which means that the subatomic particles produced by radioactive decay have enough energy to break chemical bonds and damage molecules they may hit. Changes in molecules can damage cells, tissues, and organs, and cause mutation to the DNA molecule, leading to cancer. Nuclear power produces large amounts of ionizing radiation because it is a concentrated and accelerated process of radioactive decay.

But the decay particles vary, and some are more dangerous than others. The subatomic pieces of an atom emitted as it decays are called alpha or beta particles, or gamma rays. Alpha particles emitted by radioactive decay are too small and weak to penetrate even skin. Beta particles can penetrate the body only a fraction of an inch. So the only way these particles can get inside us and do any harm is if they are inhaled or swallowed. Gamma rays, however, have more penetrating power, and only thick layers of concrete, metal, or liquids like water or oil can stop them.

~

It is also valuable to understand the basic way that the 104 operational commercial power reactors and 37 small academic or research reactors in the United States work. First the fuel is shaped into pellets about the diameter of a dime. A row of pellets is piled up to form a rod several feet long. Each rod of fuel pellets is held inside a metal tube. The tubes are then grouped in square bundles of between 36 and 289 each. In between the bundles of fuel rods, control rods made of material that can capture and absorb those little flying neutrons keep them from going on to hit and break apart other uranium nuclei. By absorbing the neutrons, control rods can limit, or stop, the chain reaction.

The fuel bundles (from several dozen up to several hundred, depending on the desired power of the reactor) are inserted into a roughly cylinder-shaped frame, the core of a reactor. The core hangs down inside the reactor vessel, a 6-to-8-inch-thick stainless-steel shell usually 10 to 15 feet wide and 30 to 40 feet high. Inside the reactor vessel, the core is submerged in water. When the operators slide out the control rods, the natural decay of some of the U235 starts a self-sustaining reaction. But that reaction is controlled by keeping enough control rods partly in place, and adding other neutron-absorbing chemicals to the water, so only the right number of neutrons are allowed to fly around. This check restricts how many atoms are splitting at any one time and controls the temperature of the core so it doesn't get too hot. The water around the core is usually about 570°F. The heated water, channeled through pipes, creates steam that drives a turbine and produces electricity.

The radioactive components of nuclear plants in the United States are enclosed in a steel-and-concrete-walled inner primary containment structure. Beyond that, the secondary containment structure forms the outer shell of nuclear plants. It's designed to capture contaminated water, steam, or air that escapes from the primary containment (which is what happened at Three Mile Island in Pennsylvania). The secondary containment structure must be able to withstand earthquakes, hurricanes, and other destructive forces. Nuclear plants must also have systems for cooling down massive amounts of steam if a major pipe breaks. Some keep millions of gallons of cool water in storage tanks inside the containment building. Some keep ice, as much as two million pounds.

~

But radioactive emissions from nuclear power plants are only one form of potential hazard. Another is the radioactive waste that's generated. The spent fuel is called high-level waste. Spent fuel is intensely radioactive. Standing right next to a spent fuel rod, you'd be exposed to four times as much radiation as it takes to cause death from radiation poisoning within one week. It's called "spent" because so much of the U235 has decayed that the fuel can no longer sustain an ongoing reaction. But though the U235 is mostly gone, many of the atoms that are left are also unstable. These atoms are radioactive byproducts of fission, the new atoms formed when the U235 atoms break apart. They are sometimes called "daughter atoms." They decay too and give off radioactive energy for periods of between months and thousands of years. This is where that familiar term "half-life" comes into play. That's the time it takes for half of the atoms to decay into a stable form in which they stop breaking down and emitting radioactive particles.

Some radioactive elements decay pretty quickly. Iodine 131 has a half-life of just 8 days. But plutonium 239 atoms take 24,000 years to get to the point at which half of them have decayed into stable daughter atoms and are therefore no longer radioactive. After another 24,000 years, half of the remaining plutonium is no longer radioactive, and so on. The shorter the half-life, the more the atom is giving off its energy in a concentrated time, so the more radioactively energetic and dangerous it is. Isotopes with long half-lives are, by definition, less energetic and less hazardous.

~

Most high-level radioactive waste is contained inside the primary containment buildings at nuclear plants, deep enough inside tanks of purified water, at least 20 feet deep, so the radioactive alpha and beta particles and gamma rays can't penetrate to the surface. The water in the storage pools also heats up from the continued radioactive decay of the fission byproducts. Since the federal government is decades behind in fulfilling its obligation to build a permanent storage facility, some older nuclear plants have started to run out of room in their spent fuel pools. They have begun to store spent fuel in special concrete containers designed to be thick enough so radiation cannot escape. Heat from the fuel dissipates into the surrounding air, but the radioactive particles can't get out of the container. These containers are usually stored outside.

As of the end of 2001, roughly 26,000 tons of spent nuclear fuel were stored in the United States. Nearly all of it was stored at reactor sites. The rest was in small off-site facilities licensed to handle such waste. A high-level waste storage facility under construction at Yucca Mountain near Las Vegas is not due to come online, at the earliest, until some time after 2010. As of the beginning of 2002, Yucca Mountain had yet to win final approval, but no other site was being developed.

There are a lot of other radioactive wastes generated at a nuclear power plant, such as clothing worn by plant workers, rags used for cleaning, and resins collected in filters. These low-level wastes are shipped to three disposal sites, in South Carolina, Washington, and Utah. There the waste is buried in specially lined and monitored landfills. Waste from filters that cleaned water from the nuclear core requires packing in lead and concrete containers, but most low-level waste requires no special handling.

THE RANGE OF CONSEQUENCES

Radiation is believed to be a relatively weak carcinogen. This is based largely on a study of survivors of the atomic weapons used in Japan in World War II. Approximately 90,000 survivors exposed to radiation from the explosions over Hiroshima and Nagasaki have been monitored. The normal rate of cancer from all causes, 17 to 20 percent, would produce 14,600 to 18,000 deaths in the study group over their entire lifetime. In the Hiroshima-Nagasaki study subjects, in the years since those atomic weapons were used, there have been approximately 8,000 cancer deaths from all causes. (More will undoubtedly occur because not everyone in the study group has reached the later ages at which most cancers develop.) By comparing the number of cancer deaths in the study group of people exposed to high levels to the cancer deaths of people exposed to little or no radiation, scientists estimate that approximately 500 of the cancer deaths in the Hiroshima-Nagasaki group are attributable to radiation. The low rate so far suggests that high-level radiation is indeed a carcinogen, but a relatively weak one.

While many different kinds of cancer are associated with radiation exposure, those most connected include leukemia (cancer of the blood), breast cancer, lung cancer, and cancer of the thyroid gland. Another significant concern about exposure to high doses of radiation is possible damage to the reproductive organs, and specifically to eggs in a woman's ovaries or to the cells in men which produce sperm. Mutations in these cells and organs can impair fertility. These mutations also cause birth defects in animal tests. But this effect has not been detected among the Hiroshima-Nagasaki survivors. The survivors who became pregnant *after* their exposure show no elevated rate of birth defects in their offspring. However, survivors who were already pregnant at the time of high exposure from the bombs had higher rates of babies with birth defects, particularly if the exposure occurred in the middle months of pregnancy when fetal central nervous system development occurs.

Acute effects from very high doses of radiation include sterilization, cataracts, skin damage, benign tumors, hair loss, and interrupted menstrual cycles. Even higher levels can cause death. These effects occur only from direct or very close contact with highly radioactive materials and are not the consequences that would be experienced by the public from a nuclear power accident. Nuclear plant workers or nuclear material handlers could experience these types of consequences. Several accidents at nuclear power facilities worldwide have resulted in acute radiation injury for a few dozen workers over the past few decades.

~

As potentially dangerous as an accident at a nuclear power plant might be, it is an example of how the concept of risk involves more than just probability. The likelihood of an accident that would release dangerous levels of radiation is very low. But the consequences of such an accident would be enormous. The Nuclear Regulatory Commission (NRC) estimates the chance of an accident that damages the core at a nuclear plant in this country, but does not release radiation, as 5 in 100,000 per year. There have been 11 such accidents worldwide since commercial nuclear power began in 1951. The last one in the United States was in 1989 at a commercial power reactor in Connecticut.

The NRC estimates the chance of a core damage accident that does release radiation as 5 in 1,000,000 per year. There have been three such accidents worldwide.

THE RANGE OF EXPOSURES

The Windscale nuclear facility (now called Sellafield) in the United Kingdom, built in the late 1940s to produce radioactive material for atomic weapons, experienced a fire and release of radioactivity in 1957. The British National Radiological Protection Board said in 1987 that the accident is likely to contribute to 33 premature cancer deaths over the lifetimes of the several hundred thousand people exposed to the radiation.

The second occurred at the Three Mile Island (TMI) Unit 2 reactor in March 1979. The accident was caused by a combination of mechanical, design, and operator error. Some of the water around the core was lost, and the partially exposed core grew so hot, 5,000°F, that half of it melted. Levels of hydrogen gas built up inside the containment structure and threatened an explosion. To reduce the pressure and the risk of explosion, gases that contained low levels of mostly radioactive xenon were released for several days. Two days into the event, as national news coverage focused on the ongoing incident, Pennsylvania governor Richard Thornburgh ordered the precautionary evacuation of 3,500 preschool children and pregnant women within 5 miles of the plant. Approximately 200,000 people in a much wider area fled. One Roman Catholic priest offered evacuees general absolution, the sacrament for those about to die.

Twenty years later, a state health registry that tracked 30,000 area residents was disbanded after it found no health effects. A citizen's lawsuit claiming health effects was dismissed for lack of evidence. Several large health studies found no health impacts. The NRC estimates that the maximum exposure to a person right at the edge of the TMI property line would have been 100 millirems. Average exposure from TMI was approximately 1 millirem. For context, the average U.S. resident is exposed to 360 millirems per year from natural background sources like the sun, the earth, and foods that pick up radioactivity from the soil. (See Chapter 34, "Radiation," for a chart showing the various sources of background radiation exposure.)

Massive changes in equipment, staffing, design, and other regulatory requirements were imposed on the nuclear industry in the wake of TMI. These changes dramatically increased nuclear plant safety and efficiency, but raised design, construction, and legal costs so much that it effectively ended expansion of the nuclear industry in the United States. Since TMI, plans for 34 new nuclear plants have been canceled, and 29 of those had been canceled by the time the third and most serious nuclear accident took place, in Chernobyl, in April 1986.

The plant in Ukraine had a different design from American nuclear reactors. Like the Windscale plant, Chernobyl used graphite instead of water to control the neutrons in the core. A series of mistakes by operators led to an explosion that set some of the graphite on fire. It was not a nuclear explosion, but it blew apart the building surrounding the core. Chernobyl did not have a secondary containment building as is required in the United States. As a result, the fire, which burned for days, spewed high levels of radioactive material directly into the environment. Ultimately, the wind spread this radioactive material across the entire Northern Hemisphere, though the levels high enough to be of health concern were confined to Ukraine, Belarus, Russia, and parts of northern Europe, including sections of Sweden and Norway.

Of those who worked to bring the catastrophe under control, 31 died from acute radiation exposure; 400 plant staff, firemen, and medical personnel suffered short-term radiation poisoning; and 800,000 "liquidators" who worked to clean up the site and build the concrete shell around it to contain dispersion of radioactive dust and debris received significant doses of radiation. More than 100,000 people living near the site had to be evacuated, and most of them received doses of radiation above naturally occurring background levels. Nearly a quarter of a million people were permanently relocated. A region 30 kilometers around the plant is still contaminated by levels of radiation too high to allow human occupation. Some locations in Europe farther from the site, including some bodies of water in Sweden, still register mildly elevated levels of radiation.

Although dozens of health studies have examined the effects of Chernobyl, the results still aren't clear. One problem is that there wasn't much reliable baseline public health data about disease rates before the accident. Epidemiologists say another problem has been the questionable methodology used in many of the Chernobyl studies. A major United Nations review of these studies in the year 2000 found that 1,800 cases of childhood thyroid cancer (fortunately, a treatable form of cancer) can be associated with the accident. There do not appear to be increases in other long-term physical health effects, notably in the rate of leukemia, a cancer strongly associated with radiation exposure which generally shows up between 2 and 5 years after exposure but which can occur as much as 40 or 50 years later.

Nor have there been confirmed long-term, chronic radiation-related health effects among the liquidators. But since cancer is largely a disease of the elderly, those exposed at Chernobyl will have to be monitored for decades before we know with more certainty the health effects of that nuclear accident. The full effects may not have shown up yet.

One impact on public health from Chernobyl is clear. There is a high occurrence of suicide among the liquidators, which epidemiologists say may well be associated with Chernobyl, and a number of psychological problems among the tens of thousands who had to evacuate their homes because of the accident.

Under normal operating conditions, nuclear plants emit gaseous radiation from vent stacks. The rate of release in this country is controlled by the U.S. government and cannot exceed 100 millirems per year, or 2 millirems per hour. These levels are widely considered safe based on the experience of atomic bomb survivors.

The area near a nuclear power plant most exposed to risk from a core damage/radioactive release accident is the zone within a 10-mile radius of the plant. Towns within this area must have alert systems to notify residents in case of an accident. People living within this zone are supposed to receive an annual mailing from state emergency management officials describing what actions to take in case of an accident. This procedure is similar to civil defense planning for other disasters, like hurricanes. In such an event, residents are instructed to tune to local emergency notification radio stations for instructions. Those downwind of a plume escaping from the plant will be instructed either to stay indoors, with windows and doors shut, or to evacuate their homes. Those upwind will be told to stay tuned for further instructions.

Each nuclear plant is required to have a detailed evacuation plan worked out with local and state emergency officials. Each plan is tested every two years with a simulated accident that requires public safety officials to initiate all the communications systems they would need in case of an actual evacuation.

Many city and state emergency officials keep a supply of potassium iodine tablets on hand for distribution in the evacuation zone. The iodine is taken up by the thyroid gland and essentially fills it, so radioactive iodine from a release can't be absorbed. More states are considering this step in the wake of evidence released by the U.S. government in early 2002 that terrorists had been considering attacks on nuclear power plants. (Many public health officials think a lot of the thyroid cancers caused by Chernobyl could have been prevented had Soviet officials admitted the scope of the accident earlier and quickly dispensed potassium iodine pills to people in the region.)

A wider 50-mile radius around the plant is the area in which radioactive material could significantly contaminate soil, water, or food. In the event of an accident, people in this area are notified but do not have to take immediate precautionary action.

REDUCING YOUR RISK

People living within the evacuation zone of a nuclear power plant should take the same precautions that people living in flood-prone or hurricane-prone areas should take in case they need to evacuate their homes. Have a kit of basic supplies ready. Be familiar with the evacuation route that applies to your location. Those routes should be described in your annual mailing from the state government. Ask for it if you haven't received it. This information is often printed on the back few pages of a calendar, to make it handy to post and keep accessible.

In addition, people in these areas should know how to access their emergency notification radio station. Having a radio that runs on batteries is recommended too. (Keep a supply of fresh batteries.) The emergency broadcast system will be used following a nuclear plant accident. It is vital that people listen to the warnings issued by local authorities. Depending on the conditions, it may be safer to stay inside while the radioactive cloud passes over than to go outside and try to drive away. Local authorities will provide guidance on the best way to avoid prolonged exposure to radiation.

FOR MORE INFORMATION

A website put together by a former nuclear reactor operator provides a detailed look at how nuclear plants operate, and more on the industry: www.nucleartourist.com.

U.S. Department of Energy
Argonne National Laboratory
www.insc.anl.gov

National Science Foundation
whyfiles.org/020radiation (health effects)

U.S. Nuclear Regulatory Commission
Office of Public Affairs
www.nrc.gov
One White Flint North
11555 Rockville Pike
Rockville, MD 20852-2738
Washington, DC 20555
(800) 368-5642

24-Hour Incident Response Operations Center: (301) 816-5100
Toll-free Safety Hotline: (800) 695-7403

Union of Concerned Scientists (advocacy group founded by
    former nuclear physicist and Nobel Prize winner Henry
    Kendall)
www.ucsusa.org
2 Brattle Square
Cambridge, MA 02238-9105
(617) 547-5552
Fax: (617) 864-9405

*This chapter was reviewed by Dr. John Little, Chairman of the Harvard School of
Public Health, Department of Cancer Cell Biology, and head of the Center for Radiation Sciences and Environmental Health; by David Lochbaum, Nuclear Scientist at the Union of Concerned Scientists; and by Gilbert Brown, coordinator of the
nuclear engineering program at the University of Massachusetts, Lowell.*

32. OZONE DEPLETION

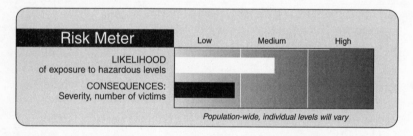

DEPLETION of the ozone layer in the stratosphere is an interesting les-
son in risk. This hazardous process is actually the result of chemicals
that were supposed to be more environmentally friendly than the dangerous ones they were replacing. But in eliminating some risks, ozone-
depleting chemicals created others.

In 1921, the American engineer and chemist Thomas Midgley, Jr., discovered the first of what would become a family of chemicals known as chlorofluorocarbons, or CFCs. The trade name of Midgley's chemical was Freon. Midgley's discovery was great news for the environment at the time, because CFCs are nontoxic, noncorrosive, and nonflammable. Because they are effective coolants, these chemicals became popular in refrigerators, and later in air conditioners. And because these gases are so environmentally benign, industry used them as propellants in aerosol spray cans and as solvents to clean electronic components like circuit boards after manufacture, replacing more hazardous materials that had been used. And because CFCs trap heat so well, they were used to blow the tiny bubbles into foam insulation and foam beverage containers.

But CFCs had several hidden properties that scientists later realized might spell trouble for a vital part of the biosphere, and for human health.

THE HAZARD

In the stratosphere, a layer of the atmosphere between 6 and 25 miles up (Mount Everest is about 5 miles high), a chemical called ozone is mixed in with the nitrogen, oxygen, carbon dioxide, and other natural components of the air.

Ozone is a molecule made up of three oxygen atoms. It is constantly being created and broken down by chemical reactions in the atmosphere, so under natural conditions the ozone concentration stays pretty much the same. There is never very much ozone in the stratosphere. For every 10,000,000 molecules of air, only 3 are ozone.

But that incredibly low level is still vital to life on earth. Ozone absorbs a form of ultraviolet radiation known as UVB (see Chapter 36, "Solar Radiation"), which damages the living tissues of animals and plants in a number of ways. Without ozone in the stratosphere, life as we know it would be impossible.

Because of some unique characteristics, CFCs destroy stratospheric ozone. First, CFCs are very stable. That means the atoms that CFC molecules are made of are strongly attached to one another. As a result, when CFCs get into the air, they last a long time, long enough to survive as the winds slowly lift them up to the stratosphere. Only the powerful radiation from the sun as it hits the upper levels of the atmosphere is strong enough to break these molecules apart.

Second, one of the Cs in CFCs stands for chlorine. This reactive element has a particularly destructive effect on the ozone molecule. So when CFCs get up to the stratosphere, and solar radiation breaks them apart and releases the chlorine atoms, each atom of chlorine can help rip apart tens of thousands of ozone molecules.

Third, because they are such stable molecules, CFCs survive in the stratosphere a long time once they get there. They don't break apart right away, even in the presence of intense solar radiation. Some last for several decades before they break down and release their destructive chlorine atoms. Consequently, CFCs release their chlorine atoms slowly, over time. Even though the production of CFCs is being phased out, because of their persistence, the molecules already released will continue to damage stratospheric ozone for several more decades.

Other chemicals that have similar characteristics and that also contribute to ozone depletion include: halons (used in fire extinguishers), the cleaning solvent carbon tetrachloride, and the fumigant methyl bromide.

THE RANGE OF CONSEQUENCES

Since the loss of stratospheric ozone leads to greater exposure to UVB radiation, the consequences of ozone depletion are the same as those discussed in Chapter 36 on solar radiation. They include two kinds of skin cancer: basal and squamous cell cancers. (No clear relationship exists between ozone depletion and higher incidence of malignant melanoma, the third type of skin cancer caused by exposure to solar radiation.) UVB also causes skin damage, cataracts, and some impairment of the immune system.

The Environmental Protection Agency estimates that every 1 percent decline in stratospheric ozone concentration results in a 1.5 to 2 percent increase in exposure to UVB radiation. It estimates that for every 2 percent increase in UVB exposure, there is an increase of 2 to 6 percent in cases of basal and squamous cell skin cancers, the types that are the most common, and if detected early, are most easily treatable.

THE RANGE OF EXPOSURES

The ozone layer is like a blanket that wraps around the entire earth. But ozone depletion is not the same everywhere. Perhaps you've heard of the ozone hole. This significant drop in stratospheric ozone concentra-

NORTH POLE

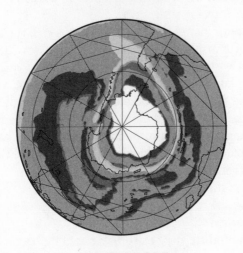

SOUTH POLE

Total ozone for October 6, 2000

tion occurs seasonally in an area around the South Pole. Unique conditions of wind, temperature, and humidity develop in this region from September through November and accelerate the breakdown of ozone. The winds cause this breakdown to occur in a confined area, which gives rise to a distinct "hole."

But it's not really a hole, just a significant depletion in ozone concentrations. In the year 2000, concentrations dropped by up to 70 percent over an area of 17.7 million square miles. (North America covers 21 million square miles.) The ozone depletion zone in 2000 extended over some populated areas in southern South America. The size of the hole and the level of ozone loss vary from year to year, and season to season, depending on climate conditions.

Another reason that the ozone depletion zone is called a hole is that at the altitudes where ozone is most abundant, 15 to 20 kilometers up, nearly all of it is destroyed, while not much happens to the ozone concentrations at higher or lower altitudes.

In the Northern Hemisphere, ozone depletion doesn't form such a distinct hole. There is still a seasonal effect, because meteorological conditions in the winter and spring favor ozone destruction more than in summer and fall. But depletion of ozone over the Northern Hemisphere is spread out, or thinned out, largely because winds don't trap the chemicals and moisture in the atmosphere in as distinct an area as occurs around the South Pole.

Satellite measurements indicate that as of 1998, the ozone loss over the Northern Hemisphere in the middle latitudes (between 30 and 60 degrees north—which covers most of North America, Europe, and much of Asia) was 6 percent in the winter and spring, and 3 percent in the summer and fall. That resulted in UVB increases of 9 to 12 percent in the winter and spring, and 4.5 to 6 percent in the summer and fall. (These percentage decreases in ozone concentration, and increases in UVB, are compared to what scientists have calculated for the level of ozone without the presence of CFCs.)

REDUCING YOUR RISK

The best way to reduce the health risk posed by ozone depletion is by controlling your exposure to sunlight, as outlined in Chapter 36. But the problem of ozone depletion is also being tackled on a global scale, and represents one of the world's most successful international efforts to deal with a global environmental problem.

Based on scientific findings about ozone depletion by Paul Crutzen, Mario Molina, and F. Sherwood Rowland (findings that won them the Nobel Prize in Chemistry in 1995), a massive global research program in the 1970s and 1980s convinced nearly all the nations of the world to sign the Montreal Protocol. That agreement called for a freeze on the production and use of CFCs. Subsequent amendments have strengthened the Montreal Protocol to accelerate the phaseout of ozone-destroying chemicals and to add newly recognized chemicals to the phaseout list.

As of 2001, the production of CFCs was completely banned. Halon production and use has been severely limited. Carbon tetrachloride has been banned. As of 2005, methyl bromide will also be banned. As a result of these steps, the global consumption of CFCs fell from 1.1 million tons in 1986 to 160,000 tons in 1996. The presence of ozone-depleting chemicals in the stratosphere has been declining since 1994. Scientists estimate that roughly 50 years from now, the natural process of ozone creation and destruction will be back in balance and the risk from man-made ozone depletion will be gone.

FOR MORE INFORMATION

Environmental Protection Agency
www.epa.gov/docs/ozone
Ariel Rios Building
1200 Pennsylvania Avenue NW
Washington, DC 20460
Stratospheric Ozone Protection Hotline: (800) 296-1996

The National Academy of Sciences has a fascinating account of the ozone depletion issue in an article, with graphic illustrations, based on the account of F. Sherwood Rowland, who shared the 1995 Nobel Prize for theorizing ozone depletion with Mario Molina and Paul Crutzen.
www4.nationalacademies.org/beyond/beyonddiscovery.nsf/web/ozone

National Aeronautics and Space Administration
Advanced Supercomputing Division
www.nas.nasa.gov/About/Education/Ozone

This chapter was reviewed by Mario Molina, who shared the 1995 Nobel Prize for his contributions to the theory of ozone depletion. Professor Molina is the Lee and Geraldine Martin Professor of Environmental Sciences in MIT's Department of Earth, Atmospheric and Planetary Sciences.

33. PESTICIDES

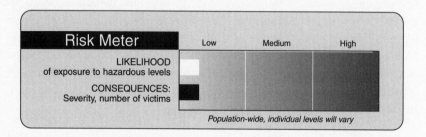

Risk Meter	Low	Medium	High
LIKELIHOOD of exposure to hazardous levels			
CONSEQUENCES: Severity, number of victims			

Population-wide, individual levels will vary

PESTS HAVE BEEN EATING our food, crowding out our crops, and spreading disease for thousands of years. People have battled back with anything they could find. Arsenic was first mentioned as a pesticide 2,000 years ago. The Egyptians, Greeks, and Romans used sulfur, lye, and other compounds that smelled or tasted foul to battle mosquitoes, wasps, lice, and fleas. Farmers used concoctions that included vinegar, cow dung, fish oil, brine, and urine to control bugs or molds and fungi, anything that threatened their crops. It was a low-tech war between humans and pests which seemed to follow one rule: "The fouler smelling-tasting-looking, the better."

But with the advent of chemical synthesis, the battle against pests took a significant turn. Some of these new chemicals were developed as far back as the late 1800s, but general advances in chemical engineering that began in World War II and have accelerated since then have given rise to a whole new age of technologically produced chemicals. Some of these chemicals were discovered to have molecular properties that could kill unwanted insects, like crop-threatening bugs or disease-carrying mosquitoes. One of the first of these was DDT, and the discovery that DDT could kill insects won Paul Müller the 1948 Nobel Prize in Medicine and Physiology. (See Chapter 24, "DDT.")

Pesticides have always been a public health conundrum. On one hand, they are designed to be lethal, killing bugs, weeds, and other damaging pests by acting on biological systems that are similar to those found in humans. But by controlling those pests, these chemicals have increased food production and dramatically reduced disease worldwide. As recently as 1970, the U.S. National Academy of Sciences lauded DDT. But in 1962, Rachel Carson's book *Silent Spring* had already raised concern that DDT and other pesticides were harming the environment, in ways that might also threaten human health.

As a result of growing concerns about possible adverse effects, pesticides are the most actively regulated class of chemicals in commerce. Dozens of uses have been banned after tests found potential danger to humans. And aggressive standards are set to make sure that those pesticides that are allowed can be used only at doses that won't cause harm to people. Still, concerns persist.

THE HAZARD

Most people think of pesticides as chemicals that kill insects. Around the home these might be ants, wasps, aphids, or cockroaches. On the farm they might be more exotic-sounding insects like the European corn borer, wireworms, stinkbugs, or leaf rollers. But those are only *insecticides*. Pesticides kill anything we don't want around. *Herbicides* kill plants (weeds on farms, or on your lawn or in your garden). *Fungicides* kill fungi. *Rodenticides* kill rats and other unwanted pest rodents. There are *algicides* for algae (commonly used in swimming pools), *miticides* for lice and mites (including on schoolchildren), *biocides* and *disinfectants* (like chlorine that kills microbes as it cleans most of America's drinking water), *nematicides* that kill microscopic worms that eat plant roots, even *slimicides,* used to kill microbes in large industrial air conditioners. Most of these work by interfering with biochemical processes in the target species and are included in the category "conventional" pesticides.

There is also a category, known as "other" pesticides, that includes things like *sulfur* and *oil.* These pesticides have a more directly physical effect, by clogging an insect's breathing passages, dissolving the protective wax on the shell of an insect or the leaf of a plant, or interfering with the ability of insects to lay eggs. They work more mechanically than biochemically.

Pesticide molecules are almost always synthetic; that is, they are manufactured in labs and chemical plants. But some of these synthetic products are actually modeled on natural pesticides. These are called "biopesticides," and they include nicotine (from tobacco), rotenone (extracted from plants in the pea family), and pyrethrum (taken from chrysanthemums). Biocides are used on both conventional and organic farms. (The synthetic form of these natural agents lasts long enough in

		Who Uses It? Agriculture Industry/Commercial/Government
Pesticide	Use (Millions of Pounds)	Home and Garden
Herbicides/plant growth regulators	568	Agriculture: 470 Ind/Comm/Gov: 48 Home and Garden: 49
Other pesticides (sulfur/oil)	173	Agriculture: 144 Ind/Comm/Gov: 14 Home and Garden: 15
Fumigants/nematicides	165	Agriculture: 140 Ind/Comm/Gov: 24 Home and Garden: 1
Insecticides/miticides	128	Agriculture: 82 Ind/Comm/Gov: 30 Home and Garden: 17
Fungicides	81	Agriculture: 53 Ind/Comm/Gov: 20 Home and Garden: 8

CONVENTIONAL PESTICIDES AND USAGE IN 1997

Source: EPA Office of Pesticide Programs

the environment to work more effectively. The natural forms usually break down too fast.)

America uses 17 pounds of pesticides per person each year. Nearly half of this amount is chlorine used to disinfect water. There are nearly 21,000 pesticide products that contain a total of about 890 active ingredients registered by the Environmental Protection Agency (EPA). Above is a recent table of most of the major categories of conventional pesticides.

Different conventional pesticides work in different ways. But the concern about some of them is that the biological processes they attack in the plant, the insect, the fungi, the rat, or the algae are similar to biological processes in humans, so what the pesticide does to the pest, it might also be doing to us. (Some pesticides attack processes like photosynthesis in plants, which have no human equivalent.) So to understand this risk, it's useful to understand how the major types of pesticides work.

Physical Toxicity

Some pesticides work in purely mechanical ways. As we mentioned, some oils clog the respiratory systems of insects or interfere with their egg laying. Some herbicides destroy the membranes of cells in certain plants. (The cell membranes in plants and animals are significantly different.)

Nervous System Disruptors

Most of the common insecticides and nematicides, and some rodenticides, attack a key part of the firing of nerve impulses in insects. The process they interrupt is similar in both bugs and humans. As an electrical signal travels along a nerve, it leaps from one nerve cell to the next across a gap called a synapse. Chemicals carry the message across the gap, and then those chemicals break down, so the signals stop. Insecticides work by keeping the chemical messenger from breaking down. That causes the nervous system signaling to continue. The continuous firing of the nerve impulses in the target species kills it.

Other pesticides interrupt the flow of nerve signals by disturbing the balance of chemicals along the nerve cells themselves, not in the synapses between them.

Photosynthesis Inhibitors

These pesticides block a plant from using sunlight to convert carbon dioxide and water into carbohydrates and energy.

Metabolic System Inhibitors

These block the processes by which plants or animals, including humans, make and transfer energy within and between cells. In insects and animals this blockage can cause cells to die. In plants and fungi, it can inhibit germination or growth.

Protein Synthesis and Enzyme Disruption

Proteins are the basic building blocks of all living cells, and enzymes are one type of protein. Some pesticides interfere with how proteins and enzymes work, and some break apart critical proteins, preventing enzyme production. Some, particularly fungicides, interfere with proteins that regulate cell division. Others, like rodenticides, interfere with the blood's ability to clot. These protein and enzyme processes are similar in both humans and the organisms that pesticides target.

Hormone Disruptors

Hormones are chemical messengers that are critical to regulating many biological functions in all plants and animals. Some pesticides are able to simulate the specific hormones in specific target pests. Some of these fake hormones trigger too much of the biological response in the organism. Others can block hormones and prevent an adequate level of that response. Such interference can inhibit growth, or the development of vital organs for survival, like part of the outer shell of some insects as they molt. Other hormone-related agents, like pheromones, flood an area with the sexual scent of female insects and confuse the male's ability to find a mate. In many cases, these pesticides target systems that have no equivalent in humans.

THE RANGE OF CONSEQUENCES

Pesticides can cause harm right away—acute poisoning from high-dose, short-term exposure—or chronic effects from low-level exposure over longer periods.

Deaths from acute pesticide poisoning are rare. According to the American Poison Control Centers in the year 2000, there were 21 poisoning deaths attributable to pesticides: 14 were suicides, 5 were unintentional or due to unknown circumstances, and 2 were in children. But poisonings that are not fatal are much more common. The EPA estimates that 80,000 children and 96,000 adults are exposed to enough pesticides each year to suffer some effects. Only about one quarter of these cases are actually registered with medical authorities. Because the symptoms are so mild, the EPA estimates that others occur but the victims and their medical providers are never aware that the cause was pesticide exposure. Perhaps of greater concern are the EPA-estimated 10,000 to 20,000 agricultural workers who suffer nonfatal pesticide poisoning each year.

Pesticide poisoning produces a number of symptoms:

- Low-level poisoning causes fatigue, headache, lightheadedness, sweating, blurry vision, cramps, nausea, diarrhea, and vomiting.
- Moderate poisoning symptoms include numbness, variable heart rate, difficulty breathing and walking, muscle tremors and twitching, and excessive salivation.
- Severe poisoning can produce convulsions, coma, and death.

Most people don't realize that the active ingredient in a pesticide is a very small proportion of the overall product. Most of the liquid or spray or powder in products that you use, or even the tiny levels of pesticide residue on food, consists of inert ingredients. These are the carriers that help deliver the active ingredient. Oils help the pesticide stick to plants, liquids help it spread in water, and powders help it spread over the soil and adhere to plants. These inert ingredients can occasionally cause acute consequences. Inert ingredients in liquid pesticides are often petroleum products that cause eye, nose, and throat irritation, and in rare cases a chemically induced pneumonia. Inert ingredients in dry pesticides include talc, clay, or ground-up plant products like corncobs.

~

Perhaps a greater concern about the consequences of pesticides are the chronic effects that may arise if low levels of exposure over a long period of time cause symptoms unrelated to those from acute poisoning. Some of the chronic effects of pesticides found in testing on laboratory animals include birth defects, nervous system disorders, damage to fertility, interference with the endocrine (hormone) system, and cancer.

Several currently approved pesticides are listed as probable or suspected carcinogens because they have been found to increase rates of certain types of cancer in rats or mice exposed to very high doses for their entire lifetime. But no pesticide is considered a known human carcinogen based on conclusive evidence in humans.

One group of pesticides, the chlorphenoxy herbicides, have received attention as potential carcinogens. The most common herbicide used in homes and gardens, 2,4-D, belongs to this group. So does its chemical cousin, 2,4 D-T. A mix of these chemicals was in Agent Orange, a defoliant used during the Vietnam War. Animal tests for cancer from chlorophenoxy herbicides have shown no increased tumor rates. But some epidemiological studies suggest an association between these compounds and cancer. Other epidemiological studies find no relationship. As of the beginning of 2002, the federal government was conducting an intense review of these pesticides to see if they have any association with cancer. These studies are complicated because dioxin, a highly toxic byproduct, is formed during the manufacture of 2,4 D-T, but it's not in 2,4-D at all.

Several fungicides are also suspected of causing cancer at high doses. They attack the biological processes in the fungi in a way that can damage human DNA. Some fungicides have been found to be carcinogenic in lab animals in very high doses.

Another class of pesticides about which cancer concerns have been raised is the biphenyl ether class, which includes roughly a dozen herbicides that have been found to cause liver tumors in rats and mice.

Given the difficulty in tracking down just which pesticides we're exposed to, at what levels, it's very difficult for epidemiologists to estimate how many cancers pesticide exposure might cause. One strong clue that the level is very low comes from an EPA estimate of cancer levels in farm workers and pesticide applicators, people exposed to much higher levels than the general public. The agency estimates the chance of an agricultural worker developing cancer over his or her entire lifetime is 1 in 10,000. They estimate that 6 farm workers a year develop cancer from exposure to pesticides. And these are conservative estimates that are intended to be precautionary but that are considered likely to overstate the true risk.

There have been very few cases of other adverse chronic effects, like infertility, associated with occupational exposure to pesticides, either among farm workers or in people who work at plants that manufacture pesticides. However, since some farm workers may choose to have children and some may not, reproductive effects are hard to monitor accurately.

~

The table opposite shows the top five pesticides in terms of usage in agriculture, and the various effects they have been found to have on lab animals given high-dose exposures. The table on page 278 shows the top five pesticides used in the home or garden and their effects.

THE RANGE OF EXPOSURES

This section may be the most important in this chapter: as the sixteenth-century scientist Paracelsus, also called the father of modern toxicology, said, "The dose makes the poison." With pesticides, it's not so much whether you're exposed. Rather, it's what level you're exposed to, for how long, and how often. Pesticides are poisons, and, as we've written, many attack systems and chemicals in insects and microorganisms that are similar to those found in humans. But in general the very low levels that are enough to work on insects and microbes are not nearly enough to have any impact on us.

COMMON INDUSTRIAL PESTICIDES AND THEIR EFFECTS

Pesticide	Acute Effects	Chronic Effects (Noncancer)	Carcinogenicity
Atrazine (herbicide)	Moderately toxic: abdominal pain, diarrhea, vomiting, eye irritation, irritation of mucous membranes, skin reactions	Same as acute effects	Inconclusive
Metolachlor (herbicide)	Slightly toxic	Skin rash Reproductive effects: Unlikely Birth defects: Unlikely	Unlikely
Metam sodium (gaseous fumigant used as a soil biocide to kill fungi, seeds of weeds, nematodes, and bacteria)	Severe irritant to mucous membranes; inflammation of respiratory tract; buildup of fluid in the lungs; coughing; bloody sputum	Birth defects: Yes	Probable
Methyl bromide (gaseous fumigant used in soil and stored grains against insects, rodents, nematodes, and fungi)	Highly toxic: Severe eye, skin, and respiratory irritation; vomiting; chest pain; dizziness; slurred speech; mental confusion. At very high levels, death.	Neurological damage; pneumonia-like syndrome; damage to heart, adrenal glands, testes	Inconclusive
Glyphosate (herbicide)	Practically nontoxic	None	No

Source: Extoxnet

COMMON HOME/GARDEN PESTICIDES AND THEIR EFFECTS

Pesticide	Acute Effects	Chronic Effects (Noncancer)	Carcinogenicity
2,4-D (herbicide)	Moderate (at high levels, respiratory irritation, muscle weakness, nausea)	Inconclusive (no in rats, yes in dogs) Reproductive effects: Unlikely Birth defects: Yes (at high doses)	Inconclusive: Several studies found effects, several did not
Glyphosphate (herbicide)	Practically nontoxic	None	No
Copper sulfate (fungicide to control plant diseases)	Caustic, severe internal damage if ingested, irritation to skin or eyes on contact	Liver disease, anemia Reproductive effects: Yes (testicular atrophy, reduced sperm production) Birth defects: Unlikely	Inconclusive
Chlorpyrifos (insecticide used on cutworms, rootworms, cockroaches, grubs, flea beetles, flies, termites, fire ants, and lice)	Moderate: Numbness, tingling sensations, lack of coordination, headache, dizziness, tremor, nausea, abdominal cramps, sweating, blurred vision, difficulty breathing or respiratory depression, slow heartbeat	Same as acute effects	No
MSMA (general-use herbicide)	Slightly toxic: Mildly irritating	Salty taste, garlicky breath, burning throat, stomach and intestinal pains Birth defects: Possible	Belongs to a group of chemicals of which some are carcinogenic, but MSMA has not been shown to be carcinogenic

Source: Extoxnet

Regulators take this dose issue into account when they consider whether a pesticide is safe, and at what level. For any chemical that might be used on products destined for direct or indirect human exposure, toxicologists subject lab animals to varying doses to determine the lowest dose at which a noncancer effect occurs. Then regulators set the permitted level of exposure at 100 to 1,000 times less than that, to make sure they protect the most sensitive populations.

Noncancer effects for which pesticides are tested include:

- Acute, single-dose effects of skin irritation, eye irritation, and neurotoxicity
- Developmental and reproductive effects
- Hormone disruption

To test pesticides for potential cancer effects, lab animals are subjected to a very high dose, to increase the likelihood that such an effect will show up. These doses are far higher for the animal subject, each day, than a person is likely to be exposed to in an entire lifetime. If cancer shows up at these high levels, it is assumed that it might show up at low levels. The results of these animal tests are then combined with epidemiological studies of people who were exposed to these chemicals (like farm workers or people who work in pesticide manufacturing) to make a fully informed judgment about whether the chemical causes cancer in humans.

Regulators then have to calculate how much of the pesticide people are likely to be exposed to if it is used as planned. Remember our general definition of risk: the consequence arises both from the hazard and the exposure. The more likely it is that people might be exposed to higher levels of the product, the more restrictive is the safety standard. When the use of a pesticide means we might be exposed to the same chemical from multiple sources, the safety limits on that product are more restrictive still. Regulators also have to take into account the cumulative effect of exposure to different pesticides that work in the same way, like different agents that all affect a specific enzyme, for example.

After all this scientific testing is done, the final decision on pesticide exposure is up to the administrator of the EPA. Under the Food Quality Protection Act (FQPA) of 1996, the administrator can establish an acceptable tolerance level for exposure only if there is a "reasonable certainty" that no harm will come from cumulative exposure to all pesticides if the one being considered is okayed at that level. The EPA is empowered to make this judgment call, so science is not the final and absolute arbiter of these levels. (The FQPA also required the EPA to reregister older pesticides to make sure they meet modern guidelines. This process is supposed to be complete by 2006.)

These acceptable levels are then used to set a "tolerance," the legal limit of a pesticide that can be used on an agricultural product. These tolerances are intended to allow safe use of a pesticide, taking into account its toxicity, the different crops on which it may be used, and consumption patterns for those foods. It also takes into account the potential for high exposure to subpopulations like children or certain ethnic groups that feature certain foods more prominently in their diet.

These levels are set to protect against chronic effects. Warning labels on consumer products still take into account the varying levels of acute toxicity of the pesticide. So on highly toxic products you will see a label that says DANGER-POISON, on mildly toxic products the label will say WARNING, and on relatively nontoxic products the label will say CAUTION. These labels also have directions for safe handling of the material, like avoiding smoking (which can cause materials on your hands to get into your mouth) and washing your hands after use.

~

We are exposed to pesticides all the time. We ingest them, inhale them, and touch them. There are numerous sources of this exposure: Pesticide residue on food is perhaps the most common exposure, though compared to possible exposures from other sources, levels on food are very low. The U.S. Department of Agriculture (USDA) tests agricultural products including fresh and processed fruits and vegetables, peanut butter, poultry, and milk for pesticide residue. It collected 10,907 samples in the year 2000 and found that about 40 percent of the samples had no detectable pesticide residues, roughly 20 percent had the remains of one pesticide, and 35 percent had more than one. Fresh produce was more likely to have detectable residues than processed food like canned tomatoes or pears. Only 19 out of the 10,907 samples were found to have levels above the safe tolerance standard.

The Food and Drug Administration (FDA) also inspects food for the presence of pesticides. In 1999, of 9,438 samples of domestic and imported food, 59 percent of the domestic products and 54 percent of the imported food samples had no detectable residues at all. Among the domestic products, fewer than 2 percent had residues exceeding safety tolerance limits. Among the imported food, 0.7 percent of the grain products exceeded tolerances; and 1.8 percent of the fruits, 3.9 percent of the vegetables, and 10.6 percent of the "other" foods sampled had residues above tolerance limits. In almost all cases in which the food had residues above tolerance limits, the actual level of the pesticide was far below the threshold known to do any harm. The violation was only because the pesticide residue was found on a product for which it had not been approved. That's why imported foods, which pass through more steps in shipping, have higher levels of these violations: there's more chance for a pesticide approved for one product to contaminate another that it's not approved for.

The FDA also tests what we actually eat in what it calls its annual Market Basket/Total Diet Study. In 1999, in 267 foods sampled in various locations across the country, the FDA found 55 different pesticides. The most common was DDT, which was banned nearly 40 years ago but doesn't readily break down in soil or water. However, the levels found were infinitesimal. In 243 of the foods, levels were below one part per million. In only two samples were levels above five parts per million. (One sample of raw peaches had one pesticide at 5.7 parts per million, and one sample of caramel candy had pesticide residue at 23.5 parts per million.)

But food isn't the only way we're exposed to pesticides. We're also exposed in the home, when we use products like lawn or garden chemicals or pool chemicals or household products like ant, roach, or wasp sprays. These levels can be much higher than food residue, especially if we do not use the product as directed. We're exposed in parks, playgrounds, public gardens, or other recreational areas. Very low levels of pesticides can occur in drinking water, after washing off from farms or lawns. Occupational exposure to pesticides occurs on farms, as the chemicals are manufactured, or as they are applied commercially at homes, businesses, or government properties.

Some pesticide exposure comes from materials no longer in use but which have left residues in the soil and water. An early class of pesticides were organochlorines, resilient chemicals that did not break down in the environment. As a result they built up to higher levels in soil and water, and stayed there for decades. In large part because of this persistence most of these pesticides, including DDT, have now been banned. But they can still be found in low levels in soil and water and at very low levels in some of the food we eat, particularly fish. The USDA testing program occasionally finds DDT and other banned pesticides in a few samples.

There is a great deal of contention about the extent of human health problems caused by pesticides, especially from pesticide residues on food. Standard risk assessment procedures, which are designed to be protective and conservative, suggest the possibility of a small rate of pesticide-induced disease. But there are few, if any, actual cases of chronic health problems directly attributable to exposure to pesticides on food. And virtually every medical organization encourages people to eat more fresh fruits and vegetables, saying the large known benefits from these foods far outweigh the hypothetical risks from pesticide residues. For example, the California Department of Environmental Protection says "nutrition experts agree that the health benefits of a diet rich in fruits and vegetables far outweigh any risk posed by the traces of pesticide residues you typically might find on produce."

As we have explained, the greater risk from pesticides appears to be the acute problems caused by exposures during pesticide use.

REDUCING YOUR RISK

Although pesticide levels on foods are low, you can take steps to reduce exposure. The California Department of Environmental Protection suggests:

• Rinse fresh fruit and vegetables thoroughly under running water. Although some pesticides are absorbed into fruits and vegetables, many residues are found on the surface. Washing will remove most surface waxes and residues, including the inert ingredients of pesticides. Peeling fruits and vegetables also removes surface residues and dirt. (But don't use soap to wash produce. Most soap is not meant to be consumed and can cause intestinal distress.)
• Throw away the outer leaves of leafy vegetables like lettuce and cabbage.
• Cooking or baking foods reduces some (but not all) pesticide residues.

Food certified as organic should have fewer pesticide residues than conventionally grown food, but don't forget that pesticides are allowed in organic farming.

FOR MORE INFORMATION

Extoxnet: The EXtension TOXicology NETwork
ace.orst.edu/info/extoxnet
U.S. Department of Agriculture

Pesticide Data Program
Agriculture Marketing Service
Science and Technology Monitoring Programs Office
www.ams.usda.gov/science/pdp/index.htm
8609 Sudley Road, Suite 206
Manassas, VA 20110
(703) 330-2300, extension 17

California Department of Environmental Protection
Department of Pesticide Regulation
www.cdpr.ca.gov/docs/factshts/factmenu.htm
1001 I Street
P.O. Box 4105
Sacramento, CA 95812-4015

Environmental Protection Agency
Citizen's Guide to Pest Control and Safety
www.epa.gov/OPPTpubs/Cit_Guide/citguide.pdf

---

*This chapter was reviewed by Carl Winter, Director of the FoodSafe program and
extension food toxicologist at the Department of Food Science and Technology, University of California, Davis who has served on several congressional advisory panels studying pesticides; and by Dr. Donald Mattison, Clinical Professor, Mailman
School of Public Health, Columbia University, former Medical Director of the
March of Dimes, and former Dean of the Graduate School of Public Health at the
University of Pittsburgh.*

34. RADIATION

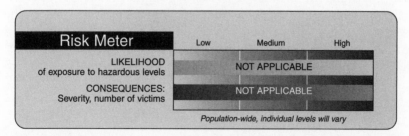

AS YOU READ this, you are being exposed to all sorts of radiation.
Some of it is hitting your eyes in what you perceive as the words

you are reading or the colors you are seeing. Radiation is hitting you if you are currently exposed to any sunlight or to the light of a fluorescent or incandescent bulb. Still more radiation is coming from radios or TV transmitters or wireless phones, from naturally occurring uranium in the earth, or even from satellites and cosmic radiation from space.

There are two basic kinds of radiation: electromagnetic, which includes radio waves, microwaves, and visible light, and nuclear radiation, which includes X rays and gamma rays. And there are two different categories in terms of their biological effects: ionizing and nonionizing. First we'll explain the two different types of radiation. Later, in "The Range of Consequences," we'll explain the different health implications of the ionizing and nonionizing forms.

THE HAZARD

Electromagnetic Radiation

This is the kind of radiation with which we're most familiar. It includes energy that moves in regular waves, like radio waves, microwaves, infrared light, visible light, ultraviolet light, X rays, and gamma rays. (X rays and gamma rays overlap into the nuclear radiation category too. More on that a bit later.)

Electromagnetic radiation has not only waves, but tiny particles called photons, which are far smaller than atoms. Photons have no mass and no electrical charge and travel at the speed of light, 186,000 miles per second.

As the photons of electromagnetic radiation travel along at the speed of light, they move in regular waves. They are measured in wavelengths, which is simply the distance between the crest of one wave and the crest of the next. The waves of electromagnetic energy that turn into sound from our radios have wavelengths in the range of a kilometer (roughly two thirds of a mile) from the crest of one wave to the crest of the next. FM radio and television signals have wavelengths of only about one meter, a little more than a yard. (Different radio and TV channels have wavelengths that vary slightly, just enough to distinguish one channel from the next.) Visible light has wavelengths in the range of only one millionth of a meter. (Different colors that we see have different wavelengths within the visible portion of the electromagnetic spectrum. Blue is blue and red is red because the size of their waves is slightly different.) X rays and gamma rays range from one billionth to one trillionth of a meter from wave crest to wave crest.

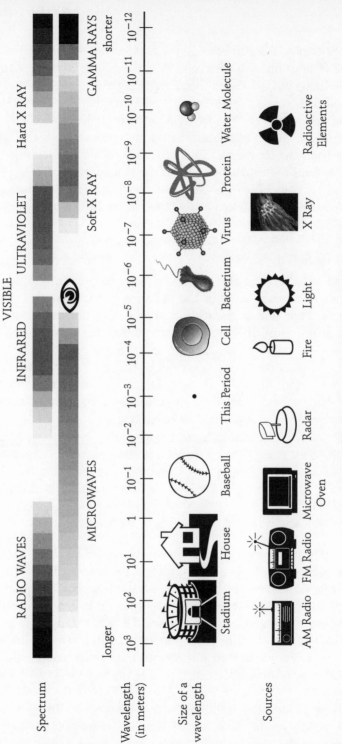

The electromagnetic radiation spectrum

Energy waves are also sometimes measured by their frequency, a measure of how many waves pass by a given spot in a given amount of time. Imagine two trains, both traveling the same speed as they pass a railroad crossing on tracks next to each other. One has very long cars, the other has very short cars. During any given stretch of time, say 30 seconds, more of the short cars will pass through the railroad crossing than the long ones. That is frequency. Frequency is a useful concept to keep in mind as we discuss the biological impacts of electromagnetic radiation because, basically, the higher the frequency, the greater the electromagnetic wave's potential to have a biological effect on you.

Nuclear Radiation

This kind of radiation is also sometimes referred to as "matter" or "particle" radiation, since most of the radiating particles in this category have mass. This kind of radiation is produced by nuclear processes inside the atom. Most of these processes involve atomic decay. Decay is the result of natural forces trying to create a balance between various parts of an atom. When things get out of balance, one way that atoms adjust is to emit either electrons, which are called beta particles, or parts of their atomic nucleii as alpha particles, protons, or neutrons. These emitted particles are nuclear radiation.

Unlike electromagnetic radiation, nuclear radiation is not described in terms of wavelengths or frequencies. Rather, nuclear radiation is measured in units of energy or absorbed dose, which is a measure of how much of the energy from the radiation particle is absorbed by what it hits.

Not all nuclear radiation is alike. There are five basic kinds.

- Alpha particles always come from the nucleus of an atom: they're made of two neutrons and two protons. Because of their relatively large mass, and because they carry two positive charges, which means they quickly interact with any atoms they hit, alpha particles have practically no penetrating power. They're stopped by a piece of paper, a fraction of an inch of water, or the outer layers of the skin. They're only a health threat in gaseous form, which makes them inhalable, or if they are attached to other particles that we inhale or ingest.
- Beta particles are electrons emitted by atoms of certain elements (tritium, strontium, phosphorus). Because beta particles are smaller and carry only a single charge, they can penetrate skin. But they

don't have much energy, so they can't penetrate more than a few centimeters. Like alpha particles, beta particles are generally only a health problem if we inhale or ingest them.

- Neutron particles are single neutrons from an atom's nucleus emitted during nuclear fission, the splitting of the atom that occurs when the nucleus of one atom is struck by a flying neutron from another atom. (See Chapter 31, "Nuclear Power.") Neutrons are the smallest of the nuclear radiation particles and have no electric charge, giving them great penetrating power. Humans absorb radiation produced by neutrons at a high rate because we're 70 percent water, which contains hydrogen. Because of the unique nature of a hydrogen atom, when a neutron hits it, it delivers all its energy into the nucleus of that atom.
- X rays aren't subatomic particles like electrons or neutrons. But they are considered nuclear radiation because they are high-energy photons produced by a nuclear process when electrons in an atom go from a high-energy state to a lower-energy state. X rays are also considered electromagnetic radiation, though, because their photons travel at fixed wavelengths and frequencies at the speed of light. (See Chapter 48, "X Rays.")
- Gamma rays are like X rays, electromagnetic energy waves produced by nuclear processes. Gamma rays often have even higher frequencies than X rays.

Radiation dose is calculated in a number of ways: the amount of radiation at a particular place, how much you're exposed to over time, how much you've absorbed throughout your whole body, or how much you've absorbed in particular organs. Bones absorb X rays differently from softer tissues, for example. These measurements are expressed all sorts of ways. There are rads (radiation absorbed doses), rems (Roentgen equivalent in man—for Wilhelm Roentgen, who discovered X rays), Svs (Sieverts), or Gys (Grays). This book uses the rem unit, which adjusts for the various types of ionizing radiation and according to which part of the body has been exposed. The rem then translates the dose to a whole-body dose equivalent. It's useful because it turns apples-and-oranges comparisons into apples-and-apples.

THE RANGE OF CONSEQUENCES

As we have stated, in terms of biological effects, what matters about radiation is whether it is ionizing or nonionizing. In the simplest terms, ionizing radiation has high enough frequencies and/or energy to damage biological tissue by changing the number of electrons in an atom, which usually leads to the breakdown of chemical bonds. Nonionizing radiation can damage tissue by raising its energy level.

~

Ionizing radiation can change an atom from electrically neutral to one that is either positively or negatively charged. These charged atoms are called ions. They tend to grab onto other nearby atoms in order to return to their original neutral electrical state. When an ion grabs onto another atom, it changes how that atom behaves, changing the structure and behavior of molecules, which can include DNA.

Ionizing radiation does its greatest harm when it alters the DNA molecule in the cell. These alterations are known as mutations. Most of the time, mutations to the DNA cause the whole cell simply to die. But sometimes the cell survives and the mutations are passed along each time the cell and its altered DNA reproduce. Such mutations can contribute to the series of genetic mutations that lead to cancer. DNA is more vulnerable to damage from ionizing radiation while the cell is reproducing, particularly in the phase of replication when the DNA is copying itself. This phenomenon explains why leukemias and lymphomas, two families of cancer associated with the blood system, are associated with ionizing radiation exposure. Cells in the blood system reproduce at a faster rate than cells in any other organ of the body. Blood cells, therefore, are more frequently in the vulnerable condition ripe for damage from ionizing radiation exposure.

But ionizing radiation is believed to be a relatively weak carcinogen. Most of what we know about the human effects of exposure to ionizing radiation comes from studies of the survivors of the Hiroshima and Nagasaki nuclear bombs. Massive studies have followed nearly 90,000 survivors of the atomic explosions. In a normal population of that size, approximately 17,000 cancer deaths would be expected for the entire lifetime of the population. In the Hiroshima-Nagasaki survivors, there have been about 8,000 so far, not an excessive number given the high levels of radiation to which these people were exposed in the moments right after the explosions. By comparing the people exposed to higher doses to those exposed to lower doses, and comparing the highly exposed population to people not exposed at all, scientists believe that only about 500 of the 8,000 cancer deaths in the atomic blast survivors can be attributed to radiation. The relatively low numbers of cancers in the Hiroshima and Nagasaki population suggests that ionizing radiation is definitely a carcinogen, but a weak one.

(The largest cancer cause of these deaths is from leukemia. Small but statistically significant radiation associations have also been observed in the atomic bomb survivors for stomach, colon, lung, breast, ovarian, urinary bladder, thyroid, liver, and nonmelanoma skin cancers.)

Scientists have also studied 3,000 children born after the atomic explosions, whose mothers were pregnant when the bombs went off. Based on their experience, it appears that in utero exposure to high levels of ionizing radiation is associated with an increased rate of smaller head size, mental retardation, lower IQ, slightly smaller body size at birth—both height and weight—and slower growth. These effects were more likely to appear in children who were at 8 to 15 weeks gestation when the bombs went off.

~

While some ionizing radiation particles don't penetrate what they hit, and some pass right through objects without hitting anything or transferring their energy to that target, *nonionizing* radiation transfers its energy to anything it hits. If the frequency and intensity of the radiation waves are high enough, they can cause what they hit to heat up. The energy intensity of the microwave, infrared, and ultraviolet radiation in the electromagnetic spectrum can be high enough to damage biological tissue.

Microwaves transmit wireless communications on devices such as mobile phones and portable phones in the home. Microwaves transmit TV signals, including signals from satellites. And of course, they're used to heat food in ovens. (See Chapter 15, "Microwave Ovens," and Chapter 8, "Cellular Telephones and Radiation.") Microwave radiation creates heat in biological tissue by causing the atoms in the tissue to vibrate so fast that friction with other atoms creates heat. The microwave radiation from cell phones, TV signals, and portable phones at home is not produced at high enough power levels to create heat, or damage, in humans. The waves simply don't have enough energy to do any harm. (Microwave ovens and portable phones use similar nonionizing frequencies. But the ovens generate more power—more waves—which is why they can heat things up and the other devices can't.)

Anyone familiar with the warmth of the sun understands the effect of *infrared* radiation. Roughly half of the sun's energy is received on earth as infrared radiation. In fact, infrared energy from the sun is the principal component in the daily local temperature and the overall temperature of the global biosphere. The flame on your stove heats a pot of boiling water with infrared radiation. Infrared radiation is what we identify as heat from anything warm or hot. Flames, radiators, our bodies, all give off infrared radiation. Since this kind of radiation comes from the motion of atoms, even ice cubes, or anything above absolute zero (459.67°F below zero, a point at which atoms and their internal parts are motionless), emits infrared radiation.

(By the way, those red heating bulbs over french fries or onion rings at a fast food shop give off infrared radiation, but the color you see is just from the visible light radiation coming through the red glass in the bulb. Infrared radiation frequencies are invisible.)

Moving up the electromagnetic spectrum, the next group of frequencies is *visible light,* roughly the other half of the energy arriving from the sun. Even though the sections of the visible light spectrum—the individual colors we see—have higher frequencies than infrared, they don't interact with atoms in a way that can cause heating or harm.

But the next section in the electromagnetic spectrum, *ultraviolet radiation,* does cause a number of health problems. (See also Chapter 36, "Solar Radiation.") There are three kinds of UV radiation: UVA, UVB, and UVC. Of the three, UVA, the most common form of UV radiation in sunlight, has the longest wavelengths and penetrates more deeply into the skin than the other two types of UV radiation. UVB has shorter wavelengths, which means higher frequencies, which means more energy. As a result, UVB is 1,000 times more effective at heating cells and causing health problems than UVA. (Good thing, then, that it is 10 to 100 times less common in sunlight.) These problems are associated with the surface of the body (skin and eyes) since UVB's shorter wavelengths can't penetrate as deeply as UVA's. Finally, there is UVC, which has the shortest wavelengths of the ultraviolet spectrum. UVC would be extremely dangerous to humans, indeed to any kind of life, but these energy waves are almost entirely absorbed by gases in the stratosphere and don't reach the earth's surface.

UV radiation is associated with a wide range of health damage. Skin cancer is the most common cancer in America, and UV radiation is the most common cause. One in seven Americans will get skin cancer over their lifetime. Nearly half of all Americans who live to at least age 65 will have skin cancer at least once. Fortunately all but a few of these cancers are easily treatable. In addition to causing skin cancer, UV radiation damages the skin. It causes sunburn, accelerates the sagging, wrinkling, and change in texture of the skin we associate with aging, and causes cataracts in the eyes, a clouding of the lens, reducing vision. It also interferes with the immune system by damaging cells in the outer layers of the skin that detect pathogens entering the body.

THE RANGE OF EXPOSURES

We're exposed to ionizing radiation all the time, from a wide range of natural and man-made sources. The bulk of our natural background exposure to ionizing radiation is radon. The average American is exposed to about one third of a rem's worth of ionizing radiation a year.

• Radon is a radioactive gas that produces alpha radiation, a byproduct of the decay of uranium in the earth (see Chapter 35).

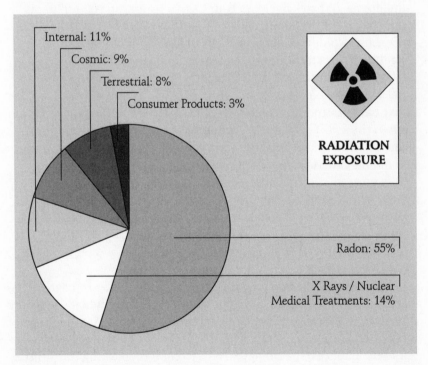

Sources of background radiation exposure for the average American, per year. *Source: EPA*

- Medical sources of ionizing radiation include X rays (dental, mammography, fluoroscopy, etc.) and nuclear medicine treatments and diagnostics. An average X ray exposes us to about 4 millirems.
- Cosmic radiation comes from the sun and other astronomical processes. Because these forms of radiation interact with the atmosphere on the way down to the earth's surface, the higher you are and the thinner the atmosphere, the more cosmic radiation you're exposed to. An average American at sea level gets 26 millirems a year of cosmic radiation. Someone at 10,000 feet elevation gets 107 millirems annually. And passengers on commercial jets flying at 39,000 feet get 0.5 millirems per hour from rays that can penetrate the outer walls of the aircraft.
- Terrestrial radiation comes from unstable forms of potassium, thorium, and uranium in the soil (not including the radon gas the uranium also produces). Exposure varies depending on where you are and what minerals are in the rocks and soil. Average annual exposure is 16 millirems on the Atlantic and Gulf coasts, 63 millirems in the Denver and Four Corners areas, and 32 millirems for the rest of the country.
- Consumer products also expose us to low levels of ionizing radiation. Cigarette smoke releases ionizing radiation that's been absorbed from the soil by tobacco. Our bodies' natural metabolism releases trace amounts of ionizing radiation from the foods we eat, like bananas and Brazil nuts, which also have absorbed the radiation from the soil. There are tiny amounts of radiation-producing materials in smoke detectors, and glow-in-the-dark watches. Some building materials include minerals that release tiny amounts of ionizing radiation. Mantle wicks for camping lanterns have tiny amounts of ionizing radiation, picked up from the soil by the silkworms that produced the material from which the mantles are made. Fertilizers that include mined phosphates are another consumer product that releases tiny amounts of ionizing radiation.

Like ionizing radiation, exposure to nonionizing electromagnetic radiation is also ubiquitous. Sunlight exposes us to the whole spectrum. Any source of heat exposes us to infrared radiation. A wide range of modern technologies, from radios to cell phones to televisions to wireless home phones to microwave ovens to baby-monitoring transmitter/receivers, expose us to radio and microwave radiation. We're also exposed to electromagnetic radiation when we are close to any wires or motors carrying electricity (see Chapter 9).

Pregnant women should pay special attention to exposure to ionizing radiation because of the sensitivity of the fetus, as demonstrated by the Hiroshima-Nagasaki studies. Also, people being X-rayed should make sure their reproductive organs are shielded. (Lead is used as the shield for the same reason the creators of Superman used it to block his X-ray vision. Lead is too thick to let X rays pass through it.)

FOR MORE INFORMATION

Uranium Information Center
www.uic.com.au/ral.htm
GPO Box 1649N
Melbourne 3001
Australia

National Safety Council
E-mail: cohend@nsc.org
Understanding Radiation Kit
Environmental Health Center
1025 Connecticut Avenue NW, Suite 1200
Washington, DC 20036
(202) 293-2270

This chapter was reviewed by Peter Valberg, who has a Ph.D. in physics, taught for 12 years at the Harvard School of Public Health, and is now an environmental consultant for the Gradient Corporation; and by Dr. John Little, Chairman Emeritus of the Harvard School of Public Health's Department of Cancer Cell Biology and head of the Center for Radiation Sciences and Environmental Health.

35. RADON

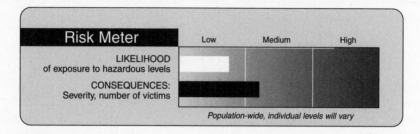

Risk Meter	Low	Medium	High
LIKELIHOOD of exposure to hazardous levels			
CONSEQUENCES: Severity, number of victims			

Population-wide, individual levels will vary

YOU DON'T KNOW it's in the air you're breathing, because radon is an odorless, colorless, tasteless gas. Nor would you think to worry much about it, since it's completely natural. But because it is also radioactive, radon is the second leading cause of lung cancer death in America. The Environmental Protection Agency (EPA) says radon kills thousands of people each year. And this hazardous form of radiation doesn't come from a nuclear power plant or an X-ray machine. The source is the earth itself.

THE HAZARD

Radon is the byproduct of uranium, an abundant element in the earth. Uranium is unstable, which means uranium atoms naturally change into a more stable form by rearranging the internal parts of the atom. But the newly created atom, radium, is also unstable and decays into radon. Each time this decay takes place, a subatomic particle is emitted. That particle carries radioactive energy. The decay of uranium and radium in the earth doesn't expose us to any of that energy because those elements are solid and stay in the earth as they decay.

But radon, which is also unstable and also decays, is a gas, which means it can seep up out of the soil into the air. Sometimes it dissolves into underground water, which can also carry it to the surface. When radon decays in the air we breathe or the water we drink, we can be exposed to the radiation it emits.

Radon presents a hazard in two ways. As it decays, it gives off energy in the form of alpha particles. Alpha particles aren't much of a threat because they have low penetration power. They can't make their way through a thick piece of paper or the surface of our skin. But if humans breathe in radon gas, the alpha particles emitted as the radon decays can penetrate the cells of the tissue in the respiratory system, especially the lungs. Energy from the alpha particles can damage DNA in these cells, which can contribute to the series of mutations that lead to cancer.

In addition to its emission of radioactive alpha particles, radon poses a risk because it decays into daughter atoms like polonium 214 and polonium 218. These daughter atoms, which are also unstable, are solid and can attach to tiny particles that we breathe in. Some of these particles stick to the layer that lines the lung. As the unstable polonium atoms next to the lung tissue decay, they emit still more radioactive energy inside us, which can contribute to cancer. A great majority of the damage done by radon comes from these progeny as they decay inside our lungs, not from the radon itself.

THE RANGE OF CONSEQUENCES

Radon exposure indoors, where levels are higher than outside, can raise your risk of lung cancer. The EPA estimates that radon is responsible for 15,000 lung cancer deaths annually in the United States. A National Research Council (NRC) study estimates the range of annual radon-related lung cancer deaths to be between 3,000 and 32,000, with the most likely range between 15,400 and 21,800. These estimates, however, are not confirmed by epidemiological studies. They are based on exposure to radon of people who worked in uranium mines.

People who smoke are at much greater risk of harm from radon. Radon and tobacco smoking apparently have a synergistic effect: that is, the effect of both together is greater than the effect of one added to the effect of the other. Studies indicate that the risk of lung cancer in people who smoke and inhale radon is higher than would be expected from the effect of smoking alone plus the effect from radon alone.

The NRC estimates that roughly 85 percent of radon-related lung cancer victims are people who smoke. They estimate that between 2,100 and 2,900 nonsmokers will die of lung cancer from radon exposure each year.

Radon is almost exclusively a lung cancer risk because most of the radiation given off by radon and its progeny are alpha particles, which have low penetrating power, so the damage they do is confined to the cells exposed immediately after we inhale them. Ingestion of water

containing radon is thought to be a low risk. A National Academy of
Sciences report concluded that drinking radon in water causes about 19
stomach cancer deaths per year in the United States.

THE RANGE OF EXPOSURES

Uranium is a common element in the earth's crust. As a result, low levels of radon are everywhere, even over open water. Levels are greater
over land, and greater still over regions where the underlying rock contains more uranium. Levels can even vary from home to home in the
same neighborhood, based on how much uranium is in the rock and
soil below, the permeability of the soil, and building characteristics,
which we will discuss in more detail. Estimates of outdoor natural levels of radon range from 0.2 to 0.4 picocuries per liter of air (pCi/L). A
picocurie is a measure of the rate of radioactive decay of radon. One
picocurie equals 2.22 radioactive decay disintegrations per minute. A liter is about half the space in a large soda bottle.

Radon is really only a health threat indoors. That's because concentrations outside are diluted in the presence of so much fresh air. Radon
gets into buildings by seeping up through cracks and seams in foundations. How much of it gets into the building depends on how much is in
the soil, how many holes or cracks there are in the building's foundation, and the relative air pressure in the soil versus the building. The gas

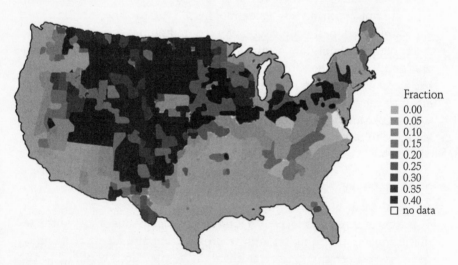

Fraction
- 0.00
- 0.05
- 0.10
- 0.15
- 0.20
- 0.25
- 0.30
- 0.35
- 0.40
- ☐ no data

Geologic radon potentials for the United States. *Source: U.S. Geologic Survey*

will flow to where the pressure is lower. So ventilating the building can actually encourage more radon to seep in from the ground below because the ventilation reduces the air pressure. (Closing off ventilation can be a problem too, causing concentrations to rise. The trick is to keep radon out of the building in the first place. See "Reducing Your Risk.")

Yet no matter what the radon levels, exposure and health risk occur only where people spend more than a few minutes in the rooms containing elevated levels. Radon in an unused basement, a crawlspace, or a rarely used room poses little direct risk because there is limited human exposure. Levels are usually higher in the basement or on the ground level because that's where the radon, in its highest concentrations, gets in. Radon levels are lower in buildings that have an open air space between the structure and the ground.

Indoor radon levels can vary from day to day. Higher atmospheric pressure outside makes it harder for the radon to seep up from the soil. Just opening doors or windows or adjusting the heat in a building also affects the rate at which radon enters, as well as the concentrations that build up. Radon levels fluctuate with the seasons because of changing soil conditions. Levels are typically higher during the winter in buildings in cold areas, where freezing leads to contraction of the loose materials in the ground, raising pressure on the radon gas that may be in that soil. That encourages the radon to seek the relatively lower pressure in the open spaces in the building.

Some of the highest radon exposures are in underground mines. In fact, the elevated rate of lung cancer in uranium miners first identified the hazard of radon, and studies of miners have provided most of the information we have about the human health effects of radon. Based on those studies, the EPA estimates that at or above 4 pCi/L, radon levels present sufficient risk to recommend remediation. (The National Council on Radiation and Protection sets the "acceptable" level at 8 pCi/L.) At the 4 pCi/L level, the EPA estimates that 1 home in 15 in the United States has radon levels high enough to warrant remedial action.

REDUCING YOUR RISK

Though radon is odorless, colorless, and tasteless, it is detectable with the proper equipment, and there are ways to reduce your exposure.

You can test for radon yourself. Hardware and home supply stores carry kits that contain a canister filled with charcoal. Left open, the charcoal passively adsorbs whatever radon might be in the air. After the specified testing period, usually 2 to 7 days, you send the kit to a lab, which analyzes the sample and mails you the results. The price for this kind of testing is roughly $20 to $25 per test. In buildings with high levels, long-term ongoing testing is usually advised.

Testing should be done on the lowest level of the building that is frequently occupied. Avoid testing in a kitchen, bathroom, laundry room, or hallway, where humid or drafty conditions can interfere with accurate results. Use a living room or family room or den, someplace that people often use. If you have a below-ground basement, even if it isn't used, test there, as well as on the next floor up. Licensed contractors can also do radon testing. They may use sampling devices that actively draw air through a measuring instrument. These tests can produce more precise information, such as radon levels at various times during the day.

If the first test detects levels of 4 pCi/L or above, the EPA suggests doing a second test to confirm the first one. If the two results yield an average of 4 pCi/L or higher, they recommend remediation. You can take a number of steps to reduce radon levels indoors. The most effective is called "sub-slab depressurization," which involves installing a pipe below the foundation. The pipe, which is open to the soil on one end, then runs above ground, usually to the roof. A fan located in an attic or outside draws air up through the pipe. That renders the surrounding air pressure in the underground portion of the pipe less than in the soil, and less than in the foundation just above the underground stretch of pipe. This zone of low pressure causes air to leak *out* of the building, instead of into it. So radon in the soil goes into the pipe, where the pressure is lower, instead of into the building. The fan draws the radon up and through the pipe, and it is discharged outside. Radon control for an existing building can cost between $800 and $2,500.

Sealing cracks in the foundation floor and in the walls of the foundation won't work. It's hard to find all the cracks and seal them completely. Settling and even small undetectable earthquakes can create new cracks. And ventilating the lowest level of your building where the radon enters is not recommended. That lowers the pressure inside the building even further and just encourages more radon to enter.

FOR MORE INFORMATION

National Radon Safety Board
www.nrsb.org

P.O. Box 426
Putnam Valley, NY 10579
(866) 329-3474

Environmental Protection Agency
www.epa.gov/iaq/radon/index.html
Indoor Air Quality Information
P.O. Box 37133
Washington, DC 20013-7133
(800) 438-4318

National Safety Council
Radon Hotline: (800) 767-7236 (low-cost do-it-yourself radon
    test kits)

---

*This chapter was reviewed by Jonathan Samet, Professor and Chairman, Depart-
ment of Epidemiology, School of Hygiene and Public Health, the Johns Hopkins
University, who has done extensive study on radon and public health; and by
William Bell, Radiation Scientist with the Massachusetts Department of Public
Health.*

36. SOLAR RADIATION

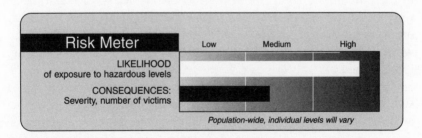

WE WROTE in the "Introduction" about the nature of our response to
risk. We subconsciously "decide" what to be afraid of and how afraid to
be as much from emotion as from reason. Few risks illustrate this point
more clearly than the risks from solar radiation. The psychology of risk

perception has found that we tend to be more afraid of something if it is man-made and less afraid if it's natural. There are several types of radiation-related hazards that evoke deep public concern and fear, all of which are man-made: nuclear power, cell phones, X rays. But far and away the most significant radiation-related risk to human health is natural, yet it gets a lot less attention and causes much less worry. In fact, it's a form of radiation that causes effects we admire, effects we describe as signs of good health, a form of radiation we actually pay a lot of money to be exposed to. It's radiation from the sun, and it is one of the biggest health risks in this book.

THE HAZARD

Radiation from the sun provides the energy for all life on the surface of the earth. But it also causes an estimated 1.3 million cases of skin cancer each year. As many as half of all Americans who live to age 65 will have skin cancer at some point over their lifetime, caused by exposure to the sun. One form of skin cancer, melanoma, kills about 7,800 people a year. Solar radiation contributes significantly to the sagging and wrinkling of our skin that we describe as signs of aging. It can damage our eyes and our immune system. Yet as we fret over man-made forms of radiation, which pose far lower dangers, most of us disregard this natural but much more serious radiation-related risk.

∼

The electromagnetic spectrum of energy waves is divided into sections defined by the size of the waves, measured from the peak of one to the peak of the next. (See the chart in Chapter 34, "Radiation.") The waves in the part of the spectrum known as ultraviolet (UV) radiation are tiny, just a few billionths of an inch wide. They are smaller than the waves in the visible part of the spectrum, and much smaller than microwaves, TV, or radio waves. (A single AM radio wave can be as long as a football field.)

The shorter the waves, the higher their frequency, which is a measure of how many waves pass by a given length of space in a given amount of time. Short waves, with high frequencies, have more energy than longer waves with lower frequencies. UV radiation has waves with frequencies so high that when they hit biological tissue, they raise the energy level in the atoms in that tissue. They literally heat it up.

There are actually three different kinds of UV waves, defined by their wavelengths, which we measure in nanometers (nm), or billionths of a meter. They each have specific effects.

- UVA (320–400 nm): These waves are tiny. But within the UV portion of the spectrum, these are the longest rays, and because of that they penetrate more deeply into the skin. They remain relatively constant year-round. There is 10 to 100 times more UVA in sunlight than UVB.
- UVB (290–320 nm): These waves are shorter, so they don't penetrate very deeply. They are less abundant than UVA, but their shorter length and higher frequency means they have more energy. UVB waves are 1,000 times more capable of producing skin redness and sunburn than UVA. They tend to be strongest in the summer.
- UVC (10–290 nm): The shortest-wave rays, they are very dangerous to plants and animals. But gases in the upper atmosphere absorb them before they reach the earth's surface. Many evolutionary biologists believe that it wasn't until billions of years ago, when chemicals evaporating from the ocean formed a protective layer in the stratosphere, that life on the surface of the earth became possible, largely because of protection from UVC radiation.

THE RANGE OF CONSEQUENCES

Exposure to UV radiation causes three kinds of skin cancer, several problems in the eyes, and general forms of skin damage. It also interferes with the immune system, key components of which are in the cells of the epidermis, the outermost layer of our skin.

Skin cancer is the most common type of cancer in the United States. It is so common, in fact, that when national cancer statistics are given they don't include skin cancer because this single category would overwhelm all the others: 1 in 7 Americans will get some form of skin cancer over the course of their lifetimes.

Basal cell carcinoma is the least aggressive of all the cancers of the skin. It begins in basal cells, which are small, round cells found in the lower part (or base) of the epidermis. In its most common form, it is characterized by a pearly, waxy-looking nodule that may ulcerate after a period of time. Basal cell carcinoma accounts for more than 90 percent of the 1.3 million skin cancer cases in the United States in 2001. It is a slow-growing cancer that seldom spreads to other parts of the body and is readily treatable.

Squamous cell carcinoma, on the other hand, if allowed to grow, can spread to the nearby lymph glands and internal organs. Fortunately, these cancers, which occur primarily on the sun-exposed areas of the face, ears, neck, and hands, are much more rare than the basal cell type. Squamous cell carcinoma accounts for approximately 6 percent of the skin cancer cases in the United States. These carcinomas begin in squamous cells, which are thin, flat cells resembling fish scales. They are found in the tissues that form the surface of the skin, the lining of the hollow organs of the body, and the passages of the respiratory and digestive tracts. This cancer usually appears as bumps or as red, scaly patches. Squamous cell carcinoma also rarely spreads, but it does so more often than basal cell carcinoma. It is also treatable and most of the time is not a dangerous tumor.

Melanoma is the most lethal of the skin cancers because it is the most likely to metastasize—to spread to other parts of the body. Melanoma accounts for about 4 percent of skin cancer cases but causes about 80 percent of skin cancer deaths. While conservative estimates predict that 7,800 people in the United States will die of melanomas during 2002, some researchers say this figure could be as high as 10,000 because of unreported cases. Melanoma begins in melanocyte cells. Melanocytes are found throughout the lower part of the epidermis, the outer layer of our skin. They produce melanin, the pigment that gives skin its natural color. Melanocytes form the noncancerous growths we call moles. Melanoma occurs when mutations in the DNA of melanocyte cells allow them to experience uncontrolled growth. (For much more on the processes and causes of cancer, see Chapter 40, "Cancer.") Since 1973, the mortality rate for melanoma (the number of deaths from melanoma per 100,000 people each year) has risen by about 44 percent, from 1.6 to 2.3. Mortality rates from melanoma are rising most rapidly among white men ages 50 and above.

~

More common than skin cancer are the noncancerous effects of UV radiation on our skin. Most of the physical changes we generically refer to as "aging"—sagging, wrinkling, leathery texture, blotchy discolorations—are largely caused by chronic UV exposure, which can accelerate the natural changes that take place in our skin as we get older.

If you've ever felt the sting of sunburn you know the power of UV radiation to heat the skin. The redness of sunburn comes from the dilation of blood vessels in the skin as they try to carry away the extra heat. Sunburn also produces unique "sunburn cells," which may be associated with the epidemiological evidence that two or more severe sunburns before the age of 18 raises your risk of skin cancer later in life.

Tanning is the increased production of melanin pigmentation in melanocyte cells. It is believed to be a way the skin tries to protect itself from some of the effects of UV exposure. There is an immediate pigment darkening, within the first couple of days, and a delayed response of further darkening about three days after the exposure. The degree of protection this extra melanin affords the skin is unclear, but it is thought to be minimal.

But skin damage isn't the only harm that comes from UV radiation. It also has the following effects on the eyes:

- Cataracts: A cloudiness in the lens due to a "clumping" of proteins, which blocks some of the light before it can pass through to the retina in the back of the eye. Cataracts are most common in people over 60 and are attributed to "the aging process," some of which is believed to include years of exposure to solar UV radiation. Excessive UV exposure can accelerate cataract formation. Prolonged exposure to UVA is also associated with malfunction of the lens, which helps the eye to focus.
- Photokeratitis: Sunburn of the cornea, the outermost layer of the eye. It results in discomfort, blurred vision, light sensitivity, and temporary vision loss (also called "snow blindness").
- Cancer: Squamous cell carcinoma of the eye.

Finally, UV exposure interferes with the immune system. Cells very close to the surface of the skin, called Langerhans cells (named for the nineteenth-century German anatomist Paul Langerhans), are part of the body's early warning system about invading microorganisms. UV radiation damages Langerhans cells and diminishes the immune system's ability to respond to these invaders. Other cells in the skin associated with immune system protection are also damaged by UV radiation.

THE RANGE OF EXPOSURES

Sunlight is by far the most common form of UV radiation. Tanning beds also use UV radiation (mostly only UVA).

UV exposure varies according to several factors. Location matters. UV radiation increases the closer you are to the equator, where the sun is more directly overhead, which means it has a shorter path through the atmosphere, so less of its radiation is absorbed or reflected back into

space. Texans have a greater risk from solar radiation than Minneso-
tans. You are also exposed to more solar radiation the higher up you
live, where the thinner atmosphere absorbs less UV radiation before it
reaches the surface. Every additional 1,000 feet above sea level represents a compounded 4 percent increase in UV radiation exposure.

The weather matters too. Clouds don't completely block solar radiation, but they do reduce it. Thicker clouds block more than thinner
ones. Also, the concentration of ozone in the ozone layer fluctuates
over short time scales of days and weeks, depending on solar activity.
Solar flares (sunspots) decrease ozone concentration and can vary the
amount of UVB hitting the earth's surface by as much as 400 percent.

The time of day also matters to UV exposure. The higher the sun is
in the sky, the more it can cut straight down through the atmosphere.
During earlier or later hours, the sun's shallower angle means its radiation has to pass through more of the atmosphere, which reduces how
much of that radiation gets to you. An hour outside between 11 A.M.
and noon exposes you to more UV radiation than an hour between 3
and 4 P.M.

Finally, there is the issue of long-term depletion of the ozone layer
from human industrial activity. The ozone layer is an effective filter for
certain UV wavelengths, particularly UVB. The less ozone there is in
the stratosphere, the more UVB we're exposed to. Scientists estimate
that every 1 percent decrease in ozone results in a compounded 1.5 to 2
percent increase in UVB reaching the earth's surface and a corresponding 1 to 3 percent increase in nonmelanoma skin cancer cases.

International agreements have all but eliminated use of the chemicals that cause ozone depletion. But they are persistent chemicals,
which means they don't readily break down. Emissions to date will
continue to damage the ozone layer for several decades, producing a
general thinning of ozone concentrations over the Northern Hemisphere and a more distinct ozone hole over Antarctica and parts of the
Southern Hemisphere. Scientists don't know how much thinning will
occur until the ozone layer has naturally healed itself in approximately
50 to 60 years.

~

Childhood exposure is perhaps the most important risk factor for skin
cancer. Some experts believe that the first two or three decades of life
are important in determining lifetime skin cancer risk from UV exposure. About 80 percent of the sun exposure of a lifetime takes place before age 18, yet most cases of skin cancer appear in those over 50. People

with any of the following characteristics are at greater risk for melanoma and other skin cancers due to excessive exposure to UV:

- Fair skin
- Blond or red hair
- Light-colored eyes
- Tendency to burn before tanning
- Tendency to freckle
- Two or more severe sunburns before age 18. Some experts believe that any history of severe sunburn, at any age, raises your risk.
- Family history of skin cancer
- Many moles (more than 50). Melanoma usually begins in the melanocytes of an existing mole, so having many moles increases the risk of developing this disease.
- People with HIV or who are otherwise immunocompromised

Several medications increase susceptibility to damage from UVA radiation. These products undergo chemical reactions when exposed to UVA, and those reactions release energy into the body, creating photosensitivity. Common drugs that cause this effect include topical and oral antihistamines, topical and oral antibiotics, nonsteroidal anti-inflammatory drugs used to control pain and inflammation, drugs to control hypertension, and immunosuppressive drugs. Sometimes these drugs also cause what is called "photophobia," a heightened sensitivity in the eyes to sunlight.

Sunscreens don't do much to fight photosensitivity because they mostly block only UVB radiation. In fact, sunscreens sometimes include ingredients such as bergamot, sandalwood oil, PABA (para-aminobenzoic acid), cinnamates, benzophenones, salicylates, and others that can actually cause UVA photosensitivity.

REDUCING YOUR RISK

Fortunately you can reduce the risk from UV radiation in several ways:

- Limit exposure to the sun during peak hours—11 A.M. to 2 P.M.
- If you can, stay in the shade.
- Apply broad-spectrum sunscreen and lip balm with a sun protection factor (SPF) of at least 15. Sunscreens between SPF 15 and 30 are as effective at blocking UV radiation as products with higher SPF ratings, but only if they are used properly, which means a nice thick application, uniformly applied and frequently reapplied. A lot of

people who don't use sunscreens correctly may benefit from those with higher SPF factors. Remember to reapply sunscreens regularly while outdoors.

- Shield as much skin as possible with closely woven garments.
- Wear broad-brimmed hats.
- Wear UV-blocking sunglasses. UV-blocking contact lenses are available, but they don't cover the whole eye, so they're intended to be used with sunglasses, not instead of them.
- Don't rely on fog, haze, and light cloud cover. They offer little protection against UV radiation.
- Be aware of the photosensitivity effects of drugs you may be taking and reduce direct exposure to the sun if you have to take them, either by not going outdoors, avoiding hours of peak sunshine, wearing protective clothing, or using appropriate sunscreens.
- Winter sports enthusiasts should bear in mind that fresh snow reflects up to 85 percent of UV radiation. You're getting it from above and below.
- Monitor UV by consulting the UV index, a forecast of the amount of skin-damaging UV radiation. It's calculated based on the elevation of the sun in the sky, the amount of ozone in the stratosphere, and the amount and type of clouds. The UV index can range from 0 (at night) to 15 or 16 (in the tropics at high elevations under clear skies). The higher the UV index, the shorter the time it takes for skin damage to occur. The UV index is becoming a common feature of weather forecasts. If your local paper or TV or radio station doesn't provide it, ask for it. It's important information for your health. (Remember, UV exposure happens even when the index is not at its peak.)
- Tanning beds commonly use UVA light and should be avoided, since UVA radiation penetrates deeper into the skin and is thought to augment the damaging effects of UVB.

Eyedrops designed to reduce the transmission of solar UV rays have not been shown to be effective. They are, however, somewhat effective for occupational exposure to UVC rays from welding.

~

While skin cancer will kill as many as 10,000 Americans in 2002, with early detection and proper treatment, skin cancers are curable.

Check your skin occasionally. It's best to do it after a bath or shower. Get familiar with your birthmarks, moles, and blemishes. Watch for changes, particularly on your head, face, neck, shoulders, arms, and hands, which are exposed to the most sun.

Look particularly for new growths, or a sore that doesn't seem to heal. Some skin cancers can start as a small lump that's smooth, shiny, pale, waxy, or red. Sometimes the lump will bleed or form a crust. Skin cancer can also start as a flat red spot that is rough, dry, or scaly.

The following easy-to-remember alphabetical list includes possible indicators for melanoma. If you detect any of them, see a doctor:

- Asymmetry: Part of the growth doesn't match the rest in shape or outline.
- Border irregularity: The edges of the growth are blurred, notched, or ragged.
- Color: The pigmentation of the growth is not uniform. Shades of tan, brown, or black are present. Blue, white, or red may also appear.
- Diameter: Any growth greater than 6 millimeters (about the size of a pencil eraser) is cause for concern (but not necessarily alarm, since each case needs to be taken in context by medical professionals).

FOR MORE INFORMATION

National Cancer Institute
CancerNet Service
cancernet.nci.nih.gov/wyntk_pubs/skin.htm
NCI Public Inquiries Office
Building 31, Suite 10A03
31 Center Drive, MSC 2580
Bethesda, MD 20892-2580
(800) 4-CANCER or (800) 422-6237

American Academy of Family Physicians
familydoctor.org/healthfacts/159/index.html

National Oceanic and Atmospheric Administration Climate
    Prediction Center
www.cpc.ncep.noaa.gov/products/stratosphere/uv_index/
    uv_current.html (local UV index forecasts)

This chapter was reviewed by Dr. Jerome Litt, Assistant Clinical Professor of Dermatology at Case Western Reserve University and author of a widely used reference book on cutaneous reactions to hundreds of drugs; by Dr. Ivor Caro, Director of the

International Training Program in Dermatology in the Department of Dermatology at the Harvard Medical School; and by Dr. Mark Naylor, Associate Professor of Dermatology, University of Oklahoma Health Sciences Center.

37. WATER POLLUTION

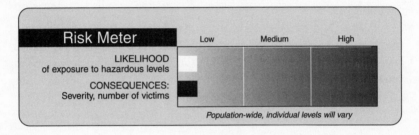

IN AUGUST 1854, the English physician John Snow investigated an outbreak of cholera in the Soho neighborhood of London. He found that people who drank from a local well had gotten sick but those who got their water elsewhere had not. The handle for the pump was removed and the cholera outbreak ended. Upon further investigation, Snow discovered that below ground, the well was being contaminated by sewage.

Snow's work is widely acclaimed as the first proof of the power of epidemiology, the study of patterns of human illness to detect hazards and exposures that might be causing those illnesses. But Snow's work did more than that. It was a pioneering step in establishing the field of public health, the idea of trying to prevent problems population-wide rather than treat them medically one patient at a time. And it was a critical advance in the realization of how important clean drinking water is to maintaining public health.

THE HAZARD

Strict safety standards have made the water supply system in the United States one of the safest in the world. The average American uses

52,500 gallons of water a year, and health problems are rare. Still, given our massive exposure to drinking water, maintaining that safety record is a constant priority. Each year about 8 percent of the nation's public water supply systems report some violation of safety standards, usually for failure to comply with monitoring schedules and government reporting requirements. Fortunately, these violations rarely result in anyone getting sick.

Most people in the United States get their drinking water from public water supply systems, defined as any water supply that serves 25 people or more. There are more than 55,000 such systems nationwide, and they are regulated by federal law. About 42 million Americans get their water from private sources, usually wells but occasionally from streams or cisterns, which are not federally regulated. In the United States, 53 percent of residents get their water from underground sources. The other 47 percent drink from surface water—lakes, rivers, and reservoirs.

Pollution in drinking water can come in a variety of forms and originate from a wide variety of sources, affecting either surface or underground supplies. Pesticides and fertilizers migrate from fields and lawns. Chemicals can be emitted from factories or migrate from disposal sites. Underground fuel storage tanks leak. The accumulated run-off from roads, streets, sidewalks, and rooftops carries a wide range of contaminants that are concentrated by storm sewers that flow into local streams and rivers. This is known as nonpoint source pollution. Household waste, including paints, solvents, and household chemicals disposed of in older landfills, can leach down into the groundwater below. (New landfills have liners to collect this leachate.) Animal wastes get carried into streams and reservoirs by rainfall or snowmelt. Unhealthful amounts of human waste are released by inadequate or malfunctioning municipal sewage systems or septic systems. Some airborne chemicals emitted from power plants, industry, or motor vehicles can be washed out of the air by rain and snow into drinking water supplies. And some contaminants are simply washed into the water as it flows through soil and rock, eroding naturally occurring minerals that in large enough amounts can become dangerous.

Against this litany of sources of contamination, two factors help keep our water safe. The first is the rigorous safety regimen of treatment and inspection that public water supplies must meet. And the second is the simple power of dilution. Even when a contaminant gets into the water, the levels remain extraordinarily low per gallon because of the millions or billions of gallons of water with which the contaminant is mixed.

The nature of water pollution isn't just a matter of which contaminant is involved. It also has to do with the unique nature of surface and underground water supplies. Rivers and streams go through a natural cleansing process as their flow eventually flushes out most contaminants. But a pollutant can get into drinking water quickly. Also, rivers and streams are exposed to sources of contamination that don't threaten underground aquifers, including spills, runoff from streets or storm sewers or lawns or agricultural land, or chemicals precipitating down in rainfall or snow. These sources can contaminate lakes and reservoirs too, which don't naturally flush themselves as quickly as flowing rivers and streams. A full exchange of all the water in lakes and reservoirs can take years. During that time, some pollutants that are heavier than water can settle to the bottom and accumulate. (But remember the dilution factor. The level of that pollutant in a glass of water you might drink would most likely be infinitesimal.)

Underground water sources, which supply about half of the people in the United States and 95 percent of rural Americans, have their own unique circumstances. Groundwater supplies have unique sources of contamination. Underground fuel storage tanks leak. Hazardous waste sites, where wastes have been buried, can also pollute underground aquifers. Chemical or gasoline spills on the surface can quickly move through porous soil and rock into aquifers that are not far below. And if these underground water systems become polluted, the cleanup challenge is much greater than for surface water. The average recycling time for underground supplies is 1,400 years. Also, underground aquifers often extend below wide areas, so pollution in any location can eventually threaten the entire aquifer, though the spread of pollution underground is very slow. (One single aquifer in the United States, the Ogalalla aquifer, extends from Texas to South Dakota.) Further, the hydrogeology of these aquifers—the way the water is stored in soil or rock and the way it moves—is often complex, making recovery and cleanup much more difficult.

To deal with these challenges, public water systems use five basic treatment processes to ensure the safety of our drinking water. Many systems use a combination of these processes.

- The most common is disinfection, which uses either chlorine or related chemicals to kill dangerous living microorganisms: 200 million Americans drink water treated by disinfection.

- A relatively new approach to disinfection is called ozonation, in which highly reactive oxygen molecules are injected into the water to kill microbes. (This is the same highly reactive ozone molecule that we know as smog when it occurs as air pollution, and the same molecule that forms the ozone layer in the upper atmosphere. It's reactive in our lungs, which is why ground-level ozone—smog—is a harmful air pollutant. In the stratosphere ozone intercepts solar radiation and protects us from dangerous forms of ultraviolet rays.)
- Flocculation and sedimentation uses alum or iron salts or synthetic polymers to cause particles in water to clump together and settle out.
- Filtration can capture and remove silt, organic matter, metals, microbes, and residues from water.
- A few systems use ion exchange or adsorption to remove some organic compounds that make it through the other processes, and to remove minerals, including arsenic, chromium, nitrates, fluoride, and radium.

THE RANGE OF CONSEQUENCES

Drinking water poses two basic types of risk: biological and chemical. The biological risk comes from pathogenic bacteria, viruses, and parasites. These are usually controlled by disinfectant treatment such as chlorination or ozonation. Pathogens cause acute and immediate effects that are usually short-lived, though they can be life-threatening to the young, the elderly, or people with weakened immune systems (including people with HIV, people receiving chemotherapy, or people taking steroidal medication). Chemical pollutants in water generally pose the risk of more long-term, chronic health consequences, like cancer or birth defects. The greatest risk is from the biological contaminants.

Listed below are the categories that the Environmental Protection Agency (EPA) uses to classify water contaminants, and some of the major pollutants in each category.

Biological Organisms

Cryptosporidium and giardia are bacteria that can cause intense gastrointestinal illness, including diarrhea, vomiting, and abdominal cramps. They are among the most common sources of waterborne illness and usually enter the water supply from the feces of animals in or near reservoirs. These parasites are difficult to treat, but they only rarely cause illness because the vast quantities of water in most systems means that the levels of these pathogens in any given unit of water we drink are practically nonexistent.

Still, they do pose a threat. The largest single outbreak of a waterborne disease in the United States was a 1993 outbreak of *cryptosporidium,* in Milwaukee, which draws its water from Lake Michigan. The outbreak occurred during a period of unusually heavy rainfall and ineffective water treatment. Over the course of a few weeks an estimated 400,000 people got sick, about 4,000 went to the hospital, and between 50 and 100 died; the original source was never identified. Other *cryptosporidium* outbreaks have occurred in recent years in Nevada, Oregon, and Georgia.

Legionnaire's disease is a microbe found naturally in water. It multiplies in warm, stagnant tanks, and in significant concentrations can cause pneumonia. Since the disease is pulmonary, it is usually caused when the organism is inhaled after the water gets into ventilation systems, not from drinking the water.

You may have heard that levels of E. coli bacteria in water are monitored. This form of E. coli is not dangerous (unlike the strain found in contaminated food). It's a nonpathogenic species that normally lives in the human digestive system. But its presence indicates an increased likelihood of other microbial contaminants carried in fecal waste, including hepatitis A, typhoid fever, and cholera. High E. coli levels mean inadequate water treatment. Sources of this contamination include failing septic systems, inadequate sewage treatment facilities, a leak in a supply pipe downstream of the water treatment facility, or an improperly disinfected repair site. Enteric (intestinal) viruses can cause vomiting, diarrhea, and cramps. They pass in feces into rivers and streams, and sometimes survive inadequate disinfection of drinking water.

Disinfection Byproducts

Most disinfection programs use chlorine-based chemicals to kill off dangerous bacteria, viruses, and parasites in drinking water. The chlorine interacts with the chemicals and organic material in the water to form disinfection byproducts (DBPs). These include chlorite, haloacetic acids, and trihalomethanes. Laboratory tests have associated these compounds, particularly trihalomethanes, with increased cancer risks, anemia, and liver, kidney, or central nervous system problems. New research is also raising concern that DBPs could cause birth defects if a pregnant mother drinks water with high levels of these byproducts at certain critical times during the first few months of fetal development.

Levels of DBPs can be controlled by raising or lowering the amount of chlorination. But water suppliers can't drop the chlorine levels too low or they won't kill enough of the germs in the water. They also have to keep chlorine levels high enough so that some residual chlorine continues to disinfect the water as it flows to the farthest points in the distribution system, where groundwater and new sources of pathogens can invade leaky pipes. So some very low level of disinfection byproducts will almost always be present in chlorinated drinking water.

Inorganic Chemicals

These minerals come mostly from the rocks and soils that water passes over or through, though some can also come from industrial and urban nonpoint source pollution. At high enough levels, some of these chemicals can have serious effects. The regulated inorganic chemicals include:

- Arsenic: cancer risk, skin damage, circulatory problems
- Beryllium: intestinal lesions
- Cadmium: kidney damage
- Copper: long-term liver or kidney damage or short-term gastrointestinal distress
- Cyanide: nerve damage, thyroid problems
- Lead: developmental problems in children, kidney problems in adults
- Mercury: kidney damage, developmental problems
- Nitrites and nitrates: harmful interference with oxygen uptake, especially in children
- Thallium: hair loss; kidney, intestine, and liver problems; and blood changes

While we mention these chemicals because they are EPA-regulated pollutants, their presence in drinking water is rare, except in certain isolated areas. And even where they do occur, their levels are extraordinarily low.

Organic Chemicals

These 63 EPA-regulated compounds include factory discharges, chemicals found in runoff water from herbicide or pesticide use, emissions from incinerators, discharge from refineries, and nonpoint source road and urban runoff. They include dioxin, benzene, PCBs, trichloroethyl-ene, vinyl chloride, styrene, lindane, and dozens of other compounds. Their effects include nervous system problems, increased cancer risk, reproductive difficulties, and liver, kidney, stomach, or gland problems. Like other pollutants, they rarely appear in drinking water at all, and when they do, they are present at exceptionally low levels, in many cases just a few parts per billion. However, evidence suggests the pos-sibility that some specific organic pollutants can have effects at ex-tremely small doses. The risk of health problems from exposure to these chemicals in drinking water is believed to be very low.

THE RANGE OF EXPOSURES

The contaminants in your drinking water, if they exist at all, exist at levels that environmental regulators have deemed safe. But setting those safe exposure levels is tricky. One challenge is figuring out what people's exposure to drinking water is in the first place. That's tough to do. The amount of water that different people consume varies widely from one person to the next, and even varies for any given individual depending on things like the weather, diet, lifestyle, and so forth. Fur-ther, the water chemistry is significantly different in different supply systems. Minute differences in the acidity of the water, its oxygen level, or its mineral content can alter the formation of DBPs, or the way the water interacts with naturally occurring minerals in the ground or met-als in the delivery pipes.

The EPA assumes that each adult consumes about two quarts of wa-ter every day over a 70-year life span. Typically, it sets limits on any po-tential cancer-causing compounds so the assumed exposure will not in-crease the cancer risk by more than one extra case per million people. For effects other than cancer, the EPA studies lab animals, determines the lowest levels at which the animals show a negative effect, and then sets the drinking water standards several times below that level, just to be safe.

~

Of the EPA-regulated chemicals that we may actually be exposed to, the most common contaminants found in aquifers in the United States are gasoline and diesel fuels and associated chemicals such as benzene, toluene, and additives such as MTBE (methyl tertiary-butyl ether). There are an estimated 100,000 leaking underground fuel storage tanks in the United States, according to the EPA, of which approximately 18,000 are known to have contaminated groundwater. In Texas alone, 223 of 254 counties have reported such leaks.

Pesticides are also frequently found in underground water supplies. In certain farming regions, they have already begun to infiltrate many groundwater sources. In California's San Joaquin Valley, for example, tests in 1988 showed that one third of wells contained ten times the allowable level of the pesticide DBCP (dibromochloropropane). In the Midwest, $400 million a year is spent just to remove a single pesticide, atrazine, from water supplies. The National Research Council estimated that over the next 30 years, cleanup of some 300,000 sites of groundwater contamination could cost up to $1 trillion.

But in terms of exposure, the most important issue is monitoring of water after treatment. While it's always better to keep the water supply clean of contaminants, the issue for public safety is how well any contaminants can be removed before that water is actually consumed. That's why monitoring and reporting requirements are so important, and violations so significant. Monitoring ensures that the water leaving the treatment facility is safe to drink, regardless of what was in it when it first arrived for treatment.

REDUCING YOUR RISK

Under federal regulations, every public water supplier must provide consumers with an annual consumer confidence report. This report must provide information on drinking water quality, including its source, any contaminants found in the water, and suggestions of how you can help to protect the water quality. If certain contaminants are found in the local water supply and are not brought under control, the supplier has to inform you immediately and advise that you boil water intended for cooking or drinking, in order to kill microbial pathogens.

You may want to take additional steps to ensure your water quality if your water complies with safety standards but contains contaminants that could be a problem for someone in your household with a compromised immune system because of HIV/AIDS, chemotherapy, or steroid treatment. (Chemical pollution is not a risk for people with compromised immune systems.) Home filtration or distillation units can provide extra protection. These systems can be installed to treat water at your home's intake pipe, or at individual faucets. The EPA or state or local water authorities, or the agencies listed at the end of this chapter, can provide additional information about the different systems that are available.

Filters that fit on a water pitcher can remove some contaminants. Bottled water is another alternative, though in most states there are no requirements that bottled water be any cleaner than municipal tap or well water. The source of bottled water is often not listed on the label, so the quality varies significantly.

Another susceptible population group is children, who are especially sensitive to contaminants such as lead. To minimize lead exposure, use only cold water for cooking, drinking, or making formula. Cold water is less likely to release lead that has leached from supply pipes or lead-soldered joints in copper supply lines. Also, run the water for a minute before using any, to flush away any lead that has leached into the water as it sat in the pipes.

For private wells, the EPA recommends annual testing, especially for E. coli and other coliform bacteria, to catch any contamination problems early. If any specific problems are known or suspected, such as pesticide runoff or elevated contaminant levels from naturally occurring underground minerals, additional and more frequent tests should be done. Many labs can do these tests. Lists are available from state environmental agencies or health departments. Some problems can be quickly corrected; for example, high bacteria counts can often be controlled just by disinfecting the well.

FOR MORE INFORMATION

The first thing to do to find out more about your water is to contact local suppliers, since systems and conditions vary so much. Remember, they are required by law to provide you with information about the quality of your water.

Also keep in mind as you pursue more information about drinking water that there are significant differences between public and private systems.

Environmental Protection Agency
Water Resource Center
www.epa.gov/safewater
401 M Street
Washington, DC 20460
Safe Drinking Water Hotline: (800) 426-4791

National Sanitation Foundation
NSF International
www.nsf.org
P.O. Box 130140
789 N. Dixboro Road
Ann Arbor, MI 48113-0140
(800) NSF-HELP or (800) 673-4357

Farm*A*Syst/Home*A*Syst
National Farm*A*Syst/Home*A*Syst Program
www.uwex.edu/homeasyst
303 Hiram Smith Hall
1545 Observatory Drive
Madison, WI 53706
(608) 262-0024

This chapter was reviewed by Stephen Estes-Smargiassi, Director of Planning, Operating Division, Massachusetts Water Resources Authority, which supplies drinking water to 2.5 million people in Boston and eastern Massachusetts; and by John Shawcross, Director of Capital Engineering and Construction at the Massachusetts Water Resources Authority, who has overseen the concept, design, and construction of the major water projects at the MWRA since 1986.

MEDICINE

38. ANTIBIOTIC RESISTANCE

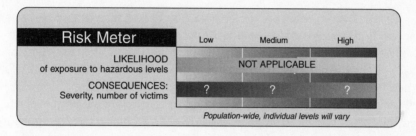

IN 1941, when penicillin first became widely available, it was hailed as a wonder drug that could control several serious bacterial infections like staph and strep. Medical experts quickly predicted that infectious diseases would become a thing of the past. But within two years, doctors began reporting cases of a common bacterium, *Staphylococcus aureus,* which somehow could resist the effects of penicillin. Almost as soon as antibiotics became available, bacteria began developing ways to resist them. Ever since, we've been in a race between the development of new and more powerful antibiotics and the ability of bacteria to adapt in ways that defeat those drugs. Many experts say the race is a dead heat. Some say the bacteria are winning.

THE HAZARD

All microscopic organisms, or microbes, secrete a variety of chemicals. Some of these secretions kill or impair competitors. Antibiotics are based on these natural microbial secretions. Penicillin, for example, is based on a secretion from the mold *Pencillium notatum,* which kills bacteria by destroying their ability to build a cell wall.

Antibiotics, also known as antimicrobials, bond to specific parts of specific bacteria, and either kill or impair them. Killing the bacteria outright, of course, eliminates the infection. But sometimes just impairing them can be enough, because that gives our natural immune system a chance to gain the upper hand in the battle against the bacteria and finish the fight. Most antibiotics bond with molecules that occur only on the surface of bacteria, which is why they don't harm human cells and why they don't kill viruses. And individual antibiotics work only on specific bacteria because their "attack" molecules bond only with specific molecules on those target bacteria. Think of the bonding as a kind of lock-and-key system. It takes a specific key (the right antibiotic molecule) to fit in a specific lock (the receptor molecule on the target bacterium).

But the bacteria that are under attack don't just take this abuse lying down. In the evolutionary competition between these attacking microbial secretions and the bacteria, the bacteria evolve ways of fighting back. They respond via the natural process of occasional changes to their DNA. These changes lead to the development of new traits, some of which allow the individual bacterium to counteract the effects of the antibiotic substance. A single bacterium that evolves a way of fighting off antibiotics has a survival advantage compared with its neighbors of the same species. It survives, and passes that resistance trait on to its offspring and they become the variation of the species that survives over time.

Bacteria are really good at reinventing themselves in this way, for two reasons. First, they can undergo changes to their DNA in several different ways. They even have the ability to share sections of their DNA between species. So a resistance trait that arises in one bacterium can be passed to other types. Second, bacteria evolve quickly because they reproduce prolifically, as frequently as once every 20 minutes. If you put a single bacterium cell in an optimal growing environment, within 12 hours you can have as many as 68,719,476,736 copies of that original cell with approximately one mutation per million copies. That creates the likelihood of tens of thousands of mutations, each of which might produce a resistance trait that helps it fight off an antibiotic.

Over time, the resistant strains proliferate while the ones that can't fight the drugs die off. The survivors spread their new, successful resistance trait to other strains within the same species, or even to other species of bacteria entirely.

The widespread use of antibiotics speeds up this process. When a microbial population of various organisms is exposed to an antibiotic, the bacteria susceptible to the antibiotic will be killed. Organisms that have some resistance survive. Without the other species around to compete for food and resources, it's easier for the resistant strains to proliferate. Those strains then pass their resistance traits to their own offspring, or share their resistance genes with other bacteria. The result is that as antibiotics kill off the bacteria they work on, they increase the prevalence of strains that can resist them.

The most significant human contribution to this accelerated an-

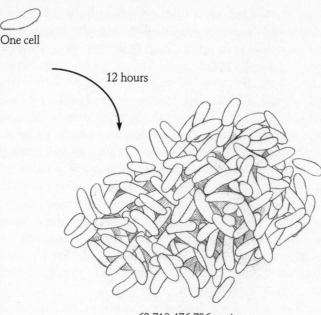

One cell

12 hours

68,719,476,736 copies
(and thousands of mutations)

Rate of bacterial reproduction

tibiotic resistance is the indiscriminate and often unnecessary use of antibiotic drugs. In the United States, between 160 and 260 million courses of antibiotics are prescribed each year. An estimated 75 percent of these are for respiratory infections, but between one third and one half are unnecessary, prescribed to people who have *viral* infections that aren't treatable with anti*bacterial* medication. Most doctors acknowledge they prescribe antibiotics to patients simply because the patients demand them. Since every application of antibiotics encourages the growth of resistant strains, misprescribed use of antibiotics accelerates the problem.

In addition, in hospitals, broad-spectrum antibiotics are often used when a more targeted drug that kills only the specific bacterium causing an infection would be enough. The advantage for the patient is that broad-spectrum antibiotics wipe out a wider range of susceptible species. But the downside is that this practice goes even further in clearing the playing field for the toughest, most resistant strains that are left behind.

Bacteria also get human help in developing antibiotic resistance when we fail to take the full course of a prescription of antibiotic medication. The weakest germs are killed within the first few days and symptoms often disappear, so we stop taking the drugs. But the stronger bacteria, which can resist the first few days of medication, survive. The full course of the medicine might have killed them off. Instead, these slightly more resistant bacteria survive and proliferate and spread the trait that helped them fight off the first few days of the drug.

Another way that humans are accelerating antimicrobial resistance is the use antibiotics in farm animals. As much as half of all the antibiotics produced for use in the United States are used on farm animals, mostly at low doses over a long period to encourage growth. But the low doses allow the more resistant strains of bacteria in animals to outsurvive the weaker ones. Some of these resistant bacteria, like strains of *Salmonella, Shigella,* and *E. coli,* can be transferred to people in improperly prepared foods. Then, when people develop bacterial illness, the strains of these bacteria are resistant to the drugs that used to control them.

Antibiotic resistance is also amplified in certain settings, like schools, hospitals, and chronic care facilities, environments where there are a lot of people with less effective immune systems carrying a lot of bacteria, so the chances of individual strains swapping their resistance genes is higher. And finally, bacterial resistance accelerates because of the global transportation system. People and goods spread microbes around the world. In many parts of the world, antibiotic drugs can be purchased without a prescription, and are often taken improperly, leading to resistant strains that spread worldwide.

THE RANGE OF CONSEQUENCES

In general, antibiotic resistance raises the likelihood that infections will have a more serious effect on a person's health. It raises the likelihood that an otherwise treatable infection might turn lethal. In developed countries like the United States, with greater access to advanced medical care and pharmaceuticals, a frequent consequence of antibiotic resistance is that a second, third, or fourth type of drug has to be used when the principal agent against a particular bacterium no longer works. These backup drugs sometimes have more side effects. They are usually in shorter supply. And they are almost always much more expensive.

It is difficult to quantify the consequences of antibiotic resistance.

Sometimes the drugs fail outright. Sometimes they merely don't work quite as well as they used to, and a patient gets sicker and stays sick longer but then recovers. Often these effects occur outside a hospital, nursing home, or other facility where accurate surveillance records can be kept. But a pattern of chilling statistics comes from a number of sources.

~

Staphylococcus aureus is a common bacterium. Most of us carry it in our noses or on our skin. It can cause minor infections, or life-threatening diseases like pneumonia. Penicillin used to kill it. But in the 1950s, less than 10 years after penicillin hit the market, *Staph aureus* had become so resistant to penicillin that healthy people going to hospitals got sick and died. Many hospital maternity wards had to close. So drug companies developed methicillin in the 1960s. By the 1980s, *Staph aureus* was resistant to methicillin. The Centers for Disease Control and Prevention (CDC) estimates that as many as 80,000 people a year get a methicillin-resistant *Staph aureus* infection after they enter the hospital. So doctors switched to the antibiotic vancomycin, a broad-spectrum drug widely thought of as the antibiotic of last resort. In 1997, the first cases of vancomycin-resistant *Staph aureus* showed up in three geographically separate locations. Many more have since been reported. In the year 2000, the first revolutionary new type of antibiotic to come out in 30 years, linezolid, was approved, offering promise in the fight against *Staph aureus* and other multidrug-resistant bacteria. It took less than a year for the first cases of linezolid-resistant *Staph aureus* to show up.

According to the CDC, each year the bacteria *Streptococcus pneumoniae* causes 100,000 to 135,000 hospitalizations for pneumonia, 6 million ear infections, and more than 60,000 cases of other invasive diseases, including 3,300 cases of meningitis. Of these, the agency estimates that at least 40 percent are caused by drug-resistant strains of *Strep pneumoniae*. Between 1993 and 1998, 45 states and the District of Columbia reported at least one case of tuberculosis that was multidrug-resistant. Health officials once thought tuberculosis had been all but wiped out in the United States. Resistant strains threaten a comeback.

The CDC estimates that 2 million people a year get so-called nosocomial infections *after* they enter the hospital. Approximately 90,000 of these people will die because of them. It is not known how many of these fatal nosocomial infections are drug-resistant, but it is believed that a significant number probably are. Between 1979 and 1987, only 0.02 percent of the pneumococcus strains infecting patients in 13 hospitals in 12 states sampled by the CDC were penicillin-resistant. By 1994, that number had risen to 6.6 percent.

In a 1999 nationwide sampling of foodborne bacteria by the National Antimicrobial Resistance Monitoring System (NARMS):

- 26 percent of the nontyphimurium *Salmonella* samples were resistant to one or more antibiotics
- 49 percent of another strain of *Salmonella, Salmonella typhimurium,* resisted one or more drugs
- 91 percent of the *Shigella* samples resisted one or more drugs
- 10 percent of the *E. coli* samples resisted one or more antibiotics
- 53 percent of the *Campylobacter* samples were resistant to one or more antibiotics

Foodborne disease outbreaks are often caused by drug-resistant strains of bacteria. In 1998, 5,000 people in the United States fell ill from *Campylobacter* caused by contaminated chicken. The strains of bacteria found in the victims were multidrug-resistant. In 1968, 12,500 people in Guatemala died in an epidemic of *Shigella*-caused diarrhea, from a strain of the bacterium that was resistant to four antibiotics. A deadly drug-resistant strain of *Salmonella typhimurium* called DT104, more virulent than other strains, appeared in the late 1990s and has killed people in Great Britain. And 28 percent of all *Salmonella typhimurium* samples tested by NARMS in 1999 had traits similar to the DT104 strain. (The Food and Drug Administration has approved a test kit for rapid detection of DT104.)

Overseas, nearly every case of gonorrhea in Southeast Asia is multidrug-resistant. In 1990, cholera bacteria in India were susceptible to common antibiotics. Just 10 years later, none of those drugs worked on cholera anymore. And our global world spreads some of these strains far and wide. Between 30,000 and 80,000 U.S. travelers returning from overseas suffer from a bacterial diarrhea that is drug-resistant. Worldwide, 2,500 travelers a year return with malaria that could not be prevented by prophylactic antibiotics that used to work. Investigators have documented the migration of one strain of multidrug-resistant *Strep pneumoniae* from Spain to the United Kingdom, the United States, South Africa, and elsewhere. Two cases of multidrug-resistant *Staph aureus* in the United States were traced to northern India. Most of the multidrug-resistant strains of typhoid found all over the world have been traced to six developing nations.

Vancomycin-resistant enterococci, normally harmless bacteria that live in the human gut, were first detected in France and England in 1987. One case appeared in New York in 1989. By 1993, 14 percent of patients in intensive care units in the United States had vancomycin-resistant enterococcus, a 20-fold increase in six years. Given the ready swapping of genes between different species of bacteria, the ability to resist vancomycin, currently the antibiotic given when others fail, could easily spread from enterococcus to other more harmful species.

One top U.S. health official said the ultimate consequence from the growing problem of antibiotic resistance "could be a return to the days before antibiotics, when common diseases were often lethal. . . . We are skating on the edge of the ice," he said.

THE RANGE OF EXPOSURES

We are exposed to bacteria constantly. There is literally no setting in which potential exposure to drug-resistant bacteria is not a concern. As we have stated, people most at risk are those who have weakened immune systems. These include people already ill from something else, infants with still developing immune systems, people taking steroidal medication, chemotherapy patients and the elderly, whose immune systems are no longer as effective as they used to be.

However, health officials are particularly worried about drug-resistant bacteria in hospitals and nursing homes, places where a combination of factors raise the risk. A significant number of people who are hospitalized come in with weakened immune systems or undergo treatments such as chemotherapy that impair their immune response. These people are at risk of more serious illness, or death, from infections that neither they nor drugs can fight. In addition, a significant percentage of people who are hospitalized are elderly, with immune systems compromised simply by age (another reason why exposure to drug-resistant bacteria is also a concern in nursing homes and assisted-living facilities). In addition, surgical patients are more susceptible to any kind of bacterial infection simply because their skin has been opened. Open wounds, or healing wounds, are another potential route of nosocomial infection. Also, hospitals are, by definition, locations where a lot of people are carrying infections. They bring various strains of bacteria in with them. There are simply more infectious bacteria around in hospitals. Inadequate hygiene by people who work in hospitals, particularly something as simple as thorough and regular hand washing, allows drug-resistant strains of bacteria to spread.

Another setting where exposure to drug-resistant bacteria is a concern is day care centers, especially for infants. Here, a combination of children with still developing immune systems, lots of direct contact among children and their caregivers, and an environment where at least one or two people are frequently sick at any given time, increases the chances of spreading bacteria, accelerating the spread of resistance traits.

REDUCING YOUR RISK

There are several steps you can take to help slow the proliferation of antibiotic-resistant bacteria.

Don't demand antibiotics any time you get sick. Remember, between one third and one half of antibiotics are misprescribed to patients who don't really need them. Don't automatically demand antibiotics for your children if they have what appears to be an ear infection. Medical authorities now think that mild cases may go away by themselves. And they say that not all ear infections are bacterial.

Pay attention to simple hygiene. Cover your nose and mouth when you sneeze to avoid spreading germs. Wash your hands frequently. Wash fruit and vegetables thoroughly. Cook beef, chicken, pork, and fish well enough to kill any germs they may contain. And don't forget that a piece of meat that you prepared on your kitchen counter might have contained a few germs. While you may kill those germs when you cook the meat, the germs are still there on the counter that you then use to prepare your salad. So wash your food preparation areas each time you finish working on one part of a meal. Sponges are a problem—throw them out or put them in the dishwasher. The Academy of General Dentistry suggests replacing your toothbrush every three to four months to avoid buildup of bacteria.

Diapers can be a source of bacteria. People handling diapers should be careful about washing their hands thoroughly. Finally, maintaining a good diet, getting half an hour of mildly aerobic exercise a few times a week, and other simple steps for staying healthy are good ways to avoid bacterial infection of any kind.

Follow the instructions for taking antibiotics. Don't stop taking them after you feel better. Some of the targeted bacteria may have just enough resistance to fight off the first few doses. They may still be lurking, ready to pass on their resistance traits. Finishing the full course of the medication will help finish off these slightly stronger germs.

FOR MORE INFORMATION

Centers for Disease Control and Prevention
www.cdc.gov/drugresistance
1600 Clifton Road
Atlanta, GA 30333
(800) 311-3435

Alliance for the Prudent Use of Antibiotics
www.healthsci.tufts.edu/apua/apua.html
75 Kneeland Street
Boston, MA 02111-1901
(617) 636-0966

This chapter was reviewed by Marc Lipsitch, Assistant Professor of Epidemiology at the Harvard School of Public Health, who has done research on antibiotic resistance; and by Dr. Don Goldman, Epidemiologist at Children's Hospital, Boston.

39. BREAST IMPLANTS

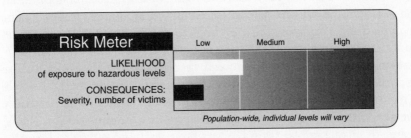

Risk Meter	Low	Medium	High
LIKELIHOOD of exposure to hazardous levels			
CONSEQUENCES: Severity, number of victims			

Population-wide, individual levels will vary

SOME RESEARCH suggests that breast size and shape are subconsciously associated with sexual attractiveness because they are associated with fertility and successful child rearing, traits that are important as males consider potential mates. Other research suggests this aspect of the female body only started getting so much attention after the Victorian era, during which completely covering the breasts became the social norm. Whatever the reason, many cultures place great emphasis on breast size and shape, and, as a result, breast augmentation has become an increasingly popular type of cosmetic surgery. Nearly 300,000 American women had breast implants in 2001. About two million American women have received implants since 1962, when implants containing silicone gel first became available.

But in the 1990s, fear of breast implants grew after widespread national attention to a study suggesting a connection between implants and cancer. Several years and hundreds of research studies later, that fear is no longer supported by the facts as we now understand them. Still, concern about implants continues.

THE HAZARD

Implants either augment the shape and size of the breasts or reconstruct breasts damaged or removed by injury or surgery. About two thirds of implants are for cosmetic augmentation, and one third are for reconstruction. Breast implants are fluid-filled sacs that contain either silicone gel or a saline solution, which is essentially salt water. The sacs are inserted under a woman's breasts, usually above the chest muscles. Silicone gel and saline water implants behave differently because the silicone gel is thicker, which gives the augmented or reconstructed breast a different shape and firmness. Silicone was the most popular type of implant because it was more like the shape and feel of the natural breast.

But a 1988 study by Dow Corning, a major manufacturer, found that the company's silicone gel caused cancer in laboratory animals. The company did not report these findings for some time, and when the results finally came to light a wave of national news coverage followed, creating widespread public concern that breast implants might cause not only cancer but a number of other illnesses. An estimated 170,000 women brought lawsuits against Dow Corning claiming a variety of ailments. In 1992 the U.S. Food and Drug Administration (FDA) banned silicone implants for breast augmentation. Dow Corning withdrew its product from the market and filed a $4.5 billion bankruptcy reorganization plan in 1995, part of which included a $3.2 billion settlement to most of the women who had sued.

But a 1999 report by the National Academy of Sciences Institute of Medicine (IOM) found that silicone gel implants do not cause cancer or other long-term systemic diseases. The panel of experts that issued that report reviewed more than 3,000 studies and medical reviews, in addition to case reports, policy papers, and other documents. The IOM review of ten years' worth of research showed that the frequency of such illnesses is no greater in women with silicone implants than it is in the general population. The IOM report found no evidence of a "novel syndrome," unique circumstances that exist only in women with implants. They also addressed the so-called human adjuvant disease, a phrase used by advocates for women suffering health problems they associated with their implants. The IOM experts agreed that such a condition does not exist and the phrase should be abandoned.

The experts found that while implants can interfere with breast-feeding—because of the effect of the device and the implantation surgery on the flow of milk to and through the nipple—no evidence exists that implants result in higher levels of silicone in breast milk (or that silicone can be passed through the placenta to the fetus). The panel found no evidence of health effects in children who had been breast-fed by women with implants. And it found that while implants can interfere with mammography because the silicone is not as transparent to X rays as natural breast tissue, there is no evidence that this problem led to diagnostic delays that in turn resulted in higher rates of cancer deaths in women with implants.

THE RANGE OF CONSEQUENCES

The IOM report did identify several health consequences from breast implants of either silicone or saline. These problems are localized to the breast area and range from mild to severe. Most are short-term, but some are permanent. They include:

Capsular Contracture

The capsule is the scar tissue that forms around the implant itself. Sometimes this scar tissue can squeeze the implant sacs and cause pain and disfigurement. Studies by manufacturers of saline implants found that 9 percent of augmentation patients and 25 to 30 percent of reconstruction patients ended up with breasts that looked abnormal, or breasts that were hard and painful, due to capsular contracture. Another study of silicone implants found that 54 percent of women with such devices developed contracture problems within 6 months of surgery.

Rupture and Leakage

Some implants rupture and leak. Women with silicone implants can experience uneven breast appearance, pain, tenderness, tingling, swelling, numbness, burning, or changes in breast sensation. Women with saline implants can experience some of those symptoms as well. A study of women who received their first silicone implants before 1988 found that a little more than half of them ruptured, and about one in five of these women had silicone gel in their breast tissue. (Several improvements in implants have been made since then.) A study of women with saline implants found that 3 percent of augmentation patients and 9 percent of reconstruction patients experienced leaking implants.

Additional Surgery

Women with implants may need more surgery at some time to correct for rupture and leakage, contracture, infection, shifting, and calcium deposits that sometimes grow around the implant. Studies of women with saline implants found that 13 to 21 percent of augmentation patients and 40 percent of reconstruction patients required additional surgery. A study of women with either type of implant found that one in four experienced problems that required additional surgery within five years after implantation.

Interference with Mammography

In addition to obscuring the passage of X rays, implants interfere with mammography in other ways. Mammography requires compression of the breast for the most accurate results, creating a small risk of rupture for women with implants. Special mammography techniques are used for these women but they can result in longer mammography sessions and more X-ray exposure. And the calcium deposits that can form in the scar tissue capsule around implants can hide indications of cancer or be misinterpreted as tumors or other abnormalities, potentially resulting in misdiagnoses and sometimes leading to unnecessary follow-up treatment, such as biopsy.

Breast implants can cause other consequences too, including:

- Changes in nipple and breast sensation
- General pain

- Infection
- Hematoma (swelling and pressure, and the appearance of a bruise, from the collection of blood inside a body cavity)
- Breast tissue atrophy
- Chest wall deformity

THE RANGE OF EXPOSURES

As we noted, the availability of silicone gel implants is restricted. The only way women can get silicone implants for reasons other than reconstruction or replacement is by taking part in special studies approved by the FDA for the purpose of gaining federal approval for the product being tested. As of the end of 2001 FDA regulators had not yet decided whether to allow silicone implants back on the market for uses other than reconstruction or replacement.

The FDA also restricts saline implants. Following seven years of study that began in 1993 and involved several implant manufacturers, the FDA in 2000 approved the saline implants of just two manufacturers—the Mentor Corporation and the McGhan Corporation.

Exposure to the risks of breast implants depends on many variables, including the surgeon's experience and competence, the surgery's purpose (augmentation, reconstruction, or replacement), placement of the implant (beneath or above the chest muscle that lies beneath the breast), type of implant (saline or silicone), the implant's surface, size, and shape (smooth or textured, large or small, round or contoured), and location of the incision (periareolar—under the nipple, inframammary—under the breast, or axillary—under the arm).

Exposure risk varies with each patient. A woman's overall health, chest structure and body shape, healing tendencies, and incidence of prior breast surgery all affect outcome. Women's goals and expectations about the results of the procedure can also vary. Medical experts recommend that women who choose breast augmentation should be realistic about what the procedure can, and cannot, accomplish.

REDUCING YOUR RISK

The IOM report said that women considering implant surgery need to be provided with more detailed information on the procedure and the possible negative side effects. In selecting a surgeon, women should ask candidates how many procedures they do in a year, how long they've been in practice, whether they are certified and by which medical groups. Women should also ask potential surgeons about the most common complications that patients of these surgeons have experienced, what the doctor's reoperation rates are, and what sorts of reoperations he most often has to perform.

After selecting a surgeon, a woman should ask the following questions about an augmentation procedure:

- What are the risks and complications? What about changes in sensation?
- How many additional operations of my implanted breast(s) can I expect over my lifetime?
- What are my choices regarding shape, size, surface texturing, incision site, and placement site? What are the benefits and drawbacks of each?
- How might my ability to breast-feed be affected?
- How can I expect my implanted breasts to look over time? After pregnancy? After breast-feeding?
- What are my options if I am dissatisfied with the cosmetic outcome of my implanted breasts?
- How will my breasts look if I choose to have the implants removed?
- Are there alternate procedures or products available if I choose not to have breast implants?
- Are there before and after photos that show each procedure?
- Can you refer me to other patients who can share their experiences?

Additional questions that should be considered by a woman thinking about breast reconstruction surgery include:

- What are my options for breast reconstruction?
- What are the risks and complications of each type of breast reconstruction surgery and how common are those complications?
- Will reconstruction interfere with my cancer treatment?
- How many steps are there in each procedure, and what are they?
- How long will it take to complete my reconstruction?
- How much pain or discomfort will I feel, and for how long?
- Are there before and after photos that show each procedure?
- What will my scars look like?
- What kind of changes in my implanted breast can I expect over time? With pregnancy?

- What are my options if I am dissatisfied with the cosmetic outcome of my implanted breast?
- Can you refer me to other patients who can share their experiences?

FOR MORE INFORMATION

U.S. Food and Drug Administration
Office of Women's Health
www.fda.gov
5600 Fishers Lane, Room 14-62
Rockville, MD 20857
(888) 463-6332

Institute of Medicine
National Academy of Sciences
www.nas.edu
Committee on the Safety of Silicone Breast Implants
2101 Constitution Avenue NW
Washington, DC 20418
(202) 334-2000

National Cancer Institute
Office of Cancer Communications
www.nci.nih.gov
Building 31, Suite 10A-24
9000 Rockville Pike
Bethesda, MD 20892
(800) 4-CANCER or (800) 422-6237

National Institutes of Health
Office of Research on Women's Health
www.nih.gov
Building 1, Suite 201
1 Center Drive
Bethesda, MD 20892
(301) 402-1770

American Cancer Society
www.cancer.org

1599 Clifton Road NE
Atlanta, GA 30329
(800) ACS-2345 or (800) 227-2345

This chapter was reviewed by Dr. Stuart Bondurant, Professor of Medicine and Dean Emeritus of the University of North Carolina School of Medicine and Chair of the IOM panel on breast implants; and by Dr. Matthew Liang, Professor of Medicine at Harvard Medical School and Professor of Health Policy and Management at the Harvard School of Public Health. Dr. Liang is Medical Director of Rehabilitation Services at Brigham and Women's Hospital in Boston.

40. CANCER

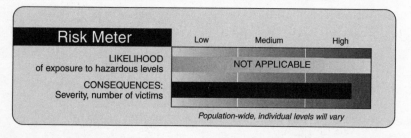

BY OUR DEFINITION, cancer is not a risk. Cancer is a consequence, an outcome, the result of processes initiated by hazards such as smoking or exposure to the sun. But since fear of this disease is so common, we devote a separate chapter to the general issues of cancer. We explain the disease itself, offer some general explanations about which external hazards can lead to various types of cancer, and we offer some general advice on how you can reduce your risk of cancer.

~

Cancer is second only to heart disease as a cause of death in the United States. One out of every two American men and one out of every three women will get cancer over the course of their lifetime. An estimated 1.3 million new cases were detected in 2001. In that year, cancer killed approximately 553,000 U.S. residents.

Cancer is one of the most widely feared health risks in the modern world. We say the "modern world" because cancer is principally a disease that occurs in people ages 50 and up and people have been living that long, on average, for only about the past 100 years. Until the early 1900s, before the advent of public health efforts like water treatment, sanitation, and the increased availability of health care, the average American lived only about 45 years. Most people simply didn't live long enough to develop cancer. The growth in the number of cancer cases over the past several decades is in large measure a result of the longer life expectancy that is a phenomenon of the last century. Three quarters of the people diagnosed with cancer are 55 or older.

HOW CANCER OCCURS

Cancer is not a single disease. There are more than 100 different kinds of cancer, each identified with the part of the body or the body system it affects. Each disease behaves differently. But almost all cancers have one thing in common. Nearly all of them result from a series of genetic mutations that build up in a single cell and interfere with that cell's ability to control how often it reproduces and makes copies of itself. This process explains why cancer is usually a disease of the later years of life. It takes many years for the several mutations necessary to cause cancer to accumulate in one cell and start the disease.

Normally, conditions inside and outside each cell control cellular reproduction. Some conditions call for more active growth, some call for less. But no matter how fast or slowly they reproduce, cells make only a certain number of copies of themselves before they either stop—a condition called senescence—or they die off altogether in a process of natural, programmed cell death. This regulated cell growth and death keeps an organism healthy. When something interferes with these controls, allowing cells to reproduce indefinitely, it leads to cancer.

The external or internal signals that trigger cells to reproduce are moderated by specific proteins, which are produced by specific genes on our DNA. Mutations to those genes either eliminate the critical growth-regulating proteins or cause them to be manufactured by the genes in a form that doesn't work properly. In either case, the natural limits on cellular reproduction don't function. Cells with these damaged genes reproduce indefinitely and can form a tumor.

DNA actually has a built-in editing system to catch these mutations and splice them out. But an additional mutation that occurs in cancer cells eliminates DNA's ability to do this internal editing. Still more

DNA mutations also give the tumor the abnormal ability to grow blood vessels for nourishment. More mutations let a few tumor cells survive even if they break away from the tumor itself. (Normally cells need to adhere to similar cells or they die.) And additional mutations to other critical genes give the tumor cells the ability to burrow through the dividing tissues that separate body organs, which is also something normal cells can't do. All of these abilities allow indefinitely reproducing cancer cells to get into the blood and lymph systems and travel to other parts of the body. It is this spreading, known as metastasis, that makes cancer so deadly. Confined to just one place in the body, most cancers could be removed or treated. But spreading as they do, they can more easily escape treatment and attack various organs and body systems, enough to be fatal.

So what causes these mutations? Sometimes they occur naturally when a cell makes a copy of itself and its DNA and doesn't get things exactly right. Once every million or 10 million times, the DNA in the daughter cell isn't a precise copy of the DNA in the mother cell. Mutations that damage genes that control cell growth are one of the steps that could lead to cancer. So the more frequently a cell reproduces, the greater the likelihood that one of these natural spontaneous mutations will occur. Consider these examples of how accelerated cell reproduction may be associated with certain forms of cancer:

The estrogen cycle in women causes monthly bursts of accelerated cell growth. Since women are reaching menstruation at an earlier age, over their lifetimes they are experiencing a total of more menstrual cycles, more of these bursts of cell growth, and therefore a greater likelihood of natural mutation. Scientists think that may contribute to higher rates of breast cancer. And heavy alcohol consumption is associated with larynx, mouth, and liver cancers because it kills cells in the mouth and throat, liver, and digestive system and prompts accelerated cellular reproduction to repair that damage.

Another source of mutation comes from chemicals called "free radicals," which our cells produce as part of our metabolism. Free radicals are molecules with an uneven number of electrons. They tear electrons away from other molecules. If they tear electrons off the DNA molecule, it can cause mutations to genes that control cell reproduction.

But external factors are by far the biggest cause of mutations to our genes. It is estimated that as many as half of all cancers are preventable simply by changes to our lifestyle. Smoking, for instance, causes roughly one third of all cancer deaths, and 90 percent of deaths from lung cancer. Diet, including excess caloric intake, is associated with approximately 30 percent of all cancer deaths. Obesity causes 10 percent of cancer deaths in men and 15 percent in women. And in addition to general lifestyle factors, there are specific external hazards, like radon, some of the chemicals in motor vehicle exhaust, asbestos, or biological agents like some viruses or bacteria, which are known as carcinogens. A carcinogen is not an agent that, by itself, causes cancer. Rather, a carcinogen is a hazard that causes any one of the mutations to genes that can contribute to cancer. (Something that can damage a gene and cause a mutation that doesn't contribute to cancer is called a mutagen.)

You may have heard that genetics plays a role in cancer. Inheritance doesn't cause mutations, of course. But a person who inherits a damaged growth-regulating gene has a greater chance of developing cancer because every cell in his or her body is already one step down the road to the several mutations necessary for cancer. Genetic predisposition is believed to be associated with 5 to 10 percent of all cancer deaths.

THE STATISTICS

Of all the types of cancer, the top four—lung, colorectal, breast, and prostate—account for more than half the cancer deaths in America each year.

The following table (pages 340–43) offers a perspective on the top ten cancers in America. Like all the numbers in this book, the table reflects conditions for the general population. Your cancer risk is higher or lower depending on several factors unique to you.

(Skin cancer is not included. It is so common that cancer statistics usually treat skin cancer separately since it would overwhelm the other categories. We discuss skin cancer in Chapter 36.)

The table on page 344 helps put in perspective the factors associated with causing cancer in the United States.

REDUCING YOUR RISK

Two thirds of the cancers in the United States each year relate to four major risk factors that you can control: tobacco use, diet, obesity, and lack of exercise. As we have stated, with modifications to these lifestyle choices, experts estimate that half of the cancers in this country are preventable. Beginning on page 344, we offer some basic steps that can help reduce your cancer risk.

LEADING CAUSES OF CANCER MORBIDITY AND MORTALITY ESTIMATED FOR 2001

Type	Cases per Year, Change 1973–1998	Deaths per Year, Change 1973–1998	Populations Most Likely to Be Affected	Modifiable Factors That Increase Risk
Lung	169,500 (27.9 percent increase)	157,400 (35.9 percent increase)	People who: • Are age 60 or older • Have a parent or sibling with lung cancer	• Cigar and cigarette smoking • Exposure to chemicals in the workplace • Exposure to radon in the home • Exposure to passive smoke • Exposure to urban air pollution *Probable* • Low intake of carotenoids (yellow and red vegetables like carrots and squash)
Colorectal	135,400 (5.8 percent decrease)	56,700 (25.9 percent decrease)	People who: • Are age 50 or older • Have certain genetic syndromes: familial adenomatous polyposis or hereditary nonpolyposis colorectal cancer • Have a parent or sibling with colorectal cancer or adenomatous polyps • Have inflammatory bowel disease	• No colorectal cancer screening • High intake of red meat • Low level of physical activity *Probable* • No multivitamin use • Greater weight • High intake of alcohol • No long-term aspirin use • No long-term use of birth control pills

| Breast | 192,200 (32.3 percent increase) | 40,200 (14.2 percent decrease) | Women who have: • Lobular carcinoma in situ • Mother or sister with breast cancer • Confirmed mutation in BRCA 1 or BRCA 2 gene • Previous breast cancer • High blood levels of estrogen • Repeated high-dose radiation to the chest in childhood or early adulthood • High breast density • Certain types of benign breast disease, including intraductal hyperplasia, atypical hyperplasia, and radial scars • Late age at menopause • Late age at giving first birth • No children • Early age for beginning menstruation | • Obesity after menopause • Substantial weight gain during adulthood • High intake of alcohol • Current or recent use of postmenopausal hormones for 5 or more years • Current or recent use of birth control pills *Probable* • Low intake of vitamin A and carotenoids • Not breast-feeding • Lack of physical activity |

Type	Cases per Year, Change 1973–1998	Deaths per Year, Change 1973–1998	Populations Most Likely to Be Affected	Modifiable Factors That Increase Risk
Prostate	198,100 (116 percent increase)	31,500 (1.6 percent increase)	Men who: • Have a father or brother with prostate cancer • Are African American	• High intake of animal fat • Low intake of tomato-based foods • Vasectomy
Pancreas	29,200 (9.2 percent decrease)	28,900 (3.5 percent decrease)	People who: • Have a parent or sibling with pancreatic cancer *Probable* • Diabetes • Chronic pancreatitis	• Cigarette smoking *Probable* • Low intake of vegetables
Non-Hodgkin's Lymphoma	56,200 (82.3 percent increase)	26,300 (47.4 percent increase)	People who: • Have HIV infection • Use immunosuppressive drugs • Have a parent or sibling with non-Hodgkin's lymphoma	
Leukemia	31,500 (6.6 percent decrease)	21,500 (28.0 percent increase)	People who have had: • Therapeutic radiation • Certain occupational exposures	• Cigarette smoking

Liver	16,200 (95.6 percent increase)	14,100 (56.3 percent increase)	People who have: • Chronic infection with hepatitis B or C • Cirrhosis • Exposure to aflatoxins • Exposure to Thorotrast • Certain occupational exposure	• Persistent, heavy alcohol use *Probable* • Cigarette smoking
Ovarian	23,400 (1.0 percent decrease)	13,900 (13.6 percent decrease)	Women who: • Have a mother or sister with ovarian cancer • Have a confirmed mutation in the BRCA 1 or 2 gene • Are older than 60 • Have fewer than 2 children	• No tubal ligation • No use of birth control pills *Probable* • No breast-feeding • No hysterectomy • Use of postmenopausal hormones
Brain	17,200 (16.8 percent increase)	13,100 (10.8 percent increase)	Very little is known about risk factors for brain cancer	

Source: Harvard Center for Cancer Prevention

RISK FACTORS FOR CANCER

Factor (types of cancer risk associated)	Percentage of Cancer Deaths
Tobacco (lung, larynx, esophagus, bladder, kidney, pancreas, liver, stomach, leukemia)	30 percent
Adult diet/obesity (colon, breast, cervix, endometrium, ovaries, uterus, prostate, lymphoma)	30 percent
Lack of exercise (colon, breast)	5 percent
Occupational exposures (various)	5 percent
Family history of cancer (breast, ovaries, colon, prostate, lung)	5 percent
Viruses (cervical, anal)	5 percent
Birth-related factors—height/weight (various)	5 percent
Reproductive/hormonal factors—adults (breast, prostate)	3 percent
Alcohol (breast, larynx, liver, mouth, pharynx, colorectal, lung)	3 percent
Socioeconomic status (various)	3 percent
Environmental pollution (various)	2 percent
Radiation (skin, lung)	2 percent
Drugs and medical treatments (various)	1 percent
Food additives/contamination (?????)	1 percent

Courtesy Harvard Center for Cancer Prevention, Kluwer Academic Publishers

Avoid Tobacco

Lung cancer is the leading type of cancer in the country, and 9 out of 10 cases are caused by tobacco use. Roughly one third of the men who die of cancer each year, and one quarter of the women, die from lung cancer. Most of these people smoked. Smoking is also associated with cancers of the esophagus, mouth, larynx, bladder, kidney, and pancreas, and recent studies suggest smoking may increase your likelihood of colon cancer, the second most common cancer in America.

Even if you don't smoke, smoking can increase your risk of cancer. An estimated 3,000 nonsmokers die of lung cancer each year because of secondhand or environmental tobacco smoke. If you don't smoke but you live with someone who does, you have a 30 percent higher chance of getting lung cancer than somebody who lives with a nonsmoker.

And you don't have to smoke or inhale secondhand smoke to face

we've learned, cancer is still a largely untreatable disease once it gets going. Regular screening exams are a powerful way to reduce your cancer risk.

FOR MORE INFORMATION

To assess your own risk, visit www.yourcancerrisk.harvard.edu, a website designed and run by the Harvard Center for Cancer Prevention which lets you enter your risk factors and determine your risk for most of the major cancers.

National Library of Medicine
MEDLINEplus
www.nlm.nih.gov/medlineplus/cancergeneral.html

National Cancer Institute
www.nci.nih.gov
Building 31, Room 10A31
31 Center Drive, MSC 2580
Bethesda, MD 20892-2580
(800) 4-CANCER or (800) 422-6237

American Cancer Society
www.cancer.org
1599 Clifton Road NE
Atlanta, GA 30333
(800) ACS-2345 or (800) 227-2345

This chapter was reviewed by Dr. Graham Colditz, Professor of Medicine at the Harvard Medical School, Professor in the Department of Epidemiology at the Harvard School of Public Health, Principal Investigator of the ongoing Nurses Health Study, and Director of Education at the Harvard Center for Cancer Prevention; and by Dr. Scott Dessain, oncologist at Massachusetts General Hospital and cancer researcher at the Whitehead Institute at the Massachusetts Institute of Technology.

41. HEART DISEASE

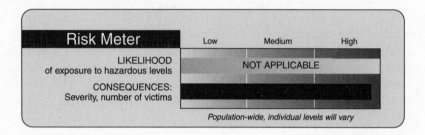

Risk Meter	Low	Medium	High
LIKELIHOOD of exposure to hazardous levels	NOT APPLICABLE		
CONSEQUENCES: Severity, number of victims			

Population-wide, individual levels will vary

LIKE CANCER, heart disease is not a hazard. It is a consequence, the outcome of exposure to hazards including poor diet, lack of exercise, smoking, and genetic factors. But we include this chapter because while heart disease may not make the headlines, as does cancer or AIDS, it is far and away the leading cause of death in the United States. Approximately 710,000 Americans died of heart disease in the year 2000. Out of every 100 people who died, of all causes, 34 of them died from heart disease. It hospitalized, but did not kill, approximately 4.5 million Americans. The economic cost of heart disease to the United States is a staggering $214 billion a year. That's one quarter of the annual cost of running every branch of the U.S. government. Slightly more than half of this cost is direct, for medical treatment. The rest is in lost life or human productivity because of disability.

～

In a sense, given the incredible amount of work the heart does, it's a wonder that heart disease is not even more prevalent. The heart is a remarkably durable, fist-sized muscle capable of pumping five quarts of blood through our bodies every minute, pumping an average of roughly 100,000 times each day. It has to push blood though a network of vessels that, laid end to end, are 60,000 miles long. Over an 80-year life span, a human heart beats 2 to 3 billion times. The vessels of the cardiovascular system are remarkable as well, built of elastic fibers and in some cases ringed by muscles, which allow them to expand with each pulse of blood and then contract back to their original size. The heart's rhythm is controlled by an intricate electrical firing mechanism that keeps heart muscle cells expanding and contracting in unison. Blood flows through four chambers, separated by valves. As that cursory explanation suggests, several systems are vital to proper cardiovascular function, which explains why heart disease is not a single condition but a family of syndromes related to the circulatory system.

Coronary heart disease (CHD, also known as coronary artery disease or ischemic heart disease) is the most common and most frequently lethal form of heart disease. It kills roughly two thirds of all victims of heart disease, and is the cause of death of approximately one in five Americans a year. It is caused by atherosclerosis, a buildup of cholesterol in the arteries which blocks the flow of blood, leading to heart attacks, angina (chest pain), sudden death, and strokes. An estimated 12.6 million Americans have a history of at least one of the major symptoms of CHD:

Heart attack, known medically as myocardial infarction, is the death of some portion of the muscle tissue of the heart because of inadequate blood flow to that tissue. It is not, as most people think, a stoppage of the heart, though that can occur during a heart attack. An estimated 1,100,000 U.S. residents have a heart attack each year, and roughly 650,000 will be first-time attacks. An estimated 7.5 million Americans have had a heart attack.

A heart attack is not an instant event, but a dynamic process that usually evolves over 4 to 6 hours. In fact, about half of heart attack victims have warning symptoms hours, days, or even weeks in advance. One early predictor of an attack may be recurrent chest pain brought on by exertion and relieved by rest. With each passing minute during a heart attack, more heart tissue is deprived of oxygen and deteriorates or dies. Also, at any moment during a heart attack, the patient can suddenly suffer a ventricular fibrillation, an abnormal heart rhythm that starts in the bottom two chambers of the heart, causing it to stop pumping blood altogether. That's why it's so important to respond promptly to symptoms of heart attack, and why the survival rate depends on the medical treatment received in the first hour. Roughly half of heart attack victims who die do so within the first hour.

Heart attacks can be "silent" and cause no symptoms at all. When symptoms do occur, they include:

- Uncomfortable pressure, fullness, squeezing, or pain in the center of the chest, usually lasting 15 minutes or more. Often the sensation goes away and returns.
- Pain or tingling that spreads to the shoulders, neck, or arms. It may

be mild to intense; may feel like pressure, tightness, burning, or heavy weight; may be located in the chest, stomach, back, neck, jaw, or inside the arms or shoulders.
- Anxiety; nervousness; cold, sweaty skin; lightheadedness; fainting; nausea; or shortness of breath
- A knot in the stomach that feels like indigestion
- Paleness or pallor
- Increased or irregular heart rate.

Angina pectoris is pain in the chest caused by inadequate blood flow to the heart. Though much has been written about all the emotional pain we feel in our heart, the heart muscle has no physical pain sensation. We feel discomfort only because of pain-sensing nerves near the heart which can be stimulated a number of different ways. For this reason, angina is a varied sensation that feels like a pressing or squeezing pain, usually in the chest under the breastbone but sometimes in the shoulders, arms, neck, jaws, or back. Angina is usually brought on by exertion and relieved either by rest or medication. Other triggers can be emotional stress, extreme cold or heat, heavy meals, alcohol, and cigarette smoking. Angina is not tantamount to a heart attack or permanent damage to the heart muscle tissue. But it is a warning sign that blood flow to the heart is restricted, which could lead to more severe symptoms.

Angina pain is usually less severe than in a heart attack, and goes away with rest or medication, which isn't the case with heart attack. Angina pain is usually worse and lasts longer than temporary chest pains that aren't related to the heart. Approximately 6,400,000 Americans suffer angina, with roughly 400,000 new cases diagnosed each year. Roughly twice as many women suffer angina pectoris as men.

Sudden cardiac death from CHD is often called a "massive heart attack," but it's actually what's known as ventricular fibrillation, a sudden abnormal heart rhythm affecting the ventricles, the bottom two chambers of the heart, keeping the heart from pumping blood. To correct this condition, doctors use electrical shock devices to restore normal beating of the heart, a process called defibrillation. As many as 220,000 of the people who die of heart attacks each year die of ventricular fibrillation/sudden cardiac death. In 90 percent of these victims, two or more major coronary arteries are narrowed by atherosclerosis.

Congestive heart failure is another main category of CHD, a weakening of the heart to the point at which it is no longer strong enough to pump blood adequately. It killed 50,824 Americans in 1999 and is a contributing factor in roughly 225,000 other deaths per year. An estimated 4.7 million Americans are living with congestive heart failure, and approximately 550,000 new cases are diagnosed each year. Heart failure is the leading cause of hospital admissions of people over 65 years old.

Heart failure does not actually involve what it implies, that the heart fails and stops. Rather, it is a progressive condition in which the capacity of the heart to pump blood is diminished over time. The most common symptoms are shortness of breath, overwhelming fatigue, and swelling of the ankles. (The condition is called *congestive* heart failure because it usually causes fluid to build up in several organs including the lungs, giving victims a feeling of respiratory congestion.) As congestive heart failure progresses, patients lose the ability to perform even modest physical activity, even caring for themselves.

Congestive heart failure increases the risk of sudden cardiac death. About half of the people diagnosed with congestive heart failure are alive five years later. Only one quarter are alive ten years after diagnosis. But as much as we understand about heart disease, 45 to 50 percent of the severe cases of congestive heart failure that can be treated only with transplantation are "idiopathic," which means doctors are not sure what the causes are.

Another general category of CHD is *cardiac arrhythmias.* These are irregular heartbeats, either too fast, too slow, or uneven, all of which interfere with blood flow and can lead to fatigue, lightheadedness, chest pain, and even death. This category includes ventricular fibrillation, the most dangerous kind of irregular heart rhythm. It's actually no rhythm at all, just a confusing set of electrical signals that makes the heart muscle quiver rather than beat regularly. As a result, it can't pump blood through the four chambers and out to the body.

Other arrythmias include atrial fibrillation, when the top two chambers of the heart, the atria, quiver instead of beat rhythmically. The blood pools and sometimes clots, which can cause strokes if those clots travel to the brain and block blood flow. About 3 to 5 percent of U.S. residents suffer this condition, most of whom are 65 or older.

Bradycardia, in which the heartbeat is less than 60 beats per minute, slower than it needs to be for normal blood circulation through the body, produces fatigue, dizziness, lightheadedness, fainting, or near-fainting spells. And tachycardia, in which the heart beats too fast, more than 100 beats per minute, effects dizziness, lightheadedness, fainting, or near fainting.

Arrythmias are treated with pacemakers, electrical devices that stabilize heartbeat, or defibrillators, which detect dramatically abnormal rhythms and shock the heart back into its normal beating pace. An estimated 4 million U.S. residents have some kind of electrical device to help maintain normal heart rhythm. Most of these are implanted, but external portable defibrillators have also been approved. A number of drugs are also prescribed to regulate heartbeat.

Almost 45,000 deaths in the United States in 1998 were attributed to cardiac arrythmias, although, as we mentioned, 220,000 sudden deaths from CHD (so-called massive heart attacks) were probably from ventricular fibrillation.

Other serious but less common forms of CHD are *diseases of the pericardium,* the sac around the heart; *diseases of the myocardium,* the heart muscle itself; endocarditis or *valvular heart disease,* which affects the heart's valves; and *congenital heart disease,* birth defects of the heart. Congenital heart disease is the most common form of birth defect in children born in the United States, occurring in about 1 percent of live births. These defects occur in various forms and can seriously limit how the child is able to function as he or she grows up. In some cases they can be fatal. Many of these defects can be at least partially repaired through surgery.

FACTORS THAT CONTRIBUTE TO HEART DISEASE

Several underlying conditions contribute to CHD, including high blood pressure, high levels of certain kinds of blood cholesterol, obesity, lack of physical activity, and smoking. We can do something about many of these factors. A lot of people with these risk factors will eventually succumb to CHD even though they try to modify them, but those modifications are still important, because changing these factors means those people will die later in life and enjoy a much better quality of life in the meantime. Experts point out that reducing your risk of CHD requires attention to several of these factors, not just any one of them.

High Blood Pressure or Hypertension

About 50 million U.S. residents have high blood pressure and roughly one third of them don't know it. In 1999, hypertension killed nearly 43,000 people directly, and contributed to the deaths of another 227,000 people.

Blood pressure is a combination of two factors: the strength of the pumping of the heart as it pushes fluid through the arteries and veins, and the force on the blood from the elastic arteries and veins pushing back as the fluid flows through them. When the pressure is high, the heart has to work harder to move blood around the body. This stress can cause the heart muscle to grow, just as any muscle will grow with excessive exercise. High blood pressure can also cause the arteries and veins to lose some of their elasticity and to scar. Both of these conditions then make the pressure grow even higher, stressing the heart further and contributing to CHD.

About 95 percent of the time, high blood pressure has no known cause. This is called essential hypertension. Secondary hypertension, for which causes are understood, is triggered by such things as kidney problems, defects in the aorta, or a tumor in the adrenal glands.

Your doctor usually gives you your blood pressure as two numbers. The first—systolic—measures the pressure when the heart is beating. The second—diastolic—measures the pressure when the heart is resting between beats. In adults, if the systolic number is 140 or above, or the diastolic number is over 90, blood pressure is considered high. And even a small rise in blood pressure can be worrisome. Two thirds of heart attack victims have only mildly high blood pressure. Optimal adult levels are 120/80 or below, but levels are affected by a number of factors and there is a wide range of what is considered healthy. (If levels are too much *below* 120/80, check with your medical professional for other possible problems.) Recommended blood pressure levels for children vary because they are still growing, and at different rates. Check with your pediatrician for what's recommended for your child.

Though science hasn't determined the cause of most cases of hypertension, researchers have identified several factors that seem to increase the likelihood that high blood pressure might develop:

- Heredity
- Race: African Americans are more likely to have hypertension, to develop it earlier in life, and to have higher blood pressures.
- Gender: Hypertension is more common in males, until age 55. After age 75, it's more common in women.
- Rising age
- Obesity
- Sedentary lifestyle
- Alcohol consumption beyond one or two drinks a day
- Diabetes, gout, and kidney disease

- Pregnancy (blood pressure will return to normal once the pregnancy is over)
- Certain medications (many over-the-counter cold and flu products that contain decongestants)
- Excessive salt consumption (in some people)
- Educational and income levels: high blood pressure tends to be more common in people with lower educational or income levels.

The best way to lower blood pressure is to alter the lifestyle factors under your control. But if that doesn't work, a range of medications can help many hypertension sufferers. Establishing the right levels of these medications takes time, and some have side effects. And the drugs work only while you're taking them; they don't cure the hypertension, only control it.

Cholesterol Levels

Much has been written about cholesterol, giving this substance a dubious reputation. But your body naturally produces cholesterol, a fatty substance in all animal cells. Your body needs cholesterol to produce hormones, vitamin D, and the bile acids that help to digest fat. But it takes only a little cholesterol to do these jobs. Too much can be dangerous.

Cholesterol is carried through the bloodstream by molecules called lipoproteins, either low-density lipoproteins (LDL) or high-density lipoproteins (HDL). LDL is the "bad" carrier because it releases cholesterol easily, allowing it to build up in artery walls, leading to the formation of plaque deposits. HDL is known as the "good" cholesterol carrier because it binds much more tightly to cholesterol and even collects the cholesterol released by LDL.

A buildup of cholesterol in the walls of the arteries is sometimes referred to as "hardening of the arteries," and, indeed, that's part of the problem. If the arteries are not as elastic as they should be, they can't expand and contract and help regulate normal blood pressure. But this buildup of cholesterol in the arterial walls is officially known as atherosclerosis, which also narrows the arteries, obstructing the flow of blood. If it shuts off blood supply to part of the heart muscle, atherosclerosis causes heart attack. Shutting off blood flow through a vessel that supplies the brain causes stroke.

One of the most important goals in fighting heart disease is to lower your LDL levels. That will reduce your risk more than raising your HDL levels, and lowering LDL is easier to do. The specific levels of LDL and HDL are more important predictors of atherosclerosis than the total cholesterol level.

Recommended cholesterol levels for adults based on research findings are:

Total
Desirable: Less than 200 mg/dL (milligrams per deciliter)
Borderline-high risk: 200–239 mg/dL
High risk: 240 mg/dL and above

LDL levels
Optimal: Less than 100 mg/dL
Desirable: 100–129 mg/dL
Borderline-high risk: 130–159 mg/dL
High risk: 160–189 mg/DL
Very high risk: 190 mg/dL and above

HDL levels (the more the better, so higher numbers are good)
Lowers your risk: 60 mg/dL or above
Intermediate: 41–59 mg/dL
Raises your risk: 40 mg/dL or below

One third of Americans have total cholesterol levels between 200 and 239; 20 percent have levels of 240 and above. Studies show that people with a total cholesterol reading above 240 have twice the risk of a heart attack as someone whose level is 200. One study showed that people with total cholesterol levels of 265 are four times more likely to suffer heart attacks than people with a level of 190 or lower.

Atherosclerosis often begins in childhood. (One study found that 10 percent of American children between 12 and 19 years old had total cholesterol levels of 200 or higher.) The LDL cholesterol recommendations for children 2 to 19 are:

Desirable: Less than 110 mg/dL
Borderline-high risk: 110–129 mg/dL
High risk: 130 mg/dL or above

Cholesterol levels are another CHD risk over which you have significant control, and reducing them, especially LDL levels, can really help. A 10 percent reduction in your total cholesterol levels can mean a 30 percent reduction in your chances of developing heart disease. The first step is to check your levels. Public health officials suggest that adults should have their levels checked at least every five years, starting at age 20.

The next step is to control your intake of the kinds of foods that contribute to high levels of LDL cholesterol. These are mostly *saturated fats,* which are found in foods like beef, beef fat, veal, lamb, pork, lard, poultry fat, butter, cream, milk, cheeses, and other dairy products made from whole milk, as well as coconut oil, palm oil and palm kernel oil (often called tropical oil), and cocoa butter. Health experts recommend that you not consume more than 7 percent of your total daily caloric intake in the form of saturated fats. They also recommend that you not eat more than 200 milligrams of dietary cholesterol per day. (The saturated fat and cholesterol content of food is on the nutrition label.) Foods that contain *unsaturated fats* may help lower your total LDL cholesterol level when they replace saturated fats in your diet. They include safflower, sesame, and sunflower seeds; corn and soybeans; many nuts and seeds; canola, olive, and peanut oils, and avocados.

Trans fatty acids are present in foods that are formed in food processing. They can raise total blood cholesterol levels. They are found in margarine, shortening, cooking oils, and the foods made from them or cooked in them, like most fried foods. They are a significant source of cholesterol-raising fat in the American diet.

Triglycerides are associated with cholesterol but aren't a direct cause of high or low cholesterol levels. Triglycerides are a form of fat that occurs in food and is produced in your body as your metabolism breaks down the fats you eat. People with high triglyceride levels often have high total cholesterol, high LDL cholesterol, and low HDL cholesterol levels. Several studies have shown that people with above-normal triglyceride levels have an increased risk of heart disease. Doctors treating high total cholesterol levels usually recommend dietary changes that will help with high triglyceride levels too.

Recommended triglyceride levels are:

Normal: Less than 150 mg/dL
Borderline-high: 150–199 mg/dL
High: 200–499 mg/dL
Very high: 500 mg/dL or higher

Health experts recommend that, no matter what kind of fat your food may contain, you limit your total daily fat intake to no more than 30 percent of your total caloric intake. The information you need in order to do so is right on the labels of most foods, since the fat and calorie content of most foods is now required information. Trans fatty acid content is not. As of the end of 2001, the U.S. government was accelerating an effort to include this information on food labels too. Other dietary choices that can reduce cholesterol levels include eating fruits and vegetables, cereals, breads, pasta, and other whole-grain products, and avoiding whole-milk products, opting for low-fat varieties instead.

Regular physical activity and weight loss also help keep total blood cholesterol levels down. Several prescription medications can also help.

Tobacco Smoking

The U.S. surgeon general says smoking is "the most important of the known modifiable risk factors for coronary heart disease in the United States." Roughly one third of all cases of CHD each year are attributed to the effects of smoking. The risk of heart attacks for smokers is more than twice that of nonsmokers, and smokers have between two and four times greater risk for sudden death, known as a massive heart attack.

Tobacco smoke damages the cardiovascular system in a number of ways. It stimulates blood clotting and atherosclerotic deposits. It causes scarring of the lining of the blood vessels. Nicotine increases the heart rate and blood pressure, and narrows or constricts the capillaries and arteries. Carbon monoxide in the smoke reduces the level of oxygen in the blood.

Quitting smoking has dramatic benefits for CHD risk. After one year away from smoking, the risk of CHD is cut in half. Five years after quitting, former smokers have about the same risk of heart disease as people who never smoked. Male smokers between ages 35 and 39 who stop smoking add an average of five years to their lives. Women who quit in this age group add three years to their lives. People who quit between ages 65 and 69 increase their life expectancy by a year.

Overweight and Obesity

Some 108 million American adults are overweight; 44,500,000, roughly one in six, are obese; 13 to 14 percent of children ages 6 to 19 are overweight. Being overweight or obese contributes to approximately 300,000 deaths a year in the United States from a variety of causes, including heart disease, diabetes, stroke, and cancer (see Chapter 45).

Lugging around extra pounds boosts the body's need for oxygen, making the heart work harder, which raises blood pressure. Obesity raises total blood cholesterol levels and lowers your HDL levels and can induce diabetes in some people. As you'll learn a little later in this section, diabetes makes other risk factors for heart disease much worse.

The definition of obesity, being excessively overweight, is controversial. Some define it as weighing more than 30 pounds over your ideal body weight. Medical professionals use a calculation to determine your body mass index (BMI), which assesses your weight relative to your height. To determine your BMI, multiply your weight in pounds by 704 and divide that figure by your height in inches, squared. For example, if you weigh 185 pounds, and you are 5 feet 10 inches (70 inches) tall, your BMI is (185 x 704) divided by (70 x 70) = 26.58. (See "For More Information" for a website that can help you do this calculation.)

- BMI values less than 18.5 are considered underweight.
- BMI values from 18.5 to 24.9 are healthy.
- BMI values between 25 and 29.9 are considered overweight.
- Obesity is defined as a BMI of 30 or greater, or about 30 pounds overweight. People with BMIs of 30 or more are at high risk of cardiovascular disease.

While BMIs are the way most medical professionals now assess whether a person is overweight enough to be at elevated risk for heart disease, another (albeit less precise) way is simply to measure your waistline. A high-risk waistline is more than 35 inches for women, and more than 40 inches for men. Studies indicate that extra fat around the waist is more of a risk factor for heart disease than fat on the legs, buttocks, or elsewhere on the body.

Research indicates that between 20 percent and 80 percent of people who are obese have a genetic predisposition to this condition. Still, those who are overweight can reduce their risk through proper nutrition and increased physical activity: essentially, burning more calories per day than you consume. More aggressive dieting and other treatments are necessary in some cases. Treatment may include a combination of diet, exercise, behavior modification, and weight-loss drugs. In cases of severe obesity, gastrointestinal surgery may be recommended.

Diabetes Mellitus

Diabetes is a progressive inability of your body to make or use insulin to convert sugar, starch, and other foods into energy. People with diabetes have abnormally high blood sugar levels. Symptoms include hunger, thirst, frequent urination, weight loss, fatigue, blurry vision, frequent infections, and slow healing of wounds or sores. The most common form of diabetes is type 2 or adult-onset disease, which more often appears in adults over age 40. Approximately 10,600,000 Americans have been diagnosed with diabetes. An estimated 4,600,000 more have it but don't know it.

Diabetes not only kills directly (68,662 in the year 2000—the sixth leading cause of death in America), but it contributes to other deaths, an estimated 190,000 per year, largely from CHD. A person with diabetes but without established CHD is considered to be at the same level of risk for a heart attack as a person with CHD. Diabetes causes nerve damage to the heart. It lowers HDL levels and raises LDL levels. Type 2 diabetes is highly associated with obesity, another contributing cause of heart disease. And 80 to 90 percent of type 2 diabetics are overweight or obese.

In most cases, people can keep diabetes, and the resulting higher risk for heart disease, in check with proper diet, physical exercise, weight management, and in some cases medication.

Physical Activity

You can raise your risk of heart disease by just lying around. Inactivity does as much to raise your risk of CHD as high cholesterol or high blood pressure. Less active people have a 30 to 50 percent greater chance of developing high blood pressure.

Regular aerobic exercise strengthens the heart muscle, improves blood circulation, boosts HDL, and helps break down and inhibit blood clots. Exercise helps prevent and manage high blood pressure. It helps keep weight under control, both by burning calories during the exercise, and by raising overall basal metabolism and producing a higher ongoing supply of fat-burning enzymes. Large studies have repeatedly shown that active people have half as much risk of heart disease as sedentary ones. Other studies have shown that it's how often you exercise, not how hard, that matters most. Thus, running seven miles once a week will have less long-term benefit for the cardiovascular system than walking one mile seven days a week. Health scientists recommend at least moderate exercise for a total of 30 to 60 minutes a day, at least four days a week.

Here's a table listing the calories that a 150-pound person would burn doing some common physical activities. Smaller people will burn less; larger people will burn more.

CALORIES BURNED, BY ACTIVITY	
Activity	Calories Burned per Hour
Bicycling 6 mph	240
Bicycling 12 mph	410
Cross-country skiing	700
Jogging 5½ mph	740
Jogging 7 mph	920
Jumping rope	750
Running in place	650
Running 10 mph	1,280
Swimming 25 yds/min	275
Swimming 50 yds/min	500
Tennis (singles)	400
Walking 2 mph	240
Walking 3 mph	320
Walking 4½ mph	440

Personality

Clinicians have long suspected a link between the ability to cope with stress and all sorts of health problems, including CHD, although the hard scientific evidence for such a link has not been established. Stress is not an officially recognized risk factor for CHD according to the National Institutes of Health or the American Heart Association (AHA). But several epidemiological studies, some now more than 40 years old, have shown that people who overreact even to minor stresses, who are driven by a heightened sense of time urgency, or who are aggressive or hostile have an above-average incidence of CHD. One hypothesis is that these people experience higher blood adrenaline levels. Chronic exposure to higher-than-normal levels of adrenaline accelerates the heartbeat, raises blood pressure, damages arteries, increases clotting, and also suppresses the immune system. One study showed that the stress of performing difficult mathematical problems can cause arteries to constrict. Other studies have found that chronic stress raises blood levels of an amino acid that is associated with CHD. Statistical associations have been shown between heart attacks and stressful events immediately preceding them.

Risk Factors We Can't Control

There are several risk factors for heart disease we can't control. Still, being aware of them helps. People with these risk factors should do more to work on the things they *can* control in order to minimize the chance of CHD.

- Age is the biggest risk factor: 85 percent of those who die from CHD are 65 or older.
- Gender: Chronic CHD rates in postmenopausal women are two to three times higher than in men or younger women, apparently because of decreased estrogen levels. While total deaths for CHD are about the same for both genders, men have a greater risk of heart attack and have attacks earlier in life. The average age of a person having a first heart attack is 65.8 for men and 70.4 for women. The lifetime risk of developing CHD after age 40 is 49 percent for men and 32 percent for women. Possibly because they have heart attacks later in life when they are more frail, women who suffer heart attacks are 50 percent more likely to be dead within a year than men.
- Race: African Americans develop high blood pressure earlier in life, and the levels are higher than in hypertensive Caucasians. African Americans have a 1.5 times higher rate of CHD death than Caucasians. Death rates from diabetes are twice as high in African-American men, and 138 percent higher in African-American women than in Caucasian women.

REDUCING YOUR RISK

The only people who can accurately guide you in reducing your risk of CHD are medical professionals. But you can reduce the risk of fatality if someone around you suffers a heart attack or sudden cardiac arrest. Time is critical. Brain death starts just 4 to 6 minutes after the heart stops circulating blood and oxygen. Cardiac arrest can be reversed if cardiopulmonary resuscitation (CPR) or an electrical shock to the heart from an automated external defibrillator (AED) are given within 7 to 10 minutes. Few resuscitation efforts succeed after that.

The AHA suggests that since heart attack and cardiac arrest are so common, many lives could be saved if everyone took the following steps:

- Know the warning signs of heart attack and cardiac arrest.
- Don't delay more than five minutes in calling for emergency assistance.
- Know CPR and start it right away for people suffering cardiac arrest. CPR helps keep the victim alive, and reduces the chance of brain damage, until trained medical help can arrive.

FOR MORE INFORMATION

American Heart Association National Center
www.americanheart.org
7272 Greenville Avenue
Dallas, TX 75231
(214) 750-5300 or (800) AHA-USA-1 or (800) 242-8721

National Library of Medicine
National Institutes of Health
www.medlineplus.gov

National Heart, Lung, and Blood Institute
National Institutes of Health
NHLBI Health Information Center
www.nhlbi.nih.gov/chd/index.htm (newest recommendations for avoiding heart disease or reducing the risks)
hin.nhlbi.nih.gov/atpiii/calculator.asp?usertype=prof (calculate your 10-year risk of developing heart disease)
www.nhlbi.nih.gov/actintime/index.htm (heart attacks)
www.nhblisupport.com/bmi/?htf=5&hti=10&bmi=25.8&wt=170 (BMI online calculator)
P.O. Box 30105
Bethesda, MD 20824-0105
(301) 592-8573

This chapter was reviewed by Dr. David M. Robinson, Deputy Director of the Division of Heart and Vascular Diseases, National Heart, Lung, and Blood Institute; by Dr. Mehmet Oz, Director of the Heart Institute, Columbia Presbyterian Hospital, New York; and by Dr. Dennis L. Sprecher, Cardiologist and the head of Preventive Cardiology at the Cleveland Clinic.

42. HUMAN IMMUNODEFICIENCY VIRUS

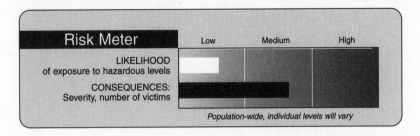

Risk Meter	Low	Medium	High
LIKELIHOOD of exposure to hazardous levels			
CONSEQUENCES: Severity, number of victims			

Population-wide, individual levels will vary

TWENTY-FIVE years ago, no one had ever heard of human immunodeficiency virus (HIV) or acquired immune deficiency syndrome (AIDS), the lethal disease that affects people with advanced HIV infection. In the short time since the virus's discovery in early 1984, however, it has become notorious as a massive global threat to human health. It has killed 22 million people worldwide since it was officially recognized, approximately 440,000 people in the United States alone.

THE HAZARD

Like other viruses, HIV can reproduce only by commandeering the genetic copying equipment of the host cells that it invades, so that every time *they* divide, more viruses are made. But what makes HIV especially heinous is that it preys on essential cells of the human immune system. As a result, the immune system cells die, weakening the body's ability to fight any kind of infection, or to combat abnormalities that can contribute to the formation of cancer, problems a healthy immune system could deal with effectively.

Thus the *immunodeficiency* in the name of the virus. HIV doesn't usually directly make us sick. It just weakens the natural immune system so much that other things that wouldn't normally be a problem can cause disease and death.

You can be infected with HIV and not have AIDS. AIDS is a designation reserved for the most advanced stages of HIV infection. In order to be officially diagnosed with AIDS, an individual must have:

- Fewer than 200 CD4-positive T cells per cubic millimeter of blood (the normal range is 500 to 1400 mm^3), and

• One or more of 26 clinical conditions, including some types of
cancer as well as opportunistic infections caused by certain bacteria,
viruses, fungi, or parasites. Pneumocystis pneumonia, for example,
is caused by a microbe that often lies dormant in the lungs of
healthy people. Normally our immune systems keep it in check. But
when the immune system has been weakened by HIV, the microbe
can proliferate and cause a severe, sometimes fatal pneumonia. And
infants and children with AIDS may suffer much more severe
attacks of ear infections, pink eye, or tonsillitis than children with
healthy immune systems.

THE RANGE OF CONSEQUENCES

The World Health Organization estimated that at the end of 2001 approximately 36.1 million people carried HIV worldwide. Every day,
HIV is transmitted to about 15,000 previously healthy people, a rate of
more than 5 million new infections per year. Since 1981, when physicians first identified AIDS as a new disease, it has killed 21.8 million
people globally.

In the United States, AIDS has claimed the lives of about 440,000
Americans over the past two decades, 15,300 in the year 2000 alone.
The Centers for Disease Control and Prevention estimates that as many
as 850,000 to 900,000 Americans are infected with HIV, that one quarter of them don't know it, and that another quarter do know but are
not receiving treatment. About 75 percent of the people with HIV in
this country are male.

The epidemic is spreading at the rate of 40,000 new infections in the
United States per year. Epidemiologists say that a disproportionate
number of these new cases are among African-American women and
young African-American men who have sex with men. Public health
surveys find that between 3 and 12 million American adults describe
themselves as participating in behaviors that put them at risk for contracting HIV.

Most people have no idea when they become infected with HIV because there are usually no initial symptoms. Some people do experience a brief, flu-like illness marked by fever, aches and pains, swollen
lymph glands, sore throat, headache, or other complaints. But these
symptoms usually disappear in a few weeks and are often attributed to
some less ominous cause. Even if there are mild symptoms at first, they
disappear and a symptom-free period then lasts for months or up to a

decade or more. The average time between infection and development of AIDS symptoms is 6 to 10 years. During this time, the virus is quietly multiplying, invading more cells, and undermining the body's defenses. When it has finally destroyed or weakened enough immune system cells, one or more opportunistic infections take hold and the infected person gets sick and begins showing the symptoms of AIDS.

Symptoms of the opportunistic infections associated with AIDS include:

- Coughing and shortness of breath
- Seizures and lack of coordination
- Difficult or painful swallowing
- Mental symptoms such as confusion and forgetfulness
- Severe and persistent diarrhea
- Fever
- Vision loss
- Nausea, abdominal cramps, and vomiting
- Weight loss and extreme fatigue
- Severe headaches
- Coma
- People with AIDS are particularly prone to developing various cancers, such as Kaposi's sarcoma, cervical cancer, or lymphomas, which are cancers of the immune system.

Most testing for HIV is initiated by blood banks, doctors, life and disability insurance programs, and the military (for recruits). Less frequently, people who don't have symptoms but have reason to believe they have contracted HIV seek out such tests. Doctors can use several different laboratory procedures to diagnose HIV infection. The first step is usually a blood test that detects HIV-specific antibodies, disease-fighting substances that the immune system generates in an effort to control the virus. But because antibodies may take several weeks to accumulate to detectable levels in the bloodstream, this type of test can yield false negative results during the first few weeks of infection. More sensitive (and expensive) tests identify specific HIV proteins and can be used to measure how much HIV is circulating in the body. These tests allow doctors to detect recent HIV infection, make predictions about the likely course of disease, or determine how well a patient is responding to therapy. Because the significance of a positive finding is so serious, most doctors will not diagnose HIV unless they've performed more than one test.

Since 1996, highly active anti-retroviral therapy (HAART), or "the AIDS cocktail" (it involves a combination of several drugs), has helped many people with HIV to live longer, healthier lives and has caused a dramatic decline in the number of AIDS deaths each year. One year after HAART was introduced in the United States, it was credited with a 47 percent decline in AIDS-related deaths. But one of the side effects of HAART has been reduced public concern about HIV because many in the press and public appear to think that since this treatment is available, the problem has been solved. This belief has actually led to a rise in the rate of infections among homosexual men. Although HAART treatments have prolonged and improved life for thousands of HIV-infected patients, in some cases the drugs stop working after a time, and other patients discontinue the medicines due to serious side effects.

At present, there is no cure for HIV/AIDS.

THE RANGE OF EXPOSURES

In the United States, approximately 75 percent of new HIV infections are acquired through sexual activity. Some forms of sex are more dangerous than others and more likely to transfer the virus. Any sexual encounter that results in bleeding is especially dangerous, apparently because it allows HIV to more readily enter the bloodstream. Researchers have found strong evidence that being the receptive partner in genital-anal intercourse carries the highest risk of HIV infection, whether the receptive partner is male or female. The risk to men who assume the insertive role is lower, but still significant for anal intercourse.

Penile-vaginal intercourse can be quite risky for a woman if the man is HIV-positive. The likelihood that an HIV-infected woman will transmit the virus to a male partner during vaginal or anal intercourse is lower. Oral-genital sex is considered less risky than intercourse, but not entirely safe. In one study of heterosexual married couples of which the husbands were HIV-positive, repeated fellatio was a risk factor for new infection among the wives. New infections have also been reported among gay men who claim to practice only oral-genital sex.

But sexual contact is only one way that people are exposed to HIV. In the United States, one fourth of new HIV infections are caused by the injection of illegal drugs. It's not the drugs, of course, but the sharing of contaminated needles among users that spreads HIV. Statistically, the only thing riskier than needle sharing for getting HIV is assuming the receptive role in anal intercourse.

The types of transmission in the United States have shifted over time. Early in the epidemic, HIV was transmitted mainly by male homosexual activity. Today, such exposure accounts for about half of all AIDS cases in this country. One quarter of new infections are attributed to intravenous drug use, and about 10 percent arise from heterosexual activity. Blood has routinely been screened for HIV since 1985 and this exposure risk has become negligible in the industrialized nations.

Transmission is more likely from people with higher levels of the virus, and these levels are usually highest right after infection, when there are no symptoms. This period of high viral burden averages six weeks. Transmission is also more likely from people with advanced AIDS, when the collapse of the immune system allows the viral levels to rise dramatically. These statistics do not mean that it's safe to have sex with an HIV-positive person at other times: Experts say that anyone who tests positive for HIV antibodies should always be considered infectious.

REDUCING YOUR RISK

Because no preventive vaccine for AIDS is available yet, the best way to avoid HIV infection is to avoid behaviors known to transmit HIV infection: having unprotected sex and sharing needles. One mathematical model finds that selecting sexual partners from high-risk groups, such as homosexual men or IV drug users, carries a 5,000-fold higher risk of HIV infection.

Although abstinence is the surest way to avoid sexual risk, it is not a workable approach for everyone. The most effective way to avoid HIV, or any sexually transmitted pathogen, is to remain sexually faithful to a partner you know to be free of infection. If that is not possible, you can reduce your risk by limiting your sexual partners and practicing safer sex. Safer sex means avoiding behaviors that are obviously dangerous, such as unprotected anal or vaginal intercourse with partners you cannot be sure are uninfected. Consistent use of condom barrier protection further reduces risk. (Diaphragms do nothing to prevent HIV transmission.) Male condoms are the best-studied method for blocking HIV transmission. According to one analysis, they are about 85 percent effective, meaning that with exclusive male condom use by an HIV-infected partner, over a year's time a woman has a 15 percent chance of becoming infected. Latex or polyurethane condoms used with water-based lubricants are best, although even these sometimes break or slip, permitting exposure to HIV-infected semen or fluids. Female condoms are now available, marketed as a means for giving women more control over reproduction, and these polyurethane pouches will probably reduce HIV transmission as well. Their efficacy has not been proved in large studies, however, and they cost more than male condoms.

To reduce exposure from IV drug use, many studies indicate that needle-exchange programs—which trade clean syringes for used ones—can significantly reduce transmission rates. The argument that distributing free needles promotes drug use, however, has limited the spread of such programs in the United States. As of 1999, 138 needle exchanges operated in most major metropolitan areas, though many were technically illegal. When clean needles are unavailable, soaking used ones in a bleach solution can reduce transmission of HIV.

Another way to reduce the risk of HIV transmission is to be aware of whether you or your sexual partner has another sexually transmitted disease. Many STD infections raise the likelihood of HIV transmission when one partner carries HIV. Early diagnosis and treatment of any STD helps reduce the risk of HIV transmission.

Finally, a person with HIV has a responsibility to reduce the risk to others. A person who carries HIV can infect others through sexual contact or by sharing needles. Knowingly engaging in an activity that can infect someone with HIV has legal consequences: since 1987, courts in the United States have found about 100 people guilty of deliberately trying to spread HIV by having sex, sharing needles, or assaulting another person. HIV-positive pregnant women also stand a 15 percent to 40 percent chance of passing the virus to their infants, a risk that can be significantly reduced with diagnosis and proper treatment—for mother and baby—around the time of delivery.

FOR MORE INFORMATION

National Institutes of Health
National Institute of Allergy and Infectious Diseases
Office of Communications
www.niaid.nih.gov/publications/aidsfact.htm
Building 31, Room 7A50, MSC 2520
Bethesda, MD 20892-2520
(301) 496-5717

University of California at San Francisco
hivinsite.ucsf.edu/InSite.jsp?page=KB

Centers for Disease Control and Prevention
Technical Information and Communications Branch
Division of HIV/AIDS Prevention
www.cdc.gov/hiv/hivinfo.htm
1600 Clifton Road NE, M/S E-49
Atlanta, GA 30333
(800) 342-2437

National Library of Medicine
www.nlm.nih.gov/medlineplus/aids.html

This chapter has been reviewed by Dr. George W. Rutherford, the Salvatore Lucia Professor of Preventive Medicine, Epidemiology and Pediatrics at the University of California, San Francisco, and a leading researcher in the epidemiology of AIDS and HIV infection; by Dr. Harvey Makadon, Associate Professor of Medicine at Harvard Medical School, Director of Harvard Medical International, Founder and Executive Director of the Boston AIDS Consortium, Harvard School of Public Health; and by Dr. Steve Morin, Associate Professor of Medicine at the University of California, San Francisco, and Director of the AIDS Policy Research Center within the UCSF AIDS Research Institute.

43. MAMMOGRAPHY

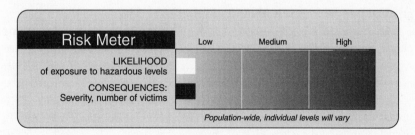

Risk Meter	Low	Medium	High
LIKELIHOOD of exposure to hazardous levels			
CONSEQUENCES: Severity, number of victims			

Population-wide, individual levels will vary

SINCE THE 1980S breast cancer has become a much more prominent health concern. It's the third leading cancer cause of death in the United States, and the number of new cases reported per year has gone up by nearly one third since President Richard Nixon declared his War on Cancer in the early 1970s. Over that same period, the percentage of breast cancer deaths has dropped nearly 15 percent. Both of those trends may well be due to efforts to identify breast cancer earlier, which makes treatment far more successful. But one of the main approaches to breast cancer detection—mammography—has raised concerns that exposure to radiation may pose a risk as well.

THE HAZARD

Mammography is a medical procedure that typically involves the use of a specially designed low-dose X-ray system and high-contrast, high-resolution film to create images of the internal tissues of the breast in order to detect changes that could be, or could become, cancer. During the procedure, a technician positions the patient's breast between two plates that compress the tissue. The uncomfortable compression is required to even out the thickness of the breast so that all of the tissue can be screened and to spread out the tissue so that small abnormalities won't be missed. It also reduces the amount of radiation that's required because the X rays have less tissue to pass through. The compression also helps to hold the breast in position to ensure a sharp, readable image.

The breast is then exposed to a small dose of radiation—usually about 0.1 to 0.3 rad (radiation absorbed dose) per picture, or about the same whole-body dose of radiation the average American receives from natural sources over six months. As with any X ray, where the tissue is more dense it absorbs some of the energy from the radiation. Less dense tissue allows more radiation to pass through. The resulting contrast on the X-ray film highlights different tissue densities that can't be detected by manual breast examination. (See Chapter 48 for more about X rays and Chapter 34 for more about radiation.)

Mammograms both screen and diagnose. Screening mammograms check women who have no signs of illness and usually involve two X rays of each breast. They can often detect lumps that cannot be felt. For women between the ages of 40 and 50, the National Cancer Institute says screening mammograms can reduce the risk of dying from breast cancer by 17 percent. For women between 50 and 69, screening mammograms lower the risk of dying from breast cancer by 25 to 30 percent. Medical experts suggest regular screening mammograms for women over 40, though there is no hard evidence on the optimal frequency. Many authorities suggest mammograms once a year or once every two years.

Diagnostic mammograms are done when a woman experiences unusual breast changes like lumps, pain, thickening, nipple discharge, or altered breast size or shape. They're also done when abnormalities turn up in screening mammograms or when screening mammograms are difficult to do because of breast implants or other special conditions. Diagnostic mammograms are more extensive and involve more X rays, each of which exposes a patient to another 0.1 to 0.3 rad, although again, that overall dose is still considered very low.

A new digital form of mammography is now coming into wider use. It converts the output of the X rays into digital information rather than projecting it onto conventional X-ray-sensitive film. The signals then produce images of the breast that can be seen on computer screens or special films. Digital mammography reduces the amount of radiation necessary to produce accurate results. It also allows radiologists to correct for overexposure and underexposure of X-ray images, which reduces the need for repeat mammograms. It allows for magnification of areas without requiring additional mammograms and allows mammograms to be electronically stored and transferred, making it easier for doctors to consult over long distances on a patient's condition. This new technology is no more or less accurate than film-based mammography.

THE RANGE OF CONSEQUENCES

Mammography creates a very low risk of cancer through exposure to ionizing radiation, a kind of radiation with high enough frequencies to break chemical bonds and therefore cause mutation in DNA molecules. Mammography poses a classic risk-benefit tradeoff. In exchange for the very slight risk of getting cancer from a mammogram, a woman gets the potential benefit of learning whether she has the disease at a time when it is more treatable, or whether she may be a candidate for treatment to remove tissues that could develop into the disease. The risk from mammography is so low, and the risk from breast cancer so much higher, that there is no question that the benefits of mammography far outweigh any risk the technology creates.

The radiation exposure from a mammogram is quite low. According to one estimate it's about the same as the dose a person receives flying from London to Australia and back (from frequencies of ionizing radiation from the sun; exposure to these frequencies is greater at high altitude, where the thinner atmosphere hasn't diffused as many of these energy waves).

But mammograms have other negative consequences. Sometimes they are inaccurate, providing false-positive results. If a woman has a mammogram every year between ages 40 and 49, she has a 30 percent chance that one of those mammograms will return a false positive. Women between the ages of 50 and 69 who have mammograms every year have a one in four chance of a false positive. Abnormal false-positive mammograms lead to further treatment that is not necessary, usually more mammograms. But 10 to 20 percent of women with abnormal mammograms go on to have biopsies. Only 20 to 40 percent of these women are found to have cancer. Other follow-up tests for women with abnormal mammograms (some of which may be false positives) can include fine-needle aspiration, ultrasound, or diagnostic mammograms. For every 1,000 women ages 40 to 49 who have mammograms, about 30 have follow-up procedures they don't need. For every 1,000 women ages 50 to 69 who have mammograms, about 12 have unnecessary follow-up treatment.

There is also the risk of false negatives: 20 percent of mammograms miss breast cancers that are present at the time of screening. This failure is more common in younger women with denser breast tissue, which contains more glands and support tissue, making cancerous tissue harder to detect.

THE RANGE OF EXPOSURES

Breast cancer is the second leading cancer cause of death for women after lung cancer. The American Cancer Society (ACS) estimates that in 2002, 203,000 women (and 1,500 men) will be diagnosed with the disease and approximately 40,000 women (and 400 men) will die from it. An average of one woman in eight will develop breast cancer in the course of her lifetime. (Remember, the risk for any given individual might be higher or lower.) So more American women are having mammograms than ever before. The National Cancer Institute reports that in 1998, two thirds of women ages 40 and older had a mammogram in the prior two years, more than twice the percentage in 1987. Poor, less educated women without health insurance were less likely to get mammograms. According to the 1998 statistics, only half of the women 40 years old or older and living below the poverty level had a mammogram in the previous two years, but two thirds of women 40 and older above the poverty line had a mammogram sometime in the previous two years.

A study released at the beginning of 2002 found that some women are not following the recommendation from medical experts that they should have regular mammograms, particularly starting at age 40. These recommendations are based on several factors. The changing nature of breast tissue makes mammography more accurate for women 40 and above. Also, the likelihood of breast cancer rises with age. The average risk of a woman developing breast cancer in a given year is:

- Age 40: 1 in 1200
- Age 50: 1 in 590
- Age 60: 1 in 420
- Age 70: 1 in 330
- Age 80: 1 in 290

Still, in a study of 60,000 Massachusetts women who had mammograms in the 1990s, nearly half waited until they were 50 to have their first screening. And only 20 percent said they had had regular screening starting at age 40. The researchers estimated that as a result of these delays, nearly half the invasive tumors found in these women were larger than they would have been had the screening been done earlier or more regularly. (The effectiveness of breast cancer treatment is largely dependent on catching the disease as early as possible.)

That study, however, followed another one by researchers in Copenhagen who questioned earlier studies supporting the effectiveness of mammography, saying they were not well done, and that newer, more rigorous studies suggest mammograms have no benefit while they raise the risk of unnecessary secondary procedures after false positive results. The consensus in the American medical community, however, is that mammograms have clear and significant benefit that far outweighs the risk, although the amount of that benefit may not be as large as thought in the past.

REDUCING YOUR RISK

Mammograms are only as effective in detecting breast cancer as the quality of the procedure and the personnel conducting it, which is why facilities that provide mammograms must meet accreditation, certification, and inspection standards enforced by the U.S. Food and Drug Administration. So make sure the facility conducting your mammography is FDA-certified. The FDA standards apply to the technicians, radiologists, medical physicists, and the equipment in the facilities as well as the overall medical organization itself. We offer information on how to check for certified mammographers in your area at the end of this chapter.

To ensure a mammogram's accuracy and reduce the risks associated with the procedure, the ACS recommends several steps.

- Women should not schedule mammograms the week before their menstrual periods if their breasts are usually tender at those times. The discomfort in having the breast compressed during mammography is unpleasant for the patient, and it makes it harder to compress the breast and achieve an accurate result.
- Before having a mammogram, women should inform their doctors or X-ray technicians if there is any possibility they might be pregnant.
- They should make sure that a lead apron is properly in place to protect the abdomen and pelvis if they are under 40 or there is any chance they may be pregnant.
- They should not wear deodorant, talcum powder, or lotion under their arms on the day of their mammograms, since these substances can appear on the X-ray film as calcium spots.
- They should describe any breast symptoms or problems to the technician performing the exam.
- They should obtain prior mammograms and make them available to the radiologist at the time of their exams.
- They should ask when they will receive the results. And they should not assume the results are normal if they do not hear from the facility.

Also, women with breast implants should inform their doctors before receiving mammograms, so that the procedure can be performed by radiologists and technicians experienced in conducting mammograms on breasts with implants. Implants change the way the natural breast tissue appears in the X rays as it is compressed during the procedure. Silicone implants can impede the accuracy of a mammogram because the silicone is not transparent to X rays and can block a clear view of the tissues behind it, especially if the implant has been placed in front of, rather than beneath, the chest muscles.

FOR MORE INFORMATION

National Library of Medicine
MEDLINEPlus
www.nlm.nih.gov/medlineplus/mammography.html

National Cancer Institute
cancernet.nci.nih.gov/cancer_types/breast_cancer.shtml
Building 31, Room 10A31
31 Center Drive, MSC 2580
Bethesda, MD 20892-2580
(301) 435-3848
Cancer Hotline: (800) 4-CANCER or (800) 422-6237

U.S. Food and Drug Administration
www.fda.gov/cdrh/mammography/certified.html
5600 Fishers Lane
Rockville, MD 20857-0001
(888) INFO-FDA or (888) 463-6332

American Cancer Society
www.cancer.org
1599 Clifton Road NE
Atlanta, GA 30333
(800) ACS-2345 or (800) 227-2345

This chapter was reviewed by Dr. Beverly Moy, Hematology-Oncology Fellow, Beth Israel Deaconess Medical Center, and Postdoctoral Clinical Researcher, Harvard School of Public Health; and by Dr. Susan Troyan, Associate Director, Breast Cancer Program, Cancer Center, Beth Israel Deaconess Medical Center, and Medical Advisor, Massachusetts Breast Cancer Coalition.

44. MEDICAL ERRORS

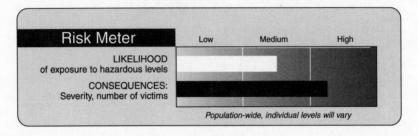

IN 400 B.C., the Greek physician Hippocrates swore, "I will follow that system of regimen which, according to my ability and judgment, I consider for the benefit of my patients, and abstain from whatever is deleterious and mischievous." Hippocrates also wrote: "As to diseases, make a habit of two things—to help, or at least, to do no harm." In fact, honoring that oath has proved surprisingly difficult. Though it's usually unintentional, health care providers and the health care system do harm far more frequently than many people realize.

A landmark 1999 report by the National Institute of Medicine (IOM), *To Err Is Human: Building a Safer Health System,* found that medical error is one of the leading causes of death and injury in America. Each year, an estimated 44,000 to 98,000 U.S. hospital patients are killed by medical errors that could have been prevented. The actual number is almost certainly higher, since many people go to outpatient clinics or surgical centers, or simply to the doctor's office, and never go to the hospital. Patients also receive care at home. They are treated in nursing homes. They get medications from both retail and online pharmacies. Medical errors occur in all of these settings too.

THE HAZARD

Medical error is tough to define. Broadly, it means the failure of a planned action to be completed as intended or the use of a wrong plan to achieve an aim. The most glaring examples—a patient having the wrong limb amputated or a surgical instrument left inside a patient's body—make headlines. But medical errors can occur anywhere in the health care system, from diagnosis to treatment to preventive care. Errors can involve medicines, surgery, equipment, reading of tests, handling of biological samples, miscommunication between health care providers, lab reports, and much more. Inherently dangerous situations include:

- The stocking of drugs in hospitals in amounts that are toxic unless diluted to normal doses
- Illegible writing on medical records
- Medical technicians operating unfamiliar equipment
- Lack of coordination when patients are treated by multiple specialists who may not know all the medicines or procedures prescribed
- Basic failure of sanitation in medical settings, such as medical providers not washing their hands after seeing one patient and before seeing the next

One of the most common forms of medical error has to do with medication mistakes. Hippocrates' oath also states, "I will give no deadly medicine to anyone if asked." But medical errors occur if patients are mistakenly prescribed the wrong drug, or the wrong level of a drug, or a drug that has dangerous interactions with other drugs they are taking, or a drug to which patients are allergic. Sometimes these errors are caused by something as simple as a pharmacist not being able to read a doctor's handwriting, or anyone in the health care system (even the patient) getting two similar-sounding drugs mixed up.

One of the more infamous cases of medical error was the 1994 death from a chemotherapy overdose of Betsy Lehman, a nationally renowned medical reporter for the *Boston Globe* who was being treated for breast cancer at the prestigious Dana-Farber Cancer Institute in Boston. She was mistakenly given four times the daily dose of a powerful anticancer drug and died of heart failure.

We cite this example because it typifies most medical errors. They are rarely what Hippocrates called "voluntary acts of mischief and corruption," recklessness that is more commonly referred to as medical negligence or misconduct. Most often medical error is accidental, caused by ignorance, oversight, lack of planning or attention to detail, communications mistakes, systems errors in medical institutions, and even by basic flaws in the way the health system is organized.

Not all errors result in harm. But research is starting to reveal the magnitude of the problem. And as baby boomers age and place more demands on the health care system, even as that system grows more complex under pressure to be economically efficient, the risk of medical errors is growing.

THE RANGE OF CONSEQUENCES

Researchers who studied hospital records in Colorado, Utah, and New York found that "adverse events," injuries resulting from a medical procedure and not from the patients' underlying condition, occurred in approximately 3 to 4 percent of hospitalized patients. In another study, the average patient in an intensive care unit was found to experience two errors per day. Extrapolating those findings to the total admissions to U.S. hospitals—an estimated 33.6 million per year—the implication is that at least 44,000 Americans die each year as a result of medical errors and that the number could be as high as 98,000.

At this rate, more Americans die in a given year as a result of medical error than die from car accidents, breast cancer, or AIDS. But it's hard to quantify this risk accurately, because disagreement remains over just what qualifies as a medical error. The dispute centers around the question of whether the negative patient outcome was caused by a medical intervention, or whether that outcome probably would have happened anyway. The numbers are open to a lot of interpretation. Another problem is that regardless of what qualifies in anyone's definition, many hospital-related errors aren't fatal so they don't get caught and don't show up in these analyses. And the errors that take place outside hospitals are even tougher to identify and count.

Medication-related mistakes, both within and outside of hospitals, are thought to account for more than 7,000 deaths annually—more than the number of Americans killed in the workplace each year. A study at two highly regarded teaching hospitals found that about 2 of every 100 hospital patients experienced a preventable medical injury from the drugs they were given and 8 patients in 100 were involved in some kind of medication-related error. The Massachusetts State Board of Registration in Pharmacy estimated that 2.4 million prescriptions are filled improperly each year just in Massachusetts. A study by the Food and Drug Administration (FDA) of 400 deaths caused by medication errors found that 16 percent were due to name mix-ups. The only other kind of medication mistake causing more deaths was giving the wrong dose.

According to a poll by the National Patient Safety Foundation, 42 percent of respondents said they had been affected by a medical error, either personally or through a friend or relative. And 32 percent of the respondents indicated that the error had a permanent negative effect on the patient's health.

Nonfatal medical errors can include:

• Allergic reactions from improper medication
• The prolongation of symptoms that the correct medication would have relieved. (Some people suffering from epilepsy, for example, continue to experience seizures when mistakenly given an anti*fungal* medication whose name resembles the name of the anti*seizure* drug.)
• Wound infections (the second most common form of medical error, according to the IOM report)
• Errors in the handling and insertion of hospital patients' catheters that lead to urinary tract infections.
• Poor sanitation by hospital workers, including failure to wash their hands frequently, which helps explain why an estimated 2 million Americans each year get infections *after* they enter the hospital. (The Centers for Disease Control and Prevention estimates that 90,000 people die from these nosocomial infections.)

One rare but high-profile kind of error is surgery on, or even amputation of, the wrong limb or organ. In a highly publicized case, the comedian Dana Carvey underwent heart bypass surgery in 1998 only to learn, after continuing chest pains, that his surgeon had bypassed the wrong artery. Carvey's blockage was cleared later through angioplasty.

REDUCING YOUR RISK

Reducing the risk of medical errors requires action by everyone in the medical care system, from patients to health care providers to the people who design and operate the systems within which health care takes place.

Institutional Changes

Instead of relying entirely on human vigilance and reflex to keep patients safe, a growing national movement is trying to create safer systems, based on the model of the aviation industry, which has demonstrated that such things as standardization, simplification, and use of protocols and checklists markedly reduce errors. Since World War II, this high-risk industry has focused extensively on building safe systems, taking the responsibility for perfect performance of pilots and engineers through monitoring instruments, feedback, automation, and redundancy of technology. By the early 1990s, the fatality rate for passengers of U.S. airlines was less than one third the rate suffered in the 1950s and 1960s. Transferring that idea to medicine, some surgery departments have created written checklists for surgical procedures that must be followed the way flight engineers and pilots go through required checks for takeoffs and landings.

The IOM report recommends that a key to reducing medical errors is to improve systems of delivering care and not to blame individual doctors, nurses, or technicians. Based on one of its recommendations, a center for patient safety has been created within the U.S. Department of Health and Human Services to set national safety goals, track progress in meeting those goals, and conduct research to learn more about preventing mistakes. The IOM report also recommended establishing a nationwide reporting system for serious medical accidents, based on systems already in place in 15 states.

Meanwhile, individual states and hospitals are testing organizational changes to minimize the likelihood of error. Hospitals are improving their record keeping to identify and track error rates, standardizing and simplifying procedures, and regulators are improving oversight of health care providers and becoming more aggressive about their relicensing inspections of hospitals.

To reduce medication errors, some hospitals and health care systems are implementing computerized prescription systems in which doctors cannot handwrite prescriptions but must use a computer entry form. Not only does this change solve the problem of a pharmacist or nurse not being able to read the doctor's handwriting, but the computer program automatically identifies dosages that may be incorrect or signals if there are any potentially dangerous drug allergies or interactions with medications the patient may already be getting. Some hospitals are assigning pharmacists to accompany doctors on hospital rounds. These two strategies have significantly reduced medication error. Some medical schools and hospitals run classes to teach their doctors more legible handwriting and to emphasize the critical importance of writing clearly. According to Harvard researchers, the poor transcription of medication orders causes 60 percent of drug-related medical errors.

The practice of anesthesia demonstrates what can be accomplished when a specialty focuses on safety. Anesthesia has adopted standardized guidelines and protocols for which drugs to administer to which patients, at what levels, and under which circumstances. They have standardized much of their equipment. As the result of these and other efforts, and advances in technology, the mortality rate from elective anesthesia dropped from 1 in 20,000 to 1 in 200,000.

The FDA is working to reduce medication mix-ups. It has issued warnings about some pairs of medications that look and sound alike, such as Lamictal (an epilepsy drug) and Lamisil (an antifungal drug). The FDA has ordered the makers of more than 30 medications to design new labels that highlight confusing names with special shading or different-colored letters to distinguish drugs whose names look and sound alike.

Individual Actions

You can do a lot to reduce your risk of being a victim of medical error. You will be safer and healthier if you become more involved in your own health care. This doesn't mean you have to do battle with your health care providers. It does mean that just because the doctor is the expert and wears that white coat doesn't mean you should stop paying attention to your own treatment.

The following advice is from the federal government's Quality Interagency Coordination Task Force.

1. Speak up if you have questions or concerns. Choose a doctor you feel comfortable talking to about your health and treatment. It's okay to ask questions and to expect answers you can understand. Think about your questions in advance and write them down. Have somebody designated to ask questions for you if you're too sick to do it yourself. A friend, family member, or other advocate is important to help you track the details of your health care. Take that person with you to the doctor's office if it will help you resolve all the items on your list.

2. Keep a list of all the medicines you take. Tell both your doctor and pharmacist about each one, including over-the-counter medicines such as aspirin, ibuprofen, and dietary supplements like vitamins and herbal products. Bring the medications with you to show your doctor, to avoid medical errors that you may cause by not getting your own drug information right. Tell your health care provider about any drug allergies you have. Make sure you can read your doctor's handwriting on the prescription. And ask your doctor to write down the reason for the medicine right on the prescription.

   Ask the pharmacist about side effects and what foods or activities to avoid while taking the medicine. When you get your medicine, read the label, including warnings, and take those warnings seriously. Make sure it's the medicine your doctor ordered, and that you know how to use it. If the medicine looks different from what you expected, ask the pharmacist about it. And take the medication as prescribed.

3. Make sure you get the results of any test or procedure. Ask your doctor or nurse when and how you will get the results. If you do not get them when expected—in person, on the phone, or in the mail—don't assume the results are fine. Call your doctor and ask for them and ask what the results mean for your care.

4. Talk with your doctor and health care team about your options if you need hospital care. If you have more than one hospital to choose from, ask your doctor which one has the best care and results for your condition. Most hospitals are well qualified to treat a wide range of problems. However, for some highly specialized procedures (such as heart bypass surgery), research shows results are often better at hospitals that do those procedures more frequently.

5. Make sure you understand what will happen when you have surgery. Make sure you, your doctor, and your surgeon all agree on exactly what will be done during the operation. Tell the surgeon, anesthesiologist, and nurses if you have allergies or have ever had a bad reaction to anesthesia.

6. Before you leave the hospital, be sure to ask about follow-up care, be sure you understand the instructions, and be sure that you follow them.

7. If you believe that an error has occurred, let somebody know. And not just a lawyer. As we said, 15 states require hospitals and/or nursing homes to report errors to a state agency, but virtually all states have agencies (usually the health department) that take patient complaints and investigate them. Many people don't realize that states have these types of programs and feel their only option is to hire a lawyer and sue. It is important to report errors so they can be investigated. The authorities may not always confirm your assessment, but bringing your concerns to their attention can often prevent further errors.

In addition to these suggestions, the American Academy of Orthopedic Surgeons "Sign Your Site" program, begun in 1998, urges member surgeons to use a skin marker to sign the operative site as part of their presurgery routine. Patients can do the same. Should you need surgery on a paired limb in particular, you can mark "NO!" on the limb that is not to be operated on.

Pay attention to your care when you are in the hospital. Check that your name is correct on your tags and charts. Every time a drug is about to be administered, even through an intravenous tube, ask what it is and what it is for. Ask a family member or friend to be at the hospital to advocate for you when you are unable to do it yourself.

Again, studies show that outcomes are better for patients who remain involved in their own care.

FOR MORE INFORMATION

"To Err Is Human: Building a Safer Health System," 1999 report
    of the Institute of Medicine. Copies available from:
National Academy Press
2101 Constitution Avenue NW
Box 285
Washington, DC 20055
(800) 624-6242 or (202) 334-3313.
Also available online at books.nap.edu/books/0309068371/html/
    index.html

U.S. Agency for Healthcare Research and Quality
www.ahrq.gov/qual/errorsix.htm
2101 E. Jefferson Street, Suite 501
Rockville, MD 20852
(301) 594-1364

This chapter has been reviewed by Dr. Lucian Leape, Adjunct Professor of Health Policy at the Harvard School of Public Health and member of the Institute of Medicine Quality of Care in America Committee. Formerly a pediatric surgeon, Dr. Leape has helped pioneer the study of medical errors in America. Nancy Ridley, Assistant Commissioner of the Massachusetts Department of Public Health, also reviewed this chapter.

45. OVERWEIGHT AND OBESITY

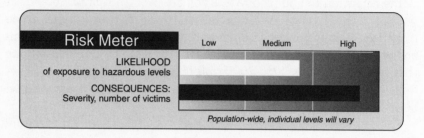

THIS BOOK is not about odds and probabilities. But there is a better than 50-50 chance that you are overweight. If so, this chapter is one of the most important you'll read: as an underlying risk factor for death, being overweight or obese is one of the leading killers in the United States. These conditions are associated with approximately 300,000 deaths a year from various forms of cancer, heart disease, stroke, and type 2 diabetes. Weighing more than is healthy for you increases your likelihood of developing arthritis, asthma, gall bladder disease, of bearing children with birth defects, and it diminishes your overall quality of life.

Yet despite the fact that awareness of these risks is as widespread as some of our waistlines, 6 Americans in 10 weigh too much for their own good. Approximately 44,500,000, roughly 1 in 6, are obese, overweight by 30 pounds or more. From 1960 to 1994, the prevalence of obesity among adults increased 60 percent. About 14 percent of children aged 6 to 19 were overweight in 1999. The prevalence of overweight adolescents has tripled since 1980.

THE HAZARD

Whether you are overweight or obese is measured in different ways. Most medical professionals currently rely on the measurement known as body mass index (BMI), which takes into account height and weight.

- BMI values of less than 18.5 are considered underweight.
- BMI values from 18.5 to 24.9 are healthy.
- BMI values between 25 and 29.9 are considered overweight.

• BMI values of 30 or greater, or about 30 pounds overweight, are considered obese.

To calculate your BMI, you can go to a website provided by the National Heart, Lung, and Blood Institute (see "For More Information" at the end of the chapter), which will do the math for you. Or you can do it yourself by multiplying your weight by 704, and dividing that number by your height in inches, squared. So if you weigh 180, multiplying that by 704 gives you 126,720. If you're 5 feet 10 inches—70 inches—tall, 70 times 70 is 4,900. Divide 4,900 into 126,720, and you get a BMI of 25.86.

A simpler if less precise way to know if you're overweight is simply to measure your waistline. A high-risk waistline is more than 35 inches for women, and more than 40 inches for men. Studies indicate that extra fat around the waist is more of a risk factor for heart disease than fat on the legs, buttocks, or elsewhere on the body.

The cause of being overweight or obese seems obvious: consuming more calories than we use. But it's more complicated than that. Several factors play a role in our tendency to gain weight. One is *genetic predisposition.* Studies of adopted children find that their body weight has little correlation with the adoptive parents they live with, but correlates closely with their biological parents. Studies of twins show that identical twins have more similar body weights than nonidentical twins. Studies of identical twins who grew up separately, with different families, show that they still have similar body weights, regardless of the environments in which they lived.

The belief that there is a genetic predisposition to being overweight or obese is further supported by the discovery of several genes, mostly in mice, which have dramatic effects on weight gain. These genes affect appetite, fat storage, and metabolic rate, key factors in how many calories we take in and how many stick.

Resting metabolism also plays an important role in body weight. Exercise and activity account for between one quarter to one third of the energy we burn. The work of digesting food accounts for another 5 to 10 percent. The rest, the majority of the calories we use, are burned as part of resting metabolism—the energy needed to keep the factory of our body going. This process accounts for 60 to 75 percent of the calories we burn, though there are variations among people by age, body type, activity level, and gender.

An average fasting, resting adult burns 73 calories per hour. Most of this energy is used by the heart, brain, liver, kidneys, and muscles. Rates of resting metabolism are higher for growing children than for the elderly, and generally higher in men than women.

But it's not just nature that predisposes us to being overweight or obese. Nurture, our *environment,* also plays a key role in body weight. About 40 percent of Americans say they don't participate in any leisure-time physical activity. Fewer than one third of Americans engage in recommended levels of physical activity, which is a minimum of 30 minutes of moderate to brisk activity per day, at least five days a week. And 43 percent of adolescents watch two or more hours of television every day. The availability of food, especially high-calorie food, is unprecedented in the United States.

THE RANGE OF CONSEQUENCES

Not long ago, obesity was a sign of affluence, admired in society, depicted in paintings as desirable. Now our culture depicts being overweight as socially and aesthetically negative. But aesthetics are the least of the reasons to be concerned about weighing too much.

Even modest weight excess, just 10 or 20 pounds for a person of average height, contributes to elevated risk of early death. People with BMIs of 30 and higher have a 50 percent to 100 percent greater risk of premature death from all causes, compared with people with BMIs between 18.5 and 24.9.

Being overweight increases your risk for heart disease (see Chapter 41). High blood pressure is twice as common in adults who are overweight. Obesity raises total blood cholesterol levels and lowers HDL, or "good," cholesterol.

Obesity can cause type 2 diabetes. Diabetes is a major killer, either directly or by contributing to heart disease. The U.S. surgeon general says a weight gain of 11 to 18 pounds doubles the risk of type 2 diabetes. Four out of five people with this dangerous medical condition are overweight or obese.

Obesity increases the risk of several cancers, including colon, gallbladder, prostate, kidney, cervical, ovarian, and endometrial (the lining of the uterus). Studies also suggest an association between obesity and cancers of the liver, pancreas, rectum, and esophagus. Women who gain more than 20 pounds in the 10 to 20 years prior to menopause double their risk of postmenopausal breast cancer, compared with women whose weight remains about the same. A study of 28,000 people in

Sweden who were monitored for 29 years found that the risk of cancers was one third higher in people who were overweight than for the population as a whole. (These associations are made by comparing disease rates in normal-weight and overweight people. Whether obesity actually caused these cancers is not understood.)

Being overweight or obese has a number of implications for women who are pregnant, and for their infants. Obesity is associated with an increased likelihood of infertility. Being obese during pregnancy increases the risk that the mother will have problems with labor and delivery, and the risk that either the mother or infant will die during pregnancy or childbirth. It raises by 10 times the risk that the pregnant mother will have high blood pressure. Infants born to obese mothers are more likely to have a high birthweight, increasing the likelihood of cesarean delivery. These infants have a higher risk of low blood sugar and of birth defects such as spina bifida.

Carrying too much weight isn't good for the bones and joints either. The surgeon general reports that once you are overweight, for every two-pound increase in weight you increase the risk of developing arthritis by 9 to 13 percent.

Respiratory problems associated with obesity include a higher prevalence of asthma and a greater prevalence of obstructive sleep apnea, a condition in which loose tissue in breathing passages causes oxygen deprivation during sleep and interrupts normal sleep cycles.

REDUCING YOUR RISK

It may seem frustrating that the tendency to be overweight has causes over which we have no control such as genetics, aging, and gender. Here's another one: when obese people dramatically reduce their caloric consumption, their bodies adjust, reduce resting metabolism, and automatically burn off calories more slowly. In other words, the effects of dramatic dieting are, to some degree, self-limiting. And many studies show that nearly all the obese people who lose weight through strenuous low-fat dieting regain it within five years.

But many things can help you lose weight and keep it off. Perhaps most important is setting reasonable expectations or goals that you can achieve and maintain. Many experts suggest shooting for weight loss of 5 to 10 percent of your starting weight, over three to six months. Even these weight reductions can significantly reduce some of the health effects of weighing too much. And probably the second most important piece of advice is to remember that weight reduction means burning more calories than you consume: both reducing caloric intake and increasing caloric output. In short, diet *and* exercise.

Diet

When it comes to dieting, the issue is both a matter of how many calories we take in, and what kind. In terms of amount, the National Institutes of Health (NIH) finds that eating 500 to 1,000 calories less than you need per day will result in the loss of one to two pounds over six months. In terms of what types of food you eat, lower-fat foods, particularly foods lower in saturated fats, are generally better if you want to lose weight. Fat is rich in calories. But so are refined sugars like those in candy and beverages. And don't forget that in terms of weight loss, all calories are the same, whether they come from jelly beans or a cheeseburger. The goal is to take in fewer calories than you burn.

To help, the NIH offers a guide for reading the nutrition labels on food, reprinted on the opposite page.

Physical Activity

Being physically active as part of weight control doesn't require running a marathon, lifting heavy weights, or swimming 100 laps in the local pool as fast as you can. Intensity is fine, but the key to physical activity for weight reduction is regularity and consistency. What counts is not how hard you work, but how often.

Just 30 to 45 minutes of moderate activity, something as simple as taking a walk most (preferably all) days of the week, is what experts suggest as an initial goal. You can even break things up and take three 10-to-15-minute walks a day. The trick, the experts say, is finding something that fits comfortably enough into your life that you can adopt it as regular behavior (30 minutes of brisk walking five days a week can add up to a weight reduction of 10 pounds in a year for many people).

Start easy if you've been sedentary for a long time. Couch potatoes don't turn into marathoners overnight. The National Heart, Lung, and Blood Institute (part of the National Institutes of Health) offers these suggestions:

• For the beginner, increase standing activities, chores like room painting, pushing a wheelchair, yard work, ironing, cooking, and playing a musical instrument.

PRODUCT: **CHECK FOR:**

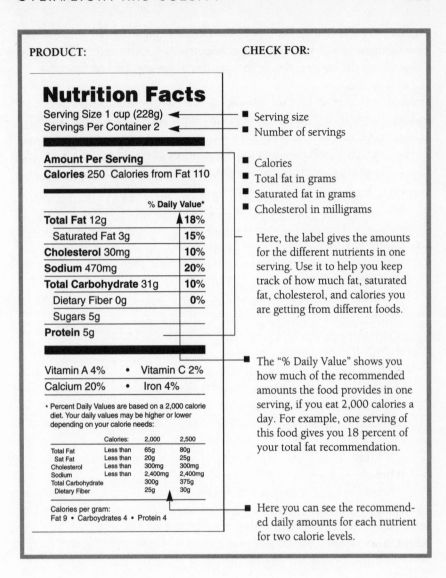

Nutrition Facts

Serving Size 1 cup (228g)
Servings Per Container 2

Amount Per Serving

Calories 250   Calories from Fat 110

	% Daily Value*
Total Fat 12g	▲18%
Saturated Fat 3g	15%
Cholesterol 30mg	10%
Sodium 470mg	20%
Total Carbohydrate 31g	10%
Dietary Fiber 0g	0%
Sugars 5g	
Protein 5g	

Vitamin A 4%   •   Vitamin C 2%

Calcium 20%   •   Iron 4%

* Percent Daily Values are based on a 2,000 calorie
diet. Your daily values may be higher or lower
depending on your calorie needs:

	Calories:	2,000	2,500
Total Fat	Less than	65g	80g
Sat Fat	Less than	20g	25g
Cholesterol	Less than	300mg	300mg
Sodium	Less than	2,400mg	2,400mg
Total Carbohydrate		300g	375g
Dietary Fiber		25g	30g

Calories per gram:
Fat 9 • Caroydrates 4 • Protein 4

■ Serving size
■ Number of servings

■ Calories
■ Total fat in grams
■ Saturated fat in grams
■ Cholesterol in milligrams

Here, the label gives the amounts
for the different nutrients in one
serving. Use it to help you keep
track of how much fat, saturated
fat, cholesterol, and calories you
are getting from different foods.

■ The "% Daily Value" shows you
how much of the recommended
amounts the food provides in one
serving, if you eat 2,000 calories a
day. For example, one serving of
this food gives you 18 percent of
your total fat recommendation.

■ Here you can see the recommended daily amounts for each nutrient
for two calorie levels.

- The next level is light activity, such as slow walking (24 minutes per mile), garage work, carpentry, housecleaning, child care, golf (walking the course, not riding in a cart), sailing, and recreational table tennis.
- Moderate activity includes walking at a pace of 15 minutes per mile, weeding and hoeing a garden, hiking while carrying a load, cycling, skiing, tennis, and dancing.
- High activity includes walking 10 minutes per mile or walking with a load uphill, climbing, heavy manual digging, basketball, or soccer.

We offer one table of activities and the calories they burn in Chapter 41, "Heart Disease." Here's another guide, this one from the U.S. surgeon general. These activities at moderate intensity burn 150 calories. Raising the intensity means you can reduce the time.

Common Chores

Washing and waxing a car for 45 to 60 minutes
Washing windows or floors for 45 to 60 minutes
Gardening for 30 to 45 minutes
Pushing a stroller 1 ½ miles in 30 minutes
Raking leaves for 30 minutes
Walking 2 miles in 30 minutes (15 minutes/mile)
Shoveling snow for 15 minutes
Climbing stairs for 15 minutes

Sporting Activities

Playing volleyball for 45 to 60 minutes
Playing touch football for 45 minutes
Bicycling 5 miles in 30 minutes
Dancing fast for 30 minutes
Water aerobics for 30 minutes
Swimming laps for 20 minutes
Basketball (just shooting baskets) for 30 minutes
Basketball (playing in a game) for 15 to 20 minutes
Bicycling 4 miles in 15 minutes
Jumping rope for 15 minutes
Running 1 ½ miles in 15 minutes (10 minutes/mile)

Remember that out of all the calories we burn, only one quarter to one third are burned by physical activity. So activity is only part of a program that also needs to include diet. You can walk for thirty minutes and burn 150 calories, but if you quench your thirst at the end of the walk with a 12-ounce soda, you put those calories right back in. A chocolate chip cookie replaces the calories burned on a 10-minute walk. You can run for two and a half hours at a 10-minute-per-mile pace, and burn 1,500 calories, but put it all right back on with a double cheeseburger, extra-large fries, and a large soft drink.

Other Weight Loss Techniques

There are many programs to help with weight loss, including diet programs and exercise regimens. Some people find it easier to lose weight using these programs than trying to go it alone. Many people find it easier to adopt an exercise routine if they join a health club. These choices are individual.

There are also drugs that can be effective in helping people lose weight. Surgical techniques include invasive surgery to staple the stomach, reducing its volume, causing a person to feel full sooner and eat less. Psychological counseling can also help since some eating behaviors are connected to a person's emotional state. All of these approaches need careful consultation with medical professionals.

FOR MORE INFORMATION

National Institutes of Health
National Heart, Lung, and Blood Institute
NHLBI Health Information Center
www.nhlbi.nih.gov/health/public/heart/obesity/lose_wt/
index.htm
www.nhlbi.nih.gov/health/public/heart/obesity/lose_wt/
lcal_fat.htm (low-fat alternatives)
www.nhblisupport.com/bmi/?htf=5&hti=10&bmi=25.8&wt=
170 (BMI online calculator)
P.O. Box 30105
Bethesda, MD 20824-0105
(301) 592-8573

Centers for Disease Control and Prevention
Division of Nutrition and Physical Activity
National Center for Chronic Disease Prevention
and Health Promotion
www.cdc.gov/nccdphp/dnpa
4770 Buford Highway NE, MS/K-24

Atlanta, GA 30341-3717
(770) 488-5820
Fax: (770) 488-5473

American Obesity Association
www.obesity.org
1250 24th Street NW, Suite 300
Washington, DC 20037
(800) 98-OBESE or (800) 986-2373

This chapter was reviewed by Dr. Edward Saltzman, Assistant Professor of Medicine at the Tufts University School of Medicine and Medical Director of the Obesity Consult Center at New England Medical Center; and by Dr. Richard Siegel, Assistant Professor of Medicine at Tufts School of Medicine, Director of the Outpatient Diabetes Clinic at New England Medical Center, and Staff Endocrinologist at the Obesity Consult Center at New England Medical Center.

46. SEXUALLY TRANSMITTED DISEASE

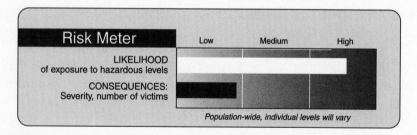

UNTIL THE LATE 1970S, diseases spread through sexual contact were called venereal diseases, derived from Venus, the Roman goddess of love. But there is little romantic about the effects of chlamydia, human papillomavirus, gonorrhea, or any of the 25 infections that are spread primarily through sexual activity: effects that range from rashes to sores to failed pregnancies to children born with birth defects, to cancer and death.

More than 65 million Americans, roughly one in five, are currently living with an incurable sexually transmitted infection. Fortunately, most of these people won't develop symptoms and active disease, and many of these cases are naturally cleared from the body in a few years without treatment. But an estimated 15 million new cases arise each year, and half are incurable. An estimated one in four Americans will contract a sexually transmitted disease at some point in their lives.

THE HAZARD

STDs are more of a threat to women than men. Women suffer more frequent infections and more serious complications from STDs, including cervical cancer (caused by the human papillomavirus), and infertility and/or death when certain diseases spread to the reproductive organs. In addition to harming the health of women, several STDs can lead to premature delivery in pregnant women, which is the number one cause of infant death and disability in America. It is estimated that 30 to 40 percent of preterm births each year are due to STDs and bacterial vaginosis, an STD-related imbalance of the bacteria in the vagina. STDs can also infect a child in utero and, if the mother is not treated early enough while still pregnant, the newborn can suffer neurological disease, eye problems, pneumonia, and death.

STDs affect people under 25 more than other age groups. Nearly two thirds of all new cases occur in people 25 and younger. Approximately one in four new cases each year is a teenager. Teens are at high risk because in addition to being in the most likely age group to have multiple sexual partners, teens are also the most likely to engage in unprotected sex. Further, younger women are biologically more susceptible to many STDs.

The presence of STDs in America is increasing, though trends vary for each specific disease. The National Institute of Allergy and Infectious Diseases, part of the National Institutes of Health, finds that this increase is caused by young people becoming sexually active at earlier ages and marrying later, thus engaging in sexual activity with a larger number of partners. In addition, the NIAID says that since divorce is now more common than it was a generation ago, more adults are engaging in sexual activity with multiple partners.

THE RANGE OF CONSEQUENCES

Each of the major sexually transmitted diseases has unique consequences.

Human Papillomavirus (HPV)

HPV diseases are the most prevalent family of STDs in the United States. They are more well known by their common names: genital warts and cervical cancer. Approximately 5.5 million Americans get a new genital HPV infection each year. An estimated 20 million people are currently infected. Most sexually active people acquire genital HPV.

More than 100 different strains of this type of virus exist, of which about 30 are sexually transmitted. The transmitted varieties divide into two groups: low risk and high risk. The low-risk strains cause genital warts. The high-risk strains lead to cellular changes that, if left untreated, can lead to cancer of the cervix, anus, or penis. Like many STDs, while HPV prevalence is high, most people with either type of HPV remain asymptomatic, but can still pass the virus to sexual partners. Approximately 90 percent of people who acquire HPV clear it from their systems without ever developing symptoms.

The most common symptom of HPV, genital warts, appears as growths or bumps, which may be flat or raised, single or multiple, small or large. They tend to be flesh-colored or whitish. They usually don't cause itching or burning. They can appear on the vulva or vagina, the penis or scrotum, in and around the anus, or in the groin, where the genital area meets the inner thigh. While most people who have HPV never develop symptoms, those who do can develop warts several weeks, months, or years after the initial infection. Some people have only one episode, after which the warts go away on their own and never recur. Some people have several recurrences. When warts are present, the virus is considered active and is more likely to spread during sexual contact.

The types of HPV that spread genital warts are usually passed on through vaginal or anal sex, and only rarely through oral sex. People do not get genital warts by touching other warts elsewhere on their bodies. There are a number of treatments for the warts caused by the low-risk strains of the virus. Warts can be frozen off with liquid nitrogen (cryotherapy), removed with an acid, injected with medication, electrically cauterized (burned off), or surgically removed either mechanically or with a laser. Prescription creams can also remove genital warts. Beyond the aesthetic reason for removal, getting rid of the warts reduces the chance of transmitting HPV to a sexual partner.

The other type of HPV, the high-risk form, can result in cervical cancer, anal cancer, and in rare cases cancer of the penis. Most people who carry these strains of HPV never develop any disease. In those who do, the first signs are cellular abnormalities, which fortunately can be detected by Pap smear tests, both for women and for men who may have high-risk types of HPV from anal sex. In mild cases these abnormal cells require no treatment and heal on their own. The HPV may still be present in these people, but in a dormant state.

People who have had these cellular abnormalities should continue to be monitored by medical professionals, who can decide whether to do follow-up tests, including a biopsy or a DNA test looking for the high-risk strains of HPV. With active monitoring, recommended for all women 18 and over (earlier if they are sexually active before age 18), potentially precancerous cell growths can be monitored and treated if necessary to prevent the development of cancer. New research findings suggest the same monitoring technologies can be used for men and women who engage in receiving anal sex. Males carrying high-risk HPV are usually asymptomatic.

In some cases, if these abnormal cells are not detected and removed, they can go on to develop into cancer. To remove the cells before that happens, doctors use either cryosurgery or loop electrosurgical excision procedure (LEEP). Approximately 15,000 women in the United States develop invasive cervical cancer, and 5,000 American women die from this disease annually. One half of the women with newly diagnosed invasive cervical cancer have never had a Pap test, and another 10 percent have not had a Pap test in the past 5 years.

HPV has also been associated with some, but not all, cases of vaginal and vulvar cancer, which are both rare forms of disease.

Chlamydia

Chlamydia is the most commonly reported bacterial-infection STD in the United States. More than 650,000 cases were reported in 1999, three quarters of which were people under 25. Underreporting is substantial because many people with chlamydia have no symptoms. An estimated 3 million U.S. residents are infected each year. Chlamydia is sometimes called the silent STD, because three quarters of infected women and half of infected men are asymptomatic. But many of those people develop symptoms much later, which can be much more severe. Chlamydia can be transmitted through all sexual activity but is most commonly transmitted through vaginal or anal sex.

When symptoms develop right away, they usually show up within one to three weeks of infection. For women, symptoms include vaginal pain or discharge, inflamed rectum, burning upon urination, abdominal pain, and conjunctivitis (eye infection, since the bacterium targets cells that include the mucous lining of the eyelid). For men, symptoms include penile discharge or pain, burning upon urination, inflamed rectum, and conjunctivitis. The newborns of mothers who had chlamydia during pregnancy can suffer conjunctivitis and pneumonia.

Chlamydia can be treated and cured with antibiotics, and it's important to know if you have this infection because it can be threatening even if it is not symptomatic. In as many as 40 percent of women with untreated chlamydia, it goes on to cause pelvic inflammatory disease (PID), a serious infection of a woman's upper genital and reproductive organs—the uterus, fallopian tubes, and ovaries. About 10 percent of women with this infection become infertile. In fact, PID is the leading cause of female infertility in the United States. PID can also cause ectopic pregnancies, in which the fertilized ovum embeds in the wall of the fallopian tube before reaching the uterus. Ectopic pregnancies can lead to infertility or death to the mother. Symptoms of PID include pain or tenderness in the lower abdomen, nausea or vomiting, bleeding between menstrual periods, or pain during sex.

Asymptomatic chlamydia in men can later cause inflamed prostate glands, scarring of the urethra, infertility, inflammation of the epididymis (tubes that carry sperm in the testes), and inflammation of the joints.

Herpes

As many as 45 million U.S. residents carry the herpes simplex virus-2 (HSV-2), the strain that causes genital herpes. Between 50 and 80 percent of the adult American population carries HSV-1, the strain that produces oral herpes or "fever blisters."

The HSV-1 and -2 viruses produce sores and blisters that cause itching and pain, and usually last from 2 to 12 days but can last up to 3 or 4 weeks. These viruses are in the family that also includes the Epstein-Barr virus, which causes mononucleosis, and the varicella-zoster virus, which causes chicken pox and shingles. Like shingles, these outbreaks can arise periodically as the virus, normally lying dormant in the nerves around the spinal cord, is stimulated into activity by a number of conditions and travels via nerve pathways to the skin. HSV-2 skin outbreaks occur in the area of the vagina, vulva, penis, scrotum, anus, groin, or buttocks. Many times these outbreaks are so mild that they are mistaken for insect bites, yeast infections, jock itch, or hemorrhoids. Outbreaks can occur once in a lifetime, or periodically. The average is four to five per year, but the frequency of symptoms varies widely from person to person. Usually there are more outbreaks in the first year, and the first is the worst. Triggers that awaken the virus include illness, poor diet, physical or emotional stress, surgical trauma, and steroidal medication, including asthma medications.

Genital herpes causes few serious physical complications, but there are serious psychological consequences from the stigma attached to this condition. Pregnant women with an active case, however, can cause potentially fatal infections in their newborns and are usually advised to have a caesarian delivery. HSV-2 can spread easily, but only to certain vulnerable tissues around the lips and genital area, and the virus dies quickly. HSV-2 spreads more effectively when sores are present, but can spread when a person is asymptomatic as well. It is easily spread by vaginal, anal, or oral sex. No cure for herpes viruses has been found, but medications are available to minimize discomfort during an outbreak and hasten recovery and healing.

Gonorrhea

Gonorrhea is a bacterial disease that infects approximately 650,000 new Americans a year. Of those cases, 20- to 24-year-old men and 15- to 19-year-old women account for 75 percent. It is transmitted by vaginal, anal, or oral sex and can infect the warm moist tissues of the genitals, the mouth and throat, and the eyes. The symptoms and risks of gonorrhea for women are similar to chlamydia. The most dangerous similarity is that there are usually no initial symptoms at all, which means that when symptoms do show up later the disease can be more serious. In women who do show early symptoms, as with chlamydia, there can be vaginal discharge, burning while urinating, and abdominal pain. As is also the case with chlamydia, undetected and untreated gonorrhea can cause pelvic inflammatory disease in women. (See the chlamydia section for more details on the serious dangers of PID.) Gonorrhea in a pregnant woman can spread to her newborn in the birth canal during delivery, and lead to blindness, joint infection, or a life-threatening blood infection.

Most men who get gonorrhea develop mild symptoms within 2 to 5 days, but up to as much as 30 days, including a burning sensation while urinating and a yellowish white discharge from the penis. Sometimes men with gonorrhea suffer painful or swollen testicles. Longer-term symptoms in men with undetected gonorrhea include epididymitis, an inflammation of the sperm-carrying tubes in the testicles, which can lead to infertility and scarring of the urethra, which in turn can cause permanently painful urination.

In both genders, untreated gonorrhea can, in rare cases, spread to the blood or joints, which can be life-threatening. Gonorrhea can be spread by people who are asymptomatic. It is curable with a range of antibiotics, though over the years the bacteria that cause gonorrhea have become resistant to several antibiotic drugs.

Hepatitis B

Hepatitis B is the most common strain of hepatitis viruses and the one of most concern as an STD, though both the B and C types can be spread through contact between mucous membranes, blood, semen, or vaginal secretions. It is estimated that of the roughly 200,000 new hepatitis B infections each year, 120,000 are caused by sexual contact. An estimated 417,000 Americans are currently living with chronic sexually acquired hepatitis B.

Hepatitis damages the liver. It usually takes about two months from the time of infection for the virus to show up in the blood. About 9 of 10 adults who get hepatitis B get rid of the virus within several months. And 30 percent of these people carry the virus and suffer no symptoms. In many other cases, the damage is mild and short-lived. Symptoms include jaundice, fatigue, abdominal pain, loss of appetite, nausea, vomiting, or joint pain. In 1 case out of 10, hepatitis B becomes a chronic, lifelong disease that often leads to serious liver damage and death. As many as 6,500 deaths occur in America each year from chronic hepatitis B–related liver disease or hepatitis B–related liver cancer.

Hepatitis B can be transmitted by vaginal, anal, or oral sexual contact. A 1996 report found that roughly 40 percent of acute hepatitis B was attributable to heterosexual activity and 18 percent was attributable to homosexual practices. Hepatitis B can be transmitted from a pregnant mother to her newborn. Approximately 20,000 infants are born with the disease each year in the United States. Without treatment at birth, 9 out of 10 of these infants will be chronic lifelong hepatitis B carriers. And 1 out of 4 of these children will likely develop liver disease or liver cancer.

Hepatitis B is the only STD preventable with a vaccine. Vaccination by a series of shots can empower the body's immune system to fight off the hepatitis B infection successfully. The vaccine is available for infants as well as older children and adults. The Centers for Disease Control and Prevention (CDC) recommends vaccination for all people ages birth to 18.

Syphilis

This notorious STD is on the decline in the United States, so much so that the CDC in 1999 launched a campaign to eradicate syphilis in this country. Still, almost 36,000 cases of syphilis were reported in 1999.

Syphilis is caused by a bacterium that produces a disease that manifests itself in three distinct stages. The primary stage appears an average of 21 days (the range is 10 to 90 days) after infection as a single sore (in rare cases, more than one can appear) called a chancre, which is almost always painless, firm, round, and small. It heals without treatment in three to six weeks. The sore appears where the bacterium entered the body. First-stage symptoms are common in most people infected with syphilis.

The second stage can show up between 17 days and 28 weeks after infection. A rash may appear anywhere on the body, including the palms of your hands and soles of your feet, which is unusual for skin rashes. This outbreak can last 2 to 6 weeks. Other symptoms may include grayish white sores in your throat or mouth and around the cervix, patchy hair loss, and a general sense of feeling ill, with fever, headache, muscle ache, and fatigue. As with the primary stage, these symptoms also fade away without treatment. Unlike the first stage, many people have no symptoms, or only mild symptoms, in the second stage.

The late stage of syphilis manifests itself from 2 to 30 or more years after infection. Of the roughly 36,000 total syphilis cases reported in the United States in 1999, nearly 29,000 were late-stage disease. At this point, the bacteria have damaged the internal organs, including the brain and nervous system, the eyes, heart, liver, blood vessels, bones, and joints. Late-stage syphilis can result in death.

The disease is most commonly transmitted while it is in the primary or secondary stages. Pregnant mothers with syphilis can transmit the disease to their fetuses and suffer stillbirth or give birth to infants who die soon after being born. Surviving newborns with undetected and untreated syphilis can suffer developmental problems, seizures, or, later, death.

Syphilis is passed through vaginal, anal, or oral sex. The fragile bacteria are not transmitted through toilet seats, doorknobs, swimming pools, hot tubs, or eating utensils. Syphilis is curable with antibiotics, most commonly with penicillin. In people with later-stage syphilis, the antibiotic arrests further damage but does not undo the damage that's already been done.

REDUCING YOUR RISK

The surest way to avoid sexually transmitted diseases is abstention, an admittedly uncommon behavior. Those who are sexually active can take several steps to reduce their risk.

- Have a mutually monogamous sexual relationship with a partner you know to be uninfected.
- Reduce the number of sexual partners you have.
- Correctly and consistently use a male condom. It does not provide absolute protection, however, since sometimes the sores or warts of various diseases exist outside the area covered by the condom, and since some forms of oral sexual activity can spread STDs. If you choose anal intercourse, condom use is even more important.
- If you engage in any activity that runs the risk of infection, you should have regular medical checkups for STDs, even if you have no symptoms. Remember, many of the most common diseases don't manifest symptoms in most people, who can still harbor and pass on the infectious pathogens. Pregnant women especially should be checked because of the risk of unknowingly passing STDs to their newborns.
- Avoid douching, since it removes some of the normal, protective bacteria in the vagina.
- If you are found to have an STD, get treatment when possible and, despite the difficulties, tell all recent sexual partners. Local health departments can actually help you do so discreetly. During treatment for an STD, avoid sexual activity.

FOR MORE INFORMATION

American Social Health Association
www.ashastd.org
P.O. Box 13827
Research Triangle Park, NC 27709
(919) 361-8400
ASHA Resource Center: (800) 230-6039 (confidential)

Centers for Disease Control and Prevention
www.cdc.gov/nchstp/dstd/dstdp.html
1600 Clifton Road
Atlanta, GA 30333
National STD Hotline: (800) 227-8922 (confidential)
National Herpes Hotline: (919) 361-8488 (confidential)

All state public health departments have information available on STDs.

This chapter was reviewed by Dr. Willard Cates, President and CEO of the Family Health Institute of Family Health International and a member of the Centers for Disease Control Advisory Committee on HIV and STD Prevention; and by Dr. Alfred DeMaria, Jr., Director of the Bureau of Communicable Disease Control, Massachusetts Department of Public Health and Massachusetts State Epidemiologist.

47. VACCINES

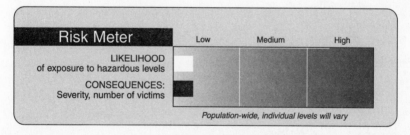

IN 1796, Edward Jenner, an English physician, made an observation that would have profound significance for human health. He noticed that dairymaids who milked cows infected with cowpox, a relatively mild disease, did not develop the much more severe disease smallpox, which had plagued humans for centuries. Jenner took the daring step of purposely exposing people to the fluid drained from cowpox sores and found that it somehow caused human resistance to smallpox. That fluid was the first vaccine.

By the beginning of the twentieth century, scientists had developed vaccines to prevent rabies, typhoid, plague, and cholera. Since then, the number of diseases preventable by vaccine has grown steadily, substantially reducing the occurrence of—or nearly eliminating—former scourges like smallpox, polio, influenza, pneumonia, whooping cough, measles, mumps, and rubella. To prevent most of these diseases, 11 vaccines are now universal for children. In addition, a total of 14 vaccines are available for children or adults to protect against hepatitis A and B, rabies, vaccinia (smallpox), influenza, yellow fever, Japanese B encephalitis, bacterial anthrax, tuberculosis, cholera, meningitis, pneumonia, Lyme disease, chicken pox and typhoid.

Vaccines have unquestionably been one of the major public health advances of the twentieth century. But as they have successfully reduced the prevalence of disease, and our fear of those diseases has gone down, concern about the negative side effects of vaccines has taken its place.

THE HAZARD

Vaccines prevent disease by preparing the immune system to react quickly in the event that a disease-causing pathogen enters the body. Without this preparation, a complete immune system response to bacteria, fungi, a virus, or a parasite can take days, and during that time the invading pathogen can multiply and cause damage. (Current vaccines combat bacteria or viruses. Vaccines against parasites and fungi are still experimental.)

Our immune system has several ways of protecting us from invading pathogens, and the first line of defense, called nonspecific immunity, works very quickly. It immediately sends special cells called phagocytes to any location where a foreign pathogen has started causing trouble. Phagocytes kill the invaders either by enveloping them and breaking them apart, or secreting chemicals that kill them. They don't have to recognize the invader in any specific way. They just have to know the pathogen is foreign.

But sometimes an invading pathogen can evade the nonspecific immune system. Then a second line of defense, the specific immune system, comes into play. Special molecules that have the unique ability to recognize molecules on the invading pathogen, called antigens, lock onto them. These special immune system molecules, called antibodies, are very specific. Each one recognizes and locks onto only the specific antigen it's designed to detect.

But this system takes longer to act than the nonspecific immune system. Sometimes when a new pathogen enters the body and starts to do its damage, there are no antibodies on hand designed to recognize it. In a sense, such a pathogen is invisible to the immune system, and can reproduce itself millions of times and do a lot of damage while the specific immune system is creating specific antibodies to recognize this new invader and start to fight back. The pathogen has a head start that can be critical to whether we live or die. Vaccines are the way we fight back in this race.

Vaccines introduce the antigens—the markers—of a harmful bacterium or virus, in a way that doesn't cause the full illness a whole healthy bacterium or virus would. Instead, the vaccine contains a version of the bacterium or virus that has been altered so that it doesn't make us sick, but it does let our immune system learn what the new antigens look like so they can make the special antibodies needed to fight it. That way, if the full unwanted pathogen shows up in the future, our immune system will have the specific antibodies it will need to recognize the invader and start to fight it off more quickly.

Here's how it worked in the case of Edward Jenner. Cowpox has some of the same antigens as smallpox, but it doesn't cause as severe a disease. So people exposed to cowpox didn't get sick, but they did get introduced to the antigens that are similar in smallpox. The immune system made antibodies to help fight these antigens, and when smallpox was introduced to these people their systems were ready to fight back much faster.

The trick, of course, is to introduce the antigens that will trigger protective antibodies in a way that won't make us sick. "Inactivated" vaccines use pathogens that have been killed but still have their antigen molecules intact. "Attenuated" or "live" vaccines use pathogens that are active and alive but have been changed so that they're too weak to cause disease. "Subunit" vaccines use just parts of the pathogen, enough to teach our system how to make the specific antibody should the whole organism ever show up.

Advances in biotechnology have made other vaccine techniques possible. For example, "recombinant" vaccines consist of pathogens from which a gene has been removed. These genetically damaged pathogens can teach our immune system to make antibodies for that pathogen, but they can't make us sick. Scientists also use genetic engineering to produce the parts of pathogens that will be used in subunit vaccines. Another new form of vaccine is called a "conjugate" vaccine, and it's designed to help babies, whose immune systems are still developing. Sometimes the infants' immature immune system can't recognize a new pathogen. Conjugate vaccines add an additional antigen to the outer wall of those pathogens, an antigen that the baby's immune system usually *will* recognize. This additional antigen marker makes an otherwise invisible pathogen visible to the baby's immune system. With this extra help, the child's system can learn to create antibodies for the potential new invader, and so be prepared to initiate an immediate response if that full healthy germ ever shows up. Conjugate vaccines have substantially reduced bacterial meningitis and drastically reduced *Haemophilus influenzae* type b in young children.

~

Although they greatly reduce the burden of disease, vaccines also pose some health risks, beyond just the sore arm you get from the injection. Adverse reactions can occur. Sometimes they're caused by contents of the drug other than the active ingredient, such as preservatives, adjuvants—which help the vaccine trigger an immune system response—or leftover proteins from the eggs some vaccines are grown in. Some of these substances may cause allergic reactions. Some may be directly toxic. Fortunately, the quantities of these foreign materials in vaccines is either very low, or zero, and the occurrence of these side effects is exceptionally rare.

THE RANGE OF CONSEQUENCES

Many of the adverse events associated with vaccines are relatively mild. Anyone who has experienced mild flu-like symptoms after getting a flu shot or whose child becomes irritable for several hours after a diphtheria, tetanus, and pertussis (DTaP) vaccination is familiar with this phenomenon. MMR (measles, mumps, rubella) shots can produce mild fever or rash. So can vaccination for chicken pox.

But more severe side effects do occasionally occur. Some are linked to specific vaccines. The table on the following pages details these effects and what is known about how often they occur.

The Centers for Disease Control and Prevention (CDC) and the Food and Drug Administration (FDA) maintain a database to track vaccine side effects. In the year 2000, doctors and families submitted more than 13,000 entries. (For perspective, there are more than 200 million vaccinations given in the United States each year.) But the list of these adverse events is only a starting point for tracking negative reactions to vaccines. Sometimes events are missed because patients don't tell doc-

DOCUMENTED ADVERSE SIDE EFFECTS OF VACCINES

Vaccine	Mild Reactions	Moderate Reactions	Severe Reactions
DTaP (diphtheria, tetanus, pertussis) (*Effects usually occur 1–3 days after shot. More common after the fourth or fifth shot in the vaccination series*)	Fussiness: Up to 1 child in 3 Fatigue: Up to 1 in 10 Mild fever: Up to 1 in 4 Vomiting: Up to 1 in 50	Nonstop crying: 1 in 1,000 Fever ≥105°F: 1 in 16,000 Seizure (jerking or staring): 1 in 14,000	Allergic reaction (difficulty breathing, shock): Fewer than 1 in 1 million
Hepatitis A (*Effects usually occur 3–5 days after shot*)	Headache: 1 in 6 adults, 1 in 20 children Loss of appetite: 1 in 12 children Fatigue: 1 in 14 adults		Serious allergic reaction: Very rare
Hepatitis B	Mild to moderate fever: 1 in 14 children, 1 in 100 adults		Serious allergic reaction: Very rare
Haemophilus influenzae (*Usually begins within a day of shot and lasts 2–3 days*)	Fever ≥101°F: 1 in 20 children		Serious allergic reaction High fever
Influenza (flu) (*Effects usually begin soon after shot and last 1–2 days*)	Fever, aches		Guillain-Barré syndrome: Fewer than 1–2 cases per million
Lyme disease	Muscle aches, joint pain, chills: 1 in 15 people		Serious allergic reaction: Very rare

Vaccine	Mild Reactions	Moderate Reactions	Severe Reactions
MMR (measles, mumps, rubella) (Symptoms occur 7-12 days after shot, less frequently after second shot in series)	Fever: 1 in 6 Mild rash: 1 in 20 Swelling of glands in cheeks or neck (rare)	Seizure caused by fever: 1 in 3,000 Temporary joint stiffness, pain: 1 in 4—mostly teenage or adult women—just from rubella vaccine Temporary low platelet count, causing bleeding disorder: 1 in 30,000 Temporary swelling of lymph nodes or salivary glands near the ear: Rare	Severe allergic reaction: Fewer than 1 in 1 million
Meningococcal disease	Fever: Small percentage		
Pneumococcal polysaccharide vaccine		Fever, muscle aches, or more severe local reactions: < 1 percent	Severe allergic reactions: Very rare
Pneumococcal conjugate vaccine	Fever: <100.5°F: 1 in 3 Fever >102.2°F: 1 in 50		
Polio (inactivated)			
Tetanus and diphtheria (for adults)			Severe allergic reaction Deep, aching pain and muscle wasting in upper arm(s) lasting many months: Rare
Varicella (chicken pox)	Fever: 1 in 10 Mild rash, up to 1 month after shot: 1 in 20 Infection of other household members (very rare)	Seizure caused by fever: 1 of 1,000	Pneumonia: Very rare

tors, or doctors don't always file a report. Sometimes an adverse event is reported, but the vaccine hasn't caused it. To help determine which side effects are caused by vaccines and which are not, the federal government's Institute of Medicine has convened an ongoing series of expert panels to evaluate the adverse event data, the results from laboratory animal studies, and basic information about how the immune system works.

One concern these experts have looked into is an association between the MMR (measles, mumps, and rubella) vaccine and the development of neurodevelopmental problems, including autism. The experts found that the evidence is "fragmentary," meaning that there were substantial elements of the hypothesis that have not been scientifically demonstrated. Another concern is that the presence of the preservative thimerosal in some vaccines might cause autism. They found that this hypothesis was "biologically plausible" but that there was no evidence that such an association actually exists. Their report on thimerosal concluded: "The hypothesis that thimerosal exposure through the recommended childhood immunization schedule has caused neurodevelopmental disorders is not supported by clinical or experimental evidence." Nonetheless, thimerosal is being phased out of vaccines.

To keep the risk of these infrequent adverse effects in perspective, remember that vaccines have dramatically decreased the incidence of many serious diseases in the United States.

- Diphtheria cases peaked in 1921 at more than 200,000. In 1997, there were only 5.
- There were approximately 214,000 cases of mumps in 1964 and just 352 in 1999.
- An estimated 500,000 cases of measles occurred each year until 1963. In 1999, there were just 86 cases.

Other diseases that vaccines have helped virtually wipe out include paralytic polio and congenital rubella syndrome. For most vaccines—for most people—the benefits vastly outweigh the risks.

THE RANGE OF EXPOSURES

Children get most vaccines by the age of six years as part of a schedule recommended by several groups of national experts including the American Academy of Pediatrics, the American Academy of Family

Physicians, and a group called the Advisory Committee on Immunization Practices, which advises the CDC. In many states such vaccinations are required as a condition for admission to school or daycare centers. They usually include the DTaP (diphtheria, tetanus, pertussis), MMR (measles, mumps, rubella), polio, Hib (*Haemophilus influenza* type b), hepatitis B, and chicken pox vaccines. Children must receive between two and five doses of each of these vaccines to be fully vaccinated. (For young children, one shot of chicken pox vaccine is usually enough.) Surveys show that, with the exception of the chicken pox vaccine, which is relatively new, about 90 percent of all children between the ages of 19 and 35 months receive these vaccines. The government recommends additional vaccines for only some children, such as influenza vaccine for children six months of age or older who have asthma.

Adolescents may need some vaccines for two reasons. First, they are exposed to pathogens that younger children are not typically exposed to (for example, meningococcal bacteria). Second, they need a "booster" to reinforce the immunity conferred by some vaccines received during early childhood. Booster shots are necessary for the chicken pox, hepatitis B, MMR, and DTaP vaccines.

Only certain subgroups of adults receive vaccinations:

- Adults whose immune systems are weakened for some reason
- Adults for whom diseases prevented by vaccines pose a greater risk (for example, women during some stages of pregnancy)
- Adults who are at greater risk for being exposed to disease (people traveling to some foreign countries, health workers, lab technicians)
- Adults who may pose a risk to other "at risk" individuals they live with

Vaccination recommendations for adults include tetanus and diphtheria, influenza, pneumococcus, MMR (measles, mumps, rubella), hepatitis, polio, chicken pox, and hepatitis A vaccines, and others, depending on various special circumstances. Older people receive at least some of these vaccines more often than younger people. For example, about one in five adults 18 to 49 years of age receive the flu vaccine each year, but close to two in three people 65 or older get an annual flu shot. Late in 2001, the CDC recommended flu shots for anyone 50 and older.

A Word on the Flu Vaccine

You may hear annual reminders to get a flu shot and may wonder why you need another one every year. Various strains of influenza virus mutate actively, and new mutants are different enough to avoid detection by the antibodies created by a previous vaccine. So health officials around the world are always on the lookout for outbreaks of new flu strains. Each year in January or February, officials figure out which strains are most likely to require new vaccines in the next flu season. Vaccine production to supply approximately 80 million shots begins months before the flu season. When officials choose accurately and order a vaccine for the right strains, the flu vaccine is more effective. If they pick wrong, or a new strain mutates and shows up after vaccine production is under way, more people get sick.

Influenza is a major public health risk, even with the success of the vaccination program. While about 75 million Americans receive flu shots each year, up from just 26 million in 1989, the CDC says flu or flu-related illness still kills approximately 20,000 Americans annually. A majority of those who succumb are elderly, people of more frail health who are more severely impacted by a respiratory disease.

REDUCING YOUR RISK

Remember, the risk from the diseases for which we can be vaccinated is far greater than the risk from the vaccines. In general, you do more to reduce your health risk by taking vaccines than by avoiding them. Still, there are some conditions under which health officials suggest that you avoid vaccination. Many of these conditions are temporary. For example, individuals who have moderate or severe illnesses are sometimes told to delay certain vaccinations. Pregnant women are told to wait until after they give birth before getting live viral vaccines like the chicken pox vaccine.

FOR MORE INFORMATION

Centers for Disease Control and Prevention
www.cdc.gov/nip
www.cdc.gov/nip/recs/contraindications.pdf (contraindications
    for childhood immunization)
1600 Clifton Road
Atlanta, GA 30333
National Immunization Hotline: (800) 323-2522

Allied Vaccine Group
www.vaccine.org

National Network for Immunization Information
www.immunizationinfo.org
66 Canal Center Plaza
Suite 600
Alexandria, VA 22314
(877) 341-6644
Fax: (703) 299-0204

If you have what you think is an adverse reaction to a vaccine,
    ask your doctor, nurse, or health department to file a Vaccine
    Adverse Event Reporting System (VAERS) form.
www.vaers.org
(800) 822-7967

---

*This chapter has been reviewed by Dr. Jerome Klein, Professor of Pediatrics at the
Boston University School of Medicine and member of the National Vaccine Advi-
sory Committee; and by Dr. Bruce G. Gellin, M.D., M.P.H., Executive Director,
National Network for Immunization Information.*

48. X RAYS

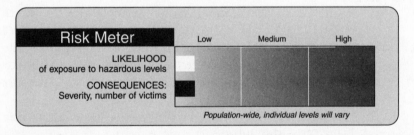

IN 1895, working in a laboratory in Germany, the physicist Wilhelm
Conrad Roentgen ran electrical currents through glass tubes to see
what kinds of emissions they would produce. At one point he held his
hand between the tube and a chemically coated screen several feet
away and watched as an image of the bones in his hand appeared. He

modestly said, "I have discovered something interesting." But Roentgen didn't know what it was. So he gave it a name befitting his uncertainty. He called the phenomenon "X ray." For his discovery, he was awarded the Nobel Prize in Physics in 1901. Roentgen had given us not only the inspiration for *The X Files* and *The X-Men,* but one of the most important diagnostic tools in medicine.

THE HAZARD

X rays, and their close cousins, gamma rays, are forms of energy. To be precise, they are photons—subatomic particles with no mass and no electrical charge—that move at the speed of light, traveling in regular waves as they go. (X-ray photons and gamma-ray photons come from different parts of the atom.)

X rays and gamma rays are part of the electromagnetic spectrum of energy waves, which is divided into sections defined by the size of the waves, measured from the peak of one to the peak of the next. (See Chapter 34, "Radiation.") X rays and gamma rays are at the high-energy end of the spectrum, the part with the smallest waves. Gamma rays are even smaller than X rays.

X-ray and gamma-ray waves vary in size, but they are all tiny. You could fit 254,000,000 X-ray waves in an inch. By comparison, the energy waves in the visible light portion of the spectrum are gigantic, averaging 50,761 per inch. A single AM radio wave can be longer than a football field.

Because X rays are so small, some of them can pass through matter like human tissue literally without hitting anything and come out, intact, on the other side. That's something that electromagnetic waves with longer wavelengths, like visible light, can't do. But X rays can't pass through everything. They revealed Roentgen's hand because while they can pass through less dense material, like the soft tissues in our bodies, they can't pass as easily through more dense tissues, like our bones.

When X rays are directed at a person during a medical exam, the ones that pass through the body without hitting anything are captured on a photographic plate or some other detector on the other side. Where X rays pass through the subject and hit the plate, the area gets darker. But some of the X rays strike atoms in the body and their energy is absorbed and doesn't get through to the detector. The more dense the material, the more likely this phenomenon is to occur. The light areas on an X-ray film represent tissues the X rays couldn't penetrate. That's why the densest material, like bone, shows up the brightest.

~

X rays can be harmful because they have high frequencies, a measure of how many waves pass by a fixed space over a fixed amount of time. X-ray waves have frequencies so high that if an X ray does hit an atom on its way through the body, it's powerful enough to knock electrons off that atom. That in turn can break apart molecules and damage biological tissue.

X rays can cause harm another way, too. Instead of breaking apart molecules, they can just break a little piece off an atom, just one electron. Atoms that are missing an electron are called "free radicals." They want to achieve their normal electron count, so they aggressively grab electrons away from other nearby atoms, ripping those atoms apart and breaking up molecules those atoms may be attached to. Radiation with high enough frequencies to do this kind of biological damage is called "ionizing radiation."

Usually when ionizing radiation damages a cell, that cell simply dies. Fortunately, organs and tissues, made up of millions of cells, can absorb a lot of this damage without any noticeable effect on the overall health of the person. The real problem from ionizing radiation like X rays and gamma rays comes when they damage atoms that are part of the DNA molecule, sometimes causing a mutation in the DNA. Sometimes the DNA can correct this mutation with a built-in self-editing process and sometimes the mutation to the DNA causes the cell to die. But if the cell survives this mutation and the DNA doesn't fix it, the mutation starts providing faulty instructions for cell function. These faulty instructions are then passed on to the cell's offspring when the cell makes copies of itself. A series of just a few such mutations to critical genes in the DNA that control cellular reproduction can lead to cancer. (For a more complete description of how DNA mutation leads to cancer, see Chapter 40.)

THE RANGE OF CONSEQUENCES

X-ray exposure is measured in several different ways. One unit of measure is the radiation absorbed dose, or rad (usually expressed in thousandths of a rad, millirads or mrads), which basically measures how much of the photon energy was absorbed by various organs of the body—in other words, photon energy that didn't get all the way through. The higher the absorbed dose, the greater the likelihood of DNA mutation (though, as you'll learn, the chances of such mutations from normal X-ray exposure are extraordinarily low).

It's important to remember that since different body tissues and organs have different densities, they will absorb X rays and gamma rays differentially.

Absorbed Dose for Some Common X-ray Procedures

Frontal Chest X Ray
Bone marrow (leukemia): 0.2 mrads
Bone: 0.7 mrads
Breast: 3.7 mrads
Lung (male): 2 mrads
Thyroid: 6.5 mrads
Mammogram
Breast: 195 mrads
Dental X Ray
Tongue: 84 mrads

What This Absorbed Dose Means for the Risk of Cancer

Lifetime Risk of Fatal Cancer per 100 Mrads
Bone marrow (leukemia): 1 in 20,000
Breast (female): 1 in 250,000
Lung: 1 in 117,647
Thyroid: 1 in 1,250,000
Liver: 1 in 666,667
Skin: 1 in 5,000,000
Increased Lifetime Cancer Risk from a Standard Frontal Chest X Ray
Bone marrow (leukemia): 1 in 10 million
Breast: 1 in 6,756,757
Lung (male): 1 in 5,882,350
Thyroid: 1 in 19,125,000

Cancer isn't the only health consequence from X-ray exposure. Very high doses of X rays can also produce acute, short-tem effects, particularly in people who are occupationally exposed to higher levels and exposed more often. Very high dose exposure is associated with skin burns, eye damage, and other tissue damage. All of these consequences are fairly immediate and short-term. People at risk of these consequences include radiologists, technicians, doctors, dentists, and nurses. However, properly trained personnel reduce this exposure by keeping at an appropriate distance, reducing exposure time, and using adequate shielding.

THE RANGE OF EXPOSURES

Most of our exposure to X rays and gamma rays comes from medical sources. There are X rays in solar radiation, but these are attenuated by the atmosphere and are substantially reduced before reaching the earth's surface.

Medical use of X rays and gamma rays is either diagnostic or therapeutic. The diagnostic use of X rays lets medical personnel see inside various parts of our bodies. Standard X ray films include dental films. But diagnostic X rays are also used in fluoroscopy and CT scans, which expose patients to higher doses. Fluoroscopy is an X-ray procedure that produces live motion pictures as opposed to a still image. Fluoroscopy is often used for assessing joint function. It is also used to guide medical personnel as they maneuver medical devices through the veins, as in catheterization or in the placement of stents inside the blood vessels. Among all the diagnostic uses of X ray, fluoroscopy often results in the highest exposure because the patient is exposed to a steady stream of X rays, focused on one part of the body for several minutes, while the procedure is under way. Also, fluoroscopy uses more X rays per unit of time in order to produce a clearer picture for the people doing the procedure. A CT scan (also known as CAT scan, for computerized axial tomography) takes multiple cross-section X rays to get a three-dimensional view. Exposure to radiation is higher with CT scan than in regular X-ray imaging because multiple pictures are taken.

Earlier we explained X rays in terms of rads. But another way of measuring ionizing radiation is in units called rems, which stands for Roentgen equivalent in man. Rems measure the potency of the radiation that hits us, not just how much actually gets absorbed. To give you an idea of the difference between a regular X ray and a CT scan, see the following table showing the different exposures for similar diagnostic testing, as measured in millirems (thousandths of a rem).

Diagnostic use of X-ray and gamma-ray photons also includes radioactive drugs (radiopharmaceuticals) that allow medical personnel

COMPARISON OF X-RAY AND CT SCAN RADIATION		
Target of Exam	Regular X Ray	CT Scan
Chest	4 mrem	780 mrem
Skull	10 mrem	180 mrem
Pelvis	111 mrem	710 mrem

to see inside our bodies. Radiopharmaceuticals, which we swallow or which are injected or surgically implanted, help medical personnel understand more about not only the structure of an organ, but about its function. An example is a radioactive isotope of iodine called I-123. In the thyroid, it gives off gamma rays that can be detected outside the body, revealing information about thyroid size and whether the organ is too active or not active enough.

X rays can also be used to treat medical problems, not just detect them. The energy from X rays and gamma rays can be used to kill unwanted tumor cells. These exposures are much higher than diagnostic X rays, but they are focused only on the unwanted cells. The patient is subjected to several X-ray beams at the same time, each entering the body from a different direction. Each source is too weak to do any acute damage by itself, so the normal tissue it passes through is exposed to low, safe doses. But where all the X rays meet—at the targeted cancer cells—they combine to have enough power to destroy those cells. X rays are also used therapeutically in the treatment of prostate cancer by implanting radioactive material, iodine-125, in the affected area. The iodine-125 emits X rays with just enough power to kill the cancer cells nearby.

~

X-ray exposure is carefully regulated, based on studies of the effects on humans from radiation exposures from the atomic bombs used in Hiroshima and Nagasaki, Japan, and from other research. Regulators have determined that an acceptable exposure to ionizing radiation for the general population is 0.1 rem per year. The average American is exposed to about 0.3 rem just from natural background sources like the sun and the earth. (A good portion of this natural exposure is from radon. See Chapter 35.) To put these numbers in perspective, remember that an average chest X ray is 4 millirems, or 0.004 rems.

RADIATION EXPOSURE FROM X RAYS, NATURE		
X-Ray Test	Typical Dose (in millirems)	Bert (time to get same dose from nature)
Dental (intra-oral)	1.5	1.8 days
Chest X-ray	3	3.7 days
Skull	15	18.3 days
Lumbar spine	200	243 days
Lower GI	600	2 years
Chest CT	780	2.6 years

Above is yet another way to put the risk of X-ray exposure in perspective. X rays can also be explained in terms of berts, which stands for background equivalent radiation time. A bert is a measure of X-ray exposure compared with the amount of time it would take to be exposed to the same amount of ionizing radiation from natural sources.

REDUCING YOUR RISK

X-ray radiography has many health-promoting benefits that generally outweigh the extraordinarily low risks. Still, many health authorities suggest that unnecessary exposure to radiation should be avoided.

You may be familiar with the leaded shields and aprons you are given to wear to protect certain body parts during an X-ray procedure to reduce exposure to your reproductive organs and to protect against the risk of damage to the DNA in your sperm or eggs. (The hereditary risk of any kind of cancer is 1 in 100,000 per 100 mrads of X-ray exposure of the gonads.) Leaded shields are also sometimes used to protect the eyes, the thyroid gland, and other organs. When receiving an X-ray exposure that is not meant to target your reproductive organs, make sure you ask for a lead-lined apron if the technician fails to provide one. Why lead? Remember how Superman, with X-ray vision, couldn't see through lead? Lead is so dense that X rays can't penetrate it.

Fetuses are particularly sensitive to radiation, since at various stages of development many of their cells are still in the stem cell phase and haven't differentiated into specific adult cells (heart, blood, brain, and so on). Damage to the DNA of those cells at these stages could lead to a number of negative developmental outcomes. So pregnant women are generally not candidates for X rays. X-ray technicians are always supposed to ask female patients if they are pregnant or breast-feeding. Any female patient scheduled to receive radiation therapy must undergo a pregnancy test. In addition, when radioactive drugs are administered, any female patient should tell her medical providers if she is pregnant or breast-feeding because a portion of the radioactive drug may concentrate in the breast milk.

Because fluoroscopy and CT scans involve more exposure to radiation than regular X rays, ask your doctor if it is absolutely necessary before you undergo this procedure. If there are lower-dose alternatives that yield the same quality of imaging, they will lower your exposure to radiation.

Recent X-ray radiography has shifted toward lower-dose techniques. The idea is to reduce the total amount of radiation exposure per X-ray picture. New materials on the panels that detect X rays are more sensitive and therefore require less radiation to produce a clear picture. New techniques have also been developed to detect the X-ray image digitally and display it on a video monitor, which allows technicians to enhance the image electronically rather than increasing the X-ray dose. Digital X ray also allows adjustments for over- or underexposure, eliminating the need for additional X-ray pictures. In medical X rays, including mammography, low-dose digital techniques have been shown to reduce radiation by 40 to 70 percent. Significant progress has also been made in the field of low-dose digital X-ray techniques in dentistry. Under some circumstances digital dental X-ray techniques can achieve 80 to 90 percent less radiation exposure than conventional X rays when taking periapical films (pictures of your roots) and about 70 percent less radiation exposure than conventional X rays when taking panoramic films (of your teeth). Digital X rays have the same accuracy as photographic film technology.

FOR MORE INFORMATION

Radiology Society of North America
American College of Radiology
www.radiologyinfo.org
P.O. Box 533
Annapolis Junction, MD 20701

National Cancer Institute
www.nci.nih.gov
Public Inquiries Office

Building 31, Room 10A31
31 Center Drive, MSC 2580
Bethesda, MD 20892-2580
(800) 4-CANCER or (800) 422-6237

This chapter has been reviewed by Dr. Frank Castronovo, Director, Health Physics and Radiopharmacology at Brigham and Women's Hospital, and Associate Professor in Radiology, Harvard Medical School; by Richard Monson, Professor of Epidemiology, Harvard School of Public Health; by Dr. John Little, Chairman of the Harvard School of Public Health, Department of Cancer Cell Biology; and by Roy Shore, Ph.D., New York University, Director, Division of Epidemiology and Biostatistics.

APPENDIX 1: VARIOUS ANNUAL
AND LIFETIME RISKS

"Fate laughs at probabilities."

—Edward Bulwer-Lytton

AS WE WROTE in the introduction, most people think of risk in terms of probability, as in "What are the chances that . . . ?" While the preceding chapters detail many of the risks you might care about, there are many risks that you're probably curious about but which don't need lengthy explanation. You just want to know what your chances are: of being hit by lightning, of dying in a plane crash, from a bee sting, in an earthquake, hurricane, or flood, and so on.

The following table will help. But be forewarned. As with all the numbers in this book, the numbers in this table will almost certainly *not* be the precise probability for you. These numbers are gross generalities. They are the simple product of dividing the population for the given year by the number of deaths from a given cause that year. To get the lifetime odds, we've divided the one-year odds by the average life expectancy for that year.

Such simple calculations leave out a lot. Time, for example. The death toll for each risk varies from year to year. Tragically, the death toll for plane crashes, firefighters, and police will be higher in 2001 than usual. These calculations also do not consider exposure. Transportation risks get much more specific when you factor in the risk per miles traveled. The odds of being killed by a shark are different for someone who swims in the ocean than for someone who doesn't. The odds of being hit by lightning are different for people who live in Florida, where lightning is more common, than they are for people in any other state. You're not at risk of dying in a plane crash if you never fly.

We haven't included demographics either. For instance, the odds of getting cancer are much higher if you're older. But we haven't calculated the rate of individuals affected by each risk in our simple table. More truck drivers died than those in any other occupation in the year 2000 (852), but since there are so many truck drivers, the rate of death per 100,000 employed is the relatively small 27.6. Only 105 timber cutters died in 2000, but at a rate of 122.1 per 100,000 employed.

And so on. The variables are so numerous that it is really impossible to come up with a table that could include them all. So what this table is really good for—besides an interesting fact or two to drop on your friends in conversation—is to put risks in some kind of relative perspective. You'll be able to see which ones are bigger or smaller, for the population as a whole, within categories.

To give you something familiar you can compare these numbers to, here are your odds of winning various lottery games:

- For a game that requires you to pick 6 numbers drawn from a total of 49 balls with the numbers 1 through 49 on them, the probability is 1 in 13,983,816.
- The U.S. Powerball game draws 5 numbers from a total of 45 balls. Then, 1 number is drawn from a separate set of balls numbered from 1 to 42. This additional step makes the odds 1 in 80,089,128.
- The U.S. Big Game, which requires you to pick 5 numbers of 50, then another from among balls numbered 1 to 36, has odds of 1 in 76,275,360.
- In a daily numbers–type lottery game in which you have to pick 4 different numbers from 0 through 9, and they have to come up in the exact order you've chosen, your odds are 1 in 10,000.

LIKELIHOOD OF DEATH OR HARM, PER YEAR, AND PER LIFETIME, FOR A UNITED STATES CITIZEN

(Statistics are based on data for the year 2001 unless otherwise noted. Some statistics, based on deaths or injuries over several years, are noted as "avg.")

CAUSE OF DEATH OR HARM	DEATHS / INCIDENTS	ONE-YEAR ODDS	LIFETIME ODDS
WEATHER			
Hurricane	17/yr (avg)	17,000,000	220,000
Tornado	94 (2000)	3,000,000	39,000
Lightning	93 (avg)	3,000,000	39,000
Flood	41 (2000)	6,900,000	89,000
While in a vehicle	20	14,000,000	180,000
Heat	381/yr (avg)	740,000	9,600
Cold	62 (1996)	4,500,000	59,000
CRIME			
Homicide	15,517 (2000)	18,000	240
By gun	10,179	28,000	360
By cutting or piercing	2,094	130,000	1,800
With blunt object	729	390,000	5,000
By intimate partner	2,000 (1996 est)	140,000	1,800
Assault (incidents)	910,744 (2000)	310	(a)
Rape (incidents)	90,186	3,100	(a)
Burglary (incidents)	2,049,946	140	(a)
Motor vehicle theft	1,165,559	240	(a)
Being arrested	14,000,000 (2000 est)	20	(a)
ANIMALS			
Shark attacks	51 attacks (2000)	5,500,000	72,000
Fatalities	1	280,000,000	3,700,000
Alligator	2 deaths (2001)	140,000,000	1,800,000
Snake, lizard, spider	5 deaths (1998)	56,000,000	730,000
Hornet, wasp, bee	46 deaths (1998)	6,100,000	80,000
Dog bite	15 deaths (1998)	19,000,000	240,000
Bear attacks	9/yr (avg; 6 by grizzlies, 3 by black bears)	31,000,000	410,000
Fatalities	3/yr (avg; 2 by grizzlies, 1 by black bear)	94,000,000	1,200,000
Other venomous animals, plants	64 deaths (1998)	4,400,000	57,000

CAUSE OF DEATH OR HARM	DEATHS / INCIDENTS	ONE-YEAR ODDS	LIFETIME ODDS
TRANSPORTATION			
Motor vehicles	41,821 (2000)	6,700	88
Trains	770 (2000)	370,000	4,800
Passengers (intercity)	4	70,000,000	920,000
People outside the train (intercity)	572	490,000	6,400
Local transit	194	1,500,000	19,000
Boats (not drowning)	801 (2000)	350,000	4,600
Recreational	701	400,000	5,200
Commercial fishing	41	6,900,000	89,000
Planes	747 (2000)	380,000	4,900
Civil aviation (public, commercial, >10 seats)	92	3,100,000	40,000
General aviation (private planes, any size)	582	480,000	6,300
Air taxis, commercial planes, <10 seats	73	3,900,000	50,000
Bus			
School	58 (1999)	4,900,000	63,000
Intercity	8 (1999)	35,000,000	460,000
Transit	32 (1999)	8,800,000	110,000
Limousines	0 (1999)	—	—
Motor homes	50 (1999)	5,600,000	73,000
All-terrain vehicles	171 (1999)	1,600,000	21,000
Snowmobiles	39 (1999)	7,200,000	94,000
RECREATION			
Amusement park rides	4/yr (avg)	70,000,000	920,000
HIGH SCHOOL, COLLEGE SPORTS			
Football	5/yr (avg)	59,000,000	770,000
Soccer	<1/yr (avg)	840,000,000	11,000,000
Basketball	<1/yr (avg)	1,700,000,000	22,000,000
Gymnastics	<1/yr (avg)	5,100,000,000	66,000,000
Ice hockey	<1/yr (avg)	2,500,000,000	33,000,000
Wrestling	<1/yr (avg)	2,500,000,000	33,000,000
Baseball	<1/yr (avg)	510,000,000	6,600,000
Lacrosse	<1/yr (avg)	1,700,000,000	22,000,000
Track	1/yr (avg)	250,000,000	3,300,000

CAUSE OF DEATH OR HARM	DEATHS / INCIDENTS	ONE-YEAR ODDS	LIFETIME ODDS
Scuba diving	90/yr (avg)	3,100,000	41,000
Skydiving	31 (1998)	9,100,000	120,000
Trampoline (injuries)	98,889 (1999)	2,800	(a)
Playground (injuries)	200,000/yr (est)	1,400	(a)
Sledding (injuries)	33,000/yr (avg)	8,500	(a)
Tobogganing (injuries)	1,500/yr (avg)	190,000	(a)
Skateboarding (injuries)	15,600/yr (est)	18,000	(a)
ACCIDENTS			
Drowning in the bathtub	337 (1998)	840,000	11,000
Suffocation from a plastic bag	27 (1998)	10,000,000	140,000
Electrocution from wiring or appliances	925 (1998)	300,000	4,000
Suffocation from being caught in an enclosed space with not enough air (refrigerator, etc.)	13 (1998)	22,000,000	280,000
Struck by a falling object	723 (1998)	390,000	5,100
Caught between objects	118 (1998)	2,400,000	31,000
Machinery	1,018 (1998)	280,000	3,600
Fireworks	16 (1999)	18,000,000	230,000
Eating poisonous plants	3 (1998)	94,000,000	1,200,000
Lawn mower (injuries)	60,000/yr (est)	4,700	(a)
Chain saw (injuries)	165,000 (est)	1,700	(a)
Elevator	30 (avg, est)	9,400	(a)
ON THE JOB (Figures are for 2000 unless otherwise specified)			
Being injured on the job	5,700,000	49	(a)
Being killed on the job	5,915	48,000	620
Deaths by industry/ profession			
Construction	1,154	240,000	3,200
Transportation	957	290,000	3,800
Service	768	370,000	4,800
Agr., forestry, fishing	720	390,000	5,100
Manufacturing	668	420,000	5,500
CEO	5 (1994)	56,000,000	730,000
Financial manager	6	47,000,000	610,000
Funeral director	11 (1992)	26,000,000	330,000

CAUSE OF DEATH OR HARM	DEATHS / INCIDENTS	ONE-YEAR ODDS	LIFETIME ODDS
Accountant	11	26,000,000	330,000
Engineer	38	7,400,000	97,000
Doctor	14	20,000,000	260,000
Nurse	17	17,000,000	220,000
Veterinarian	5 (1995)	56,000,000	730,000
Teacher (except postsecondary)	26	11,000,000	140,000
Clergy	10	28,000,000	370,000
Lawyer	8	35,000,000	460,000
Athlete	28	10,000,000	130,000
Pilot, navigator	130	2,200,000	28,000
Logging, forestry	113	2,500,000	32,000
Fisherman	52	5,400,000	71,000
Real estate	6	47,000,000	610,000
Car, boat sales	6	47,000,000	610,000
Retail sales	162	1,700,000	23,000
Door-to-door sales	8	35,000,000	460,000
Construction (nonsupervisors)	504	560,000	7,300
Cashier	68	4,100,000	54,000
Motor vehicle operator	1002	280,000	3,700
Secretary, typist, stenographer	12	23,000,000	310,000
General office clerk	6 (1999)	47,000,000	610,000
Mail carrier	10	28,000,000	370,000
Firefighter	43	6,500,000	85,000
Police	142	2,000,000	26,000
Corrections officer	8	35,000,000	460,000
Crossing guard	5	56,000,000	730,000
Bartender	12 (1999)	23,000,000	310,000
Waiter	7	40,000,000	520,000
Janitor, cleaner	60	4,700,000	61,000
Hairdresser, cosmetologist	7 (1997)	40,000,000	520,000
Amusement park attendant	9	31,000,000	410,000
Farmer	251	1,100,000	15,000
Farm laborers	159	1,800,000	23,000
Auto mechanic	49	5,700,000	75,000

CAUSE OF DEATH OR HARM	DEATHS / INCIDENTS	ONE-YEAR ODDS	LIFETIME ODDS
Heavy equipment mechanic	29	9,700,000	130,000
Electricians	53	5,300,000	69,000
Power line worker	6	47,000,000	610,000
Painter, paperhanger	45	6,300,000	82,000
Mason	15	19,000,000	240,000
Carpenter	91	3,100,000	40,000
Electrician	89	3,200,000	41,000
Roofer	65	4,300,000	56,000
Oil well driller	15	19,000,000	240,000
Butcher	6	47,000,000	610,000
Baker	5 (1993)	56,000,000	730,000
Truck driver	852	330,000	4,300
Bus driver	21	13,000,000	170,000
Cabdriver, chauffeur	70	4,000,000	52,000
Garbage collector	23	12,000,000	160,000
Deaths by cause			
Highway incidents	1,363	210,000	2,700
Falls	734	380,000	5,000
Homicide	677	420,000	5,400
HEALTH ISSUES			
Alzheimer's	49,044 (1999)	5,700	75
Anthrax	5 (2001)	56,000,000	730,000
Asthma (17,300,000 sufferers)	5,000 deaths (1995 est)	56,000	730
Cancer	551,883 (1999)	510	7
Chicken pox cases	64	4,400,000	57,000
Diabetes	68,662 (1999)	4,100	53
Flu	2,175 (1999)	130,000	1,700
Heart disease	934,110 (1999)	300	4
Hepatitis (all forms)	4,796 (1999)	59,000	770
Hernia	1,510 (1999)	190,000	2,400
Kidney diseases	37,672 (1999)	7,500	97
Liver diseases	26,219 (1999)	11,000	140
Malaria	3 (1999)	94,000,000	1,200,000
Measles	3 (1999)	94,000,000	1,200,000
Meningitis	770 (1999)	370,000	4,800
Mumps cases	352 (1999)	800,000	10,000

CAUSE OF DEATH OR HARM	DEATHS / INCIDENTS	ONE-YEAR ODDS	LIFETIME ODDS
Parkinson's	15,690 (1999)	18,000	230
Peptic ulcer	4,507 (1999)	62,000	810
Plague	15 cases/yr (est)	19,000,000	240,000
Pneumonia	64,849 (1999)	4,300	57
Pregnancy and childbirth	370 (1999)	760,000	9,900
Rheumatic fever	4,792 (1999)	59,000	770
Stroke	158,448 (1998)	1,800	23
Suicide	30,575 (1998)	9,200	120
By gun	17,424	16,000	210
By suffocation, strangulation	5,726	49,000	640
Typhoid fever	400 cases/yr (est)	700,000	9,200
Tuberculosis	16,000 cases (2000)	18,000	230
West Nile virus	7 (1999)	40,000,000	520,000
Whooping cough	5 (1998)	56,000,000	730,000
Sexually transmitted disease			
All cases	15,000,000	19	(a)
Human papillomavirus (genital warts, cervical cancer)	5,500,000	51	(a)
Chlamydia	3,000,000	94	(a)
Herpes	1,000,000	280	(a)
Gonorrhea	650,000	430	(a)

(a) Lifetime risk is not estimated for events that can affect the same person more than once in a lifetime.

Sources: American Academy of Family Physicians; Centers for Disease Control and Prevention; Consumer Product Safety Commission; Federal Emergency Management Agency; State of Florida; Stephen Herrero, Professor Emeritus of Environmental Science, University of Calgary; Michigan Association of Timbermen; National Center for Catastrophic Sports Injury Research; National Center for Health Statistics; National Center for Injury Control and Prevention; National Highway Traffic Safety Administration; National Oceanic and Atmospheric Administration; National Safety Council; National Ski Areas Association; National Transportation Safety Board; National Weather Service; University of Delaware Climate Research Center; University of North Carolina, Chapel Hill; U.S. Bureau of Labor Statistics; U.S. Department of Transportation.

APPENDIX 2: THE RISK METERS

AS WE WROTE in the Introduction, many of the issues we discuss are contentious and evoke passionate opinions. Advocates on one side or another promote their views and present the information as they interpret it, leaving the general public to sort through how much is advocacy and how much is "just the facts." We have tried to make this book a resource of advocacy-free information, as neutral as possible, to allow you to make up your own mind about risk. But as an extra piece of guidance, for use as you see fit, we offer in the simple risk meters that begin each chapter our opinions about each risk and whether it is big or small.

We explained in the Introduction what these "risk meters" try to communicate. But in the interest of transparency and openness, we offer you further explanation of our thinking behind these judgments. We repeat that these estimates are only general, for the population as a whole. Your risk may well vary. Our estimates are not precise, which is why we present most of them as a range. They are not based on any specific scientific data. They are judgments based on our analysis of the information we collected. We intend them to be consistent and to reflect the number of individuals harmed or killed across a very wide range of risks. We don't mean to play down any risks or to scare you about others, only to give a rough idea of the relative likelihood of exposure and severity of the consequences. The estimates include what we learned in general from our reviewers, but in no way do they reflect what those reviewers might say about these specific rankings. We didn't ask them. We felt that making these judgments was our risk to take.

We discussed each risk at some length, based on all the information we had gathered to prepare each chapter. We talked about what we had learned, how much we know for sure about each risk, and how much is uncertain or unknown. Between the journalist's skepticism and the scientist's caution, we challenged each other a good deal. Some of the rankings went down. Some went up.

We first judged each risk on its own, and then referred each risk against the whole list, to see if the overall perspective seemed reasonable. So, like the entire book, you can use this table in two ways. You can see what we thought about any individual risk and get a sense of whether it is low, medium, or high. Or you can look at the overall list and see where each risk stands compared to the rest, in our opinion.

OUR THOUGHTS ABOUT EACH RISK METER

Risk	Likelihood of Exposure to Hazardous Levels	Consequences (Severity plus Number of Victims)
	I. HOME, TRANSPORTATION, WORK	
Accidents	9–10: We think that there is a very high likelihood that accidents can happen to anyone, at any time. We consider any accident to be, by definition, hazardous.	6–7: Clearly, on the basis of number of victims, both fatal and nonfatal, this risk would rate a 9 or even a 10. But we think that because the consequences of many accidents are not severe, a somewhat lower number is appropriate.
Air bags	1–2: Not everyone in a car is exposed to an air bag, and not all crashes cause air bag deployment. And newer bags are replacing the older more powerful ones, further reducing likelihood that exposure to air bag deployment involves a hazard.	1–2: While there are a lot of air bag deployments each year, given the number of motor vehicle crashes that occur, few produce consequences of high severity.
Alcohol	8–9: Exposure to even a low amount of alcohol can create a real hazard, like impaired driving, loss of coordination, and loss of judgment. And millions are exposed to these levels frequently. Exposure to more dramatically hazardous levels is high in and of itself.	7–8: Millions of people suffer consequences from alcohol, a huge number of which are the most severe consequence of all: death. We temper this range only because there are also a large number of people who suffer no consequences from alcohol or consequences that aren't as severe.
Artificial sweeteners	0–1: We give this a 1 only because we don't like zeros: zero risk is practically nonexistent. We believe it's unlikely that anyone is exposed to hazardous levels.	0–1: Again, we don't think anyone is exposed so we don't think there are any consequences. But zeros make us nervous.
Bad backs, carpal tunnel syndrome, and other repetitive task injuries	7–8: The likelihood is great that we will be exposed to levels of activities that could be hazardous, both on the job and elsewhere.	2: The number of victims is large, but the severity of these consequences is fortunately temporary and relatively nonsevere in most cases.

Risk	Likelihood of Exposure to Hazardous Levels	Consequences (Severity plus Number of Victims)
Caffeine	4–6: The real question here is "What's hazardous?" We consider that effects like mood change, irritability, withdrawal headaches, and such are hazards. Moderate caffeine use is enough exposure to create these effects, but many caffeine users are never exposed to levels that will cause these problems.	1: The effects are temporary and of low severity.
Cellular telephones and driving	3–4: The use of cell phones by drivers is widespread. But not all uses automatically create a hazardous distraction.	2–3: Information is still very sketchy, but it appears that severe consequences, though they do occur, are fortunately rare. We think that in addition to these few deaths and serious injuries, there are probably a modest number of people who suffer other consequences, though in most cases, these are not severe—fender benders or minor injuries.
Cellular telephones and radiation	0: We hate zeros, but we'll stick our necks out here. There is just no science to suggest any hazard.	0: If it's not a hazard, it can't cause a negative consequence.
Electrical and magnetic fields	0–1: This exposure is about as close to zero as we can get, but the remaining uncertainty about the few studies that suggest an association with a slightly elevated rate of childhood leukemia keeps us from giving it a zero.	0–1: Without those childhood leukemia questions, which most EMF researchers think are explained by something else they haven't figured out yet, this consequence would also be a zero. Besides that question, we don't think there are any negative consequences of exposure to EMFs.

	Likelihood	Severity/Consequence
Firearms	5–6: Not counting the nearly 17,000 suicides by firearms that we found, guns killed roughly 13,000 people, injured three times that many, and were involved in an estimated 331,000 crimes. To us, that suggests high likelihood of exposure to firearms in a way that will cause a hazard.	5–6: This range is the same as the first because while the numbers of victims are high, we think in terms of physical health. The severity of being robbed is lower than serious injury. Also, we presume that while many firearms injuries were undoubtedly severe, a substantial number of those injured were thankfully less severe.
Foodborne illness	10: We rate this hazard a 10 because the likelihood of being exposed to a pathogen in your food that will make you sick is probably higher than anything in this book.	3: The consequences of foodborne illness may seem mild, not at all severe. But we note that an estimated 5,000 people a year die from this hazard (most of whom have previously weakened or suppressed immune systems), and that's probably an underestimate.
Food irradiation	0: A zero we're comfortable with. There is no science to suggest any hazard at all.	0: No hazard. No consequence.
Genetically modified food	1: Based on what we know now, though the likelihood of exposure to GM foods is very high, the likelihood that you'll be exposed to anything hazardous is very low.	1: Even if GM food exposure involves a hazard, which we don't think is likely based on what the science has found to date, the worst case appears to be the possibility of food allergies, only a small number of which would be life-threatening. So the severity of consequences is low.
Mad cow disease	0–1: Another case where we'd like to say zero, but zeros make us nervous. There are no cases in this country in animals, and no reports of human cases. But there is always the possibility of a rare case in the United States.	0–1: Any case that occurs would be fatal. But right now there aren't any, and we believe that systems to keep the disease at bay mean that even if some cases do show up, the number of victims would remain very low.

Risk	Likelihood of Exposure to Hazardous Levels	Consequences (Severity plus Number of Victims)
Microwave ovens	0: Interesting, given the lessons it holds for other new technologies as they come along. As afraid as we were when microwaves first came out, there is no exposure to hazardous levels of radiation.	0: No victims.
Motor vehicles	8–9: The presence of motor vehicles is so high, and the number of injuries so great, that the likelihood of being exposed to hazard from motor vehicles deserves this ranking.	5–6: While the number of victims is very high, fortunately the number injured is three times higher than the number killed, so the severity tempers this ranking. Note that this ranking is below that of non-motor vehicle accidents, which kill and injure more people.
School buses	1: Per vehicle mile traveled, school buses have an extraordinary safety record. As concerned as we all are for our kids, the likelihood of exposure to hazard from school buses is very low.	1: 27 children killed a year is 27 too high, of course. But population-wide, the number of victims suffering consequences from school buses (not including grumpy drivers) is thankfully very low.
Tobacco	5: This hazard gets a medium ranking even though only roughly one quarter of Americans are exposed to tobacco as direct consumers, because secondary exposure remains widespread and creates hazard.	9: The number of victims who suffer severe consequences from tobacco exposure is huge. We repeat just one statistic. One out of two of today's 47 million smokers who don't quit will die from tobacco exposure.
	II. ENVIRONMENT	
Air pollution (indoor)	4–6: Exposure to indoor air is nearly constant, but we don't believe that it's constant at *hazardous levels*, especially given the wide variety of indoor environments we're exposed to.	2–4: This range is low because we think the severity of the consequences is relatively low for most victims.

Air pollution: particulates	6–7: A large number of people are exposed to particulate pollution, and the epidemiology convinces us that sometimes the levels are hazardous. That the range isn't higher reflects our doubt about how many exposures are to hazardous levels.	6–7: The number of people who experience consequences from particulate pollution is high. And many of these consequences are severe, including many deaths. This range would be higher but for our uncertainty about the number of people affected, and questions that remain about the severity of effects of different kinds of particles.
Air pollution: smog	2–4: The likelihood of exposure is thankfully limited to sporadic, temporary episodes in only some areas. And these episodes are becoming less frequent. But when smog occurs, it's often an urban phenomenon, where large numbers are exposed.	1–3: The number of people exposed is limited by the sporadic and temporary nature of the phenomenon. More than that, we think that for the most part the consequences of smog are not severe, though recent evidence suggests it might cause asthma and have a strong association with premature mortality.
Air pollution: air toxics	1: Simply put, the exposure levels are so low that we don't think they're likely to be hazardous in most cases.	1: With levels of exposure so low, we think it's most likely that very few people suffer severe consequences from air toxics.
Asbestos	1–2: Asbestos is a classic example of how hazard alone does not make a risk. Asbestos is hazardous, and there is a lot of it around, but only rarely are we exposed to levels that are hazardous. It has to become airborne to matter, and most of the time it doesn't, at least not at levels that are deemed to be hazardous.	3–4: Deaths record earlier exposure. There are low numbers versus the whole population, but the severity of consequences for those victims is high.
Biological weapons	?: Nobody knows.	?: Nobody knows.

Risk	Likelihood of Exposure to Hazardous Levels	Consequences (Severity plus Number of Victims)
Carbon monoxide	1: Exposures to hazardous levels are rare.	1–2: While most consequences are not severe, carbon monoxide kills several hundred people each year.
DDT	1: Though it's almost as much of a cliché as a skull and crossbones, the reality appears that DDT doesn't do much to people at the infinitesimal levels at which we're most likely to be exposed. (Damage to the hormone system from low levels of DDT exposure would change our estimate, but those effects are not supported by the human evidence, so far.)	0–1: DDT has documented severe consequences for wildlife, but the number of humans who suffer *any* consequences is low, and the number of people who suffer consequences of any severity is practically zero. (Again, our estimate would change if theories about DDT interference with the human hormone system are proven. So far the evidence is against it.)
Diesel emissions	2–3: Diesel emissions contain hazardous substances, but the levels of diesel emissions alone aren't often high enough to be hazardous. They can be higher and more harmful in limited locations and at isolated times, but this rating is an average for all locations and all times.	2–4: The nature of some of diesel's consequences can be severe, including death from particles and perhaps from air toxics. But since exposure to diesel emissions occurs at relatively low levels, we think the number of people suffering consequences is modest. However, there is great uncertainty about how many victims there might be.
Environmental hormones	?: We are much too uncertain to make any judgments. The toxicology so far doesn't support the idea that low doses have dramatic developmental effects. But the wildlife evidence suggests a disturbing pattern that warrants research into possible human effects.	?: The proposed consequences range from minor to severe. The suggested number of victims range from a few to the entire human species. We don't even know if these consequences are real yet, much less how many people suffer what kinds.

Hazardous waste	1: We believe there is very little likelihood of exposure to hazardous levels of chemicals from hazardous waste sites, most of which have been identified and either fenced off or cleaned up. Exposure from groundwater contamination is also very unlikely to occur at hazardous levels.	0–1: The number of people exposed to enough hazardous waste to cause any consequences is quite low.
Incinerators	1: The likelihood of exposure to hazardous levels of pollutants from incinerators is very remote.	1: The number of people suffering any consequence from incinerator emissions is very low, and the severity of those consequences is generally low as well.
Lead	2: Lead is a hazard, but at what level? There is a lot of lead around, but we think the levels most people, including children, are exposed to produce either no effects or symptoms so mild we can't call the exposure hazardous.	2: Had we written this book several years ago when the average child blood lead level was 16 µg/dL we'd have said the severity of consequences was higher. But that severity is lower now that average levels are down to 4 µg/dL. So while there may be a large number of people exposed to some lead, and while any exposure may produce some effects, the severity to most people is low.
Mercury	1: We feel there is a very low likelihood that people are exposed to mercury at hazardous levels.	0–1: Not only is exposure to hazardous levels unlikely, but even those who are exposed fortunately suffer consequences that are not that severe.

Risk	Likelihood of Exposure to Hazardous Levels	Consequences (Severity plus Number of Victims)
Nuclear power	1: The likelihood that anyone will be exposed to hazardous levels of radiation in connection with nuclear power plants, whether through their normal operation, accidents, or from nuclear waste, is extraordinarily low.	3: This rating supposes an accident that releases radiation. Based on the findings of injury from the Chernobyl accident, and the knowledge about the effects of radiation exposure gained from following people in Hiroshima and Nagasaki, we think the number of victims suffering severe consequences would be relatively low, though a much larger number could suffer mild effects. This rating is uncertain, however, because of the wide range of possible accident scenarios. 9: We've given this risk a unique third rating. While the physical effects of an accident might not be severe for a large number of victims, the psychological, economic, and social consequences of an accident at a nuclear plant in the United States would be enormous.
Ozone depletion	4–6: Based on exposure to the sun we'd say this hazard deserves a 10. But thinning of the ozone over most of the Northern Hemisphere is not uniform. Indeed it's greater over the northern United States than in the south. And it varies from year to year. And sometimes it's not great enough to make the increased UVB exposure much more hazardous than it already is. So we give this risk a more moderate, though still relatively high rating.	3: Fortunately most of the effects of UVB exposure, including the two most common forms of skin cancer, are relatively mild. Both of these cancers are slow-growing, unlikely to metastasize, and easily treatable. Other UVB exposure consequences, such as sunburn and skin aging, we also consider mild. So though the number of victims is large, the severity is low.

Pesticides	1: There is widespread exposure to pesticides. But the evidence suggests to us that there is a very low likelihood that our exposure, either from food or from home or garden uses, occurs at hazardous levels.	1: The evidence suggests that very few people suffer either acute or chronic consequences of any severity from pesticide exposure.
Radiation	Not applicable: This chapter merely explains radiation as a background to all the other chapters where the risk is radiation-related.	Not applicable: This chapter merely explains radiation as a background to all the other chapters where the risk is radiation-related.
Radon	2–3: Radon exposure is widespread, but it is not ubiquitous and the evidence suggests that in a majority of cases, levels are not hazardous.	4: Though the range of estimated lung cancer deaths from radon exposure is wide, we rely on the middle of that range, which means that there are a significant number of victims suffering the most severe consequences.
Solar radiation	9: Exposure to the sun is almost universal, and even for short periods can be hazardous. It is also chronic. We consider skin damage a hazard. We stop short of rating this hazard a 10 only because we presume that some exposures can be at nonhazardous levels.	4–5: The number of victims of solar radiation exposure is almost certainly higher than any risk in this book. But the severity of the consequences, even including the two principal forms of skin cancer, is modest or low in nearly all cases. Nonetheless, with 7,800 melanoma deaths per year, and possibly more, and with such a huge number of other victims of minor outcomes, we think a lower ranking would be inappropriate.
Water pollution	1: Exposure to drinking water is universal. But as we noted, the safety record of the U.S. drinking water supply shows that exposure to hazardous levels of contaminants in that water is very low.	1: Population-wide, very few people suffer any severe consequences from drinking water pollution, though there are occasional localized exceptions.

Risk	Likelihood of Exposure to Hazardous Levels	Consequences (Severity plus Number of Victims)
	III. MEDICINE	
Antibiotic resistance	Not applicable: We leave the first meter blank because there is no specific hazard to which we're exposed.	?: Antibiotic resistance has many health implications, but we don't know how many people suffer them, nor how many of these consequences are severe and how many are mild. There is too little data to make an informed judgment at this time.
Breast implants	3–4: The media-fed national scare of the 1990s has some people still thinking of breast implants as a cancer risk. On that scale, it's a zero. But about 300,000 women a year get implants, and the likelihood of problems is sizable.	1–2: Though there may be many women who suffer consequences from breast implants, in most cases the effects are localized, temporary, and not severe.
Cancer	Not applicable: Cancer is not a hazard to which we're exposed. It's an outcome.	9: The consequences of cancer are severe and affect a huge number of people. We talked about giving cancer a 10, but absolutes make us nervous.
Heart disease	Not applicable: Heart disease isn't a hazard. There is no issue of exposure.	9–10: The number of people who suffer consequences from heart disease is enormous, and the severity of those consequences is too.
Human immunodeficiency virus	2: Any exposure to HIV is hazardous. But roughly only 1 in 7,000 Americans is exposed at 2002 rates, and exposure is principally in just two population subgroups, so we rate it low because our figures are for the population as a whole.	5–6: The consequences of HIV infection are severe, though new drug therapies hold out hope that it might not always be fatal. Also, the number of new infections per year, 40,000, is not as high as the number of deaths from motor vehicle crashes or deaths from non–motor vehicle accidents.

Mammography	1: As with all the low-level radiation issues in this book, we think the likelihood of exposure to a hazardous level here is very low and borders on zero. However, inaccuracies with mammography pose other risks, such as unnecessary follow-up procedures for false positives and delayed treatment for false negatives. These circumstances are rare, however.	1: Very few patients who get inaccurate results suffer severe consequences as a result.
Medical errors	4–6: So many people interact with the medical care system that we think the chances of exposure to medical error are significant. But not all of these errors lead to negative outcomes, so exposure to errors over the whole population is not always hazardous. Thus we rate this risk in the middle. The numbers on which we base our judgment are highly uncertain.	7–8: Even though many medical errors may not lead to negative consequences of any severity, those that do can be quite severe, including the possibility of tens of thousands of deaths. So we rate the consequences high, acknowledging the severity of many outcomes.
Overweight and obesity	Not applicable (6): Weighing too much is an outcome more than a hazard, but if we were rating it solely as a hazard—since it leads to consequences—we'd give it a 6, since 6 out of 10 Americans are overweight.	9: There are a huge number of victims, and the consequences of being overweight or obese are often fatal or life-threatening.

Risk	Likelihood of Exposure to Hazardous Levels	Consequences (Severity plus Number of Victims)
Sexually transmitted disease	8–9: The numbers of people who have STDs, or contract new infections each year, indicates that the likelihood of exposure to the hazard of getting one of these diseases is very high for anyone who is not celibate or engaged in a monogamous sexual relationship with a noninfected partner.	3: While the number of victims is staggering, the percentage of those suffering severe consequences is low. Still, the huge universe of victims raises the actual number of victims with severe consequences enough for us to give this ranking.
Vaccines	1: The evidence is overwhelming that though vaccines do produce side effects, the likelihood is very low.	1: Among the very small universe of people who suffer any negative consequences from vaccination, the overwhelming majority of these people suffer nonsevere effects.
X rays	1: The likelihood that X-ray exposure from man-made sources will occur at hazardous levels is very low, except for treatment uses, which are intended to eliminate much greater and more imminent threats to survival from certain cancers. The number of these X-ray exposures is also very low.	1: The number of people exposed to man-made X rays who suffer severe consequences (except those trying to save their lives with high-dose treatments) is extraordinarily low.

ACKNOWLEDGMENTS

MANY PEOPLE made vital contributions to the research and writing of this book. They include John Graham and Susan Hsia, formerly of the Harvard Center for Risk Analysis; the journalists Robert Braile, David Chandler, Richard Higgins, Vivien Marx, Janet Raloff, and Pat Thomas; Robin Herman of the Harvard School of Public Health; Terri Rutter of the Dana-Farber Cancer Institute; and Professor Jane Stevens of the University of California at Berkeley.

Joshua Cohen, Research Scientist at the Center for Risk Analysis, not only contributed to the research and writing of several chapters, but offered insightful ideas about many aspects of this work, including the risk meters and appendices.

We would also like to thank the more than 100 scientists, government officials, medical providers, and other experts who reviewed our work. Their expertise helped us to be accurate, thorough, and balanced. We are deeply grateful that they made the time in their busy schedules to help us with this work.

We owe special thanks to our editor, Laura van Dam. Her support was constant and enthusiastic. Her ideas and suggestions helped immensely. Without her dedication to and advocacy of this project, this book would not exist.

We thank Houghton Mifflin Senior Manuscript Editor Jayne Yaffe Kemp, whose thorough and thoughtful editing caught more than a few dangling participles and split infinitives. Her close attention to content caught many unanswered questions and made a truly vital contribution.

Thanks to graphic designer Elissa Traher of Traher Design for risk meters that are clear and informative. Thanks to Jennifer Fairman, Certified Medical Illustrator, for graphics that are instructive and elegant.

Finally, we thank the many professional colleagues, friends, and family members who offered their support, their ideas, their questions, and their feedback, all of which helped make this book better.

NOTES

1. Accidents

We derived most of the numbers in this chapter from the excellent reference volume of the National Safety Council "Injury Facts—2001." The NSC has been in the business of collecting these numbers for more than 70 years, using national and state databases. Most numbers for the year 2000 are round numbers, estimates based on incomplete government and hospital data, adjusted by the NSC using formulae for estimating the final numbers, which have been refined over the years. Other numbers, for earlier years, will be more precise, since the final data for those years are available. When numbers are given that come from other sources, those sources are cited.

2. Air Bags

The death count from air bags stood at 195 at the end of 2001, but 39 additional unconfirmed cases of air bag–related fatality were still under investigation. Of the total of 195 deaths, 26 of them, mostly to children, occurred in just one year, 1996. This tragic concentration got a lot of media attention and had much to do with bringing the issue of air bag risk to the public's attention.

3. Alcohol

We mention in the chapter in general terms that more men are heavy drinkers than women. The gap increases as people age. Males outnumber female heavy drinkers in the 18 to 24 age category by 3 to 1, in the 25 to 44 age group by 4.5 to 1, in the 45 to 64 age group by 5.5 to 1, and in the 65 and over age group by 9 to 1.

4. Artificial Sweeteners

The incidence of the rare genetic disorder phenylketonuria (PKU) is approximately 1 in 16,000 among Caucasian and Asian Americans. The incidence among African Americans is much smaller.

5. Bad Backs, Carpal Tunnel Syndrome, and Other Repetitive Task Injuries

Work-related repetitive stress injuries cost employers more than $20 billion a year in worker's compensation claims. A multiyear study of the problem by federal agencies led to ergonomics legislation proposed by the Clinton administration but ultimately rejected by Congress. Most observers said that the main stumbling block was the portion of the legislation that would have raised costs business and industry would have to pay to compensate injured workers, in an effort to create economic incentives to reduce the number of injuries.

6. Caffeine

Blocking adenosine is just one way that caffeine can affect cells. Caffeine can also change how our cells store and release calcium. It can inhibit an enzyme that helps blood platelets aggregate, which slows bleeding. It can block the action of another neurotransmitter, GABA (gamma-aminobutyric acid). These effects, however, are all believed to occur only at high, nearly toxic levels of caffeine intake, far greater than the levels at which most people consume caffeine.

You may be aware that chocolate is poisonous to dogs. Chocolate contains theobromine, a close relative of caffeine (trimethylxanthine) and of theophylline, an ingredient in tea. They're all part of the xanthine family of compounds. Theobromine, however, is unique in its toxicity for dogs. It takes 150 milligrams per kilogram to cause a toxic reaction. There are different levels of theobromine in different types of chocolate. Toxicity levels are:

> 1 ounce per pound of body weight for milk chocolate, which has 44 milligrams of theobromine per ounce
> 1 ounce per 3 pounds of body weight for semisweet chocolate, which has 150 milligrams per ounce
> 1 ounce per 9 pounds of body weight for unsweetened chocolate, which has 390 milligrams per ounce

Xanthines can also poison other pets, including cats and birds.

8. Cellular Telephones and Radiation

As opposition to cell phone towers grew in the late 1980s and early 1990s, the federal government passed rules that preclude local communities from simply denying approvals for such towers. Those rules also stated that local communities may not take into account any effects the radiation might have on the environment. The FCC has interpreted this stipulation to include impacts on public health. Both rulings are based on the lack of scientific evidence that radiation from these towers has any negative environmental or public health effects.

10. Firearms

Passage of the federal law banning certain semiautomatic weapons, the so-called assault weapons, has raised their prices dramatically since they can no longer be manufactured. Starting in 1998, as states began filing lawsuits against firearms manufacturers, those companies began to raise their prices 5 to 10 percent a year or more on handguns.

11. Foodborne Illness

Of the 6,285 identified outbreaks of foodborne disease in the United States from 1990 to 1998 in which groups of people became ill, officials were able to track down and identify the disease that caused only 2,060 of them.

12. Food Irradiation

The FDA is currently considering a proposal to allow irradiation as a food safety treatment for ready-to-eat foods including sprouts, seeds, and juices; frozen fruits and vegetables, such as broccoli, peas, and strawberries; cut and packaged salads; refrigerated ready-to-eat meat and poultry products, such as deli and luncheon meats and hot dogs; dried meat and poultry products, such as beef jerky and turkey jerky; and frozen meat and poultry, such as precooked beef patties and precooked fried chicken.

13. Genetically Modified Food

Many European countries require that products containing GM organisms say so on the label, and some major European food manufacturers are demanding that their suppliers in the United States set up separate food handling, storage, and processing facilities for GM and non-GM grains. But a report from Canada in late 2001 found that seeds and pollen of GM canola had spread off the farms on which they had been planted to such an extent that it has become difficult to grow non-GM canola anywhere in Canada without the likelihood of cross-fertilization with the GM strain.

One form of GM food is corn bred with a gene that contains a natural pesticide, so that while the crop is growing it requires less treatment with industrial chemicals. An unexpected side benefit is that when the corn is in storage, moths don't seem to like it as much and don't do as much damage. This happy result in turn reduces the growth of a fungus known to be dangerous to humans.

14. Mad Cow Disease

The experience of mad cow disease in Germany in early 2001 seems to confirm one of the findings of risk perception research, that we tend to be more afraid of risks when they are new than when we've lived with them for a while. After about two dozen infected animals were first discovered in Germany, a public opinion poll found that roughly 80 percent of German residents thought mad cow disease was a serious risk to human health, though no human cases had been reported. The pollsters also asked people in Great Britain, where more than 100 people had died and where hundreds of thousands of infected or potentially infected cattle had been slaughtered. In the United Kingdom, however, where the disease had been around for several years, only about 40 percent of the public rated it as a serious risk to human health.

16. Motor Vehicles

Only New Hampshire, the "Live Free or Die" state, doesn't require safety belts for people over 18. In 1998, safety belt usage there was 56 percent. New Hampshire, however, was only the sixth-lowest state for safety belt use. In South Dakota it was only 46 percent, and in North Dakota it was just 40 percent.

18. Tobacco

People in lower income groups smoke more than people in other income brackets. One third of people living in poverty smoke. Among educational groups, 44 percent of those who had earned a General Education Development (GED) high school equivalence diploma are smokers, compared with 8 percent of those with a master's, professional, or doctoral degree. Among different ethnic groups, 34 percent of

Native Americans (including Alaska natives) smoke, as well as 27 percent of African Americans, 25 percent of Caucasians, 20 percent of Hispanics, and 17 percent of Asians and Pacific Islanders.

19. Air Pollution (Indoor)

One of the first buildings where workers experienced sick building syndrome was EPA headquarters in Washington, D.C.

20. Air Pollution (Outdoor)

The standard unit for measuring a concentration of an air pollutant is a microgram per cubic meter, or $\mu g/M^3$. A microgram is one millionth of a gram, and a gram is $\frac{1}{28}$ of an ounce. A cubic meter of air is roughly the amount an average person not engaged in strenuous activity breathes in an hour. We take in about 6 to 10 liters (the size of 3 to 5 large soda bottles) of air per minute. During heavy exercise, that amount can go up to 100 liters per minute or more. Many government regulations are predicated on the assumption that we breathe 10 cubic meters of air during an 8-hour workday, and another 10 cubic meters or so during the other 16 hours of the day when we are usually less active or asleep.

The maximum lung capacity averages 5 to 10 liters per breath, but most people take in only one half to three quarters of a liter per breath. Interestingly, you can breathe in more air on the deepest breath than you can breathe out. The inspirational muscles have more capacity than the expirational muscles do, ensuring that our lungs can't completely empty and collapse between breaths.

22. Biological Weapons

Ebola and Marburg viruses were discovered accidentally in 1967 in Marburg, Germany, when 25 lab workers doing research on blood samples from African green monkeys from Uganda developed a rash on their bodies about five days later. One third of those lab workers died. The first major human outbreaks were in Zaire and Sudan in 1976. In Zaire, 88 percent of the people infected died; 53 percent of the infected population in Sudan died.

23. Carbon Monoxide

CO poisoning of a pregnant mother is a risk not only to her but to her developing fetus, since a healthy supply of oxygen is so important to normal fetal development.

24. DDT

The use of DDT worldwide remains controversial because in many countries it continues to be the most effective way to control disease-carrying mosquitoes and to reduce deaths from malaria. Supporters of its continued use in these countries say that the proven and dramatic benefits far outweigh the speculative and relatively minor health damage that opponents cite. When an international treaty was signed in May 2001 banning or phasing out several environmentally persistent pesticides, DDT was the only one that was exempted in those developing countries that rely on it to save lives, pending development of alternatives.

25. Diesel Emissions

There have been efforts to reduce the emissions of particles and nitrogen oxides (NOx) from diesels. But due to the nature of a diesel engine—the way it compresses the air and then mixes in fuel—lowering particle production tends to raise NOx, and vice versa.

26. Environmental Hormones

While the study of industrial chemicals and their impact on the endocrine system is relatively new, the FDA has studied the impact of phytoestrogens for some time. The vast majority of this research suggests that these natural plant estrogens, to which we're exposed at much higher levels than environmental industrial chemicals, appear to reduce a woman's chance of breast cancer and may reduce an individual's chance of colon cancer. They also appear to have cardiovascular and skeletal benefits. No good evidence exists, however, to explain why these effects occur.

27. Hazardous Waste

One form of hazardous waste disposal is called bioremediation, in which naturally occurring microbes are used to break the hazardous material into harmless gases and water vapor. This process requires the right temperature and soil conditions and works only on certain materials. Bioremediation is an approved hazardous waste disposal technology used by some large businesses, which spread the material out in a controlled area and let the microbes munch the waste into harmless byproducts. This technology is used on oil sludge, the gritty dirt produced in oil drilling which also carries petroleum products.

28. Incinerators

Incinerators do not emit significant amounts of two other main categories of air pollution, volatile organic compounds (VOCs) and polyaromatic hydrocarbons (PAHs).

~

The EPA reports that the greatest source of airborne dioxin emissions in the United States is the burning of household waste in residential fireplaces and back yards. Tests on back yard barrel burning of household waste conducted by the agency in 1997 indicated that as much as 1,200 grams of dioxin are emitted per year nationwide by the burning of household waste. That's roughly 100 times more dioxin than is emitted by municipal waste incinerators. Several states have made this burning illegal, but enforcement is rare.

29. Lead

Lead has a unique value in helping date very old rocks or fossils. Lead is the byproduct of the radioactive decay of other elements, including uranium. Scientists know the rate at which this decay happens, and by studying minerals like zircon, which contains both uranium and lead, they can compare the proportion of each and calculate how long the uranium-to-lead decay has been going on, and therefore, how old the rock is. The oldest rock found so far (that is not thought to have fallen from outer space) was identified by the geologist Sam Bowring, of the Massachusetts Institute of Technology. It was found in Canada's Northwest Territories. Known as Acasta gneiss, it is 4.02 billion years old. Scientists can also use this lead-dating method to figure out the age of fossils, by dating the sedimentary rocks next to which those fossils were found. Figuring out the age of the specific sedimentary layers where the fossils were found indicates the age of the fossils.

30. Mercury

Because fish feed in a natural ecosystem and accumulate both natural and anthropogenic doses of mercury, and because it bioaccumulates, fish carry levels of mercury between 10 and 100 times higher than foods like cereals, vegetables, fruits, farm-fed meat and poultry, eggs, and dairy.

31. Nuclear Power

Chain reactions don't occur naturally in the earth because U235, the kind of uranium used in nuclear power and nuclear weapons, is so rare. Only 0.7 percent of the natural uranium in the earth is U235. Most is U238, which is also unstable and decays, but which does not have a nucleus that breaks apart as readily as U235 when a neutron strikes. U235 isn't concentrated enough in any one spot in the earth to spur a self-sustaining chain reaction. But geologists have found evidence that 2 billion years ago, long before radioactive decay had broken down a lot of the U235 in the earth, a concentration of this type of uranium started a natural nuclear chain reaction in West Africa.

Among the cancers most closely associated with radiation exposures, leukemia is thought to arise because the cells in the blood system reproduce more often than in other body systems, and the ionizing radiation does its damage while the cell is reproducing, not while it's in its resting phase. Leukemia can arise with fewer mutations to the DNA than most other cancers, so fewer mutational events are necessary to trigger this form of the disease. Thyroid cancer is associated with radiation because a particular fission byproduct, radioactive iodine, is absorbed in the body by the thyroid gland.

In the past couple of years new designs for nuclear plants, which would be smaller and have less expensive and more accident-proof safety systems than existing plants, have reenergized the industry. But no formal license request to build a new nuclear plant had been made to the NRC as of the end of 2001.

The areas in and around Hiroshima and Nagasaki are safe to live in because the radiation from the atomic bombs came only from the small amount of material in those bombs. The radiation levels were high in the moments after the explosions, but the radiation from the blasts did not make other materials in the area radioactive (atomically unstable and changing structure, therefore giving off radiation). So after the source of the radiation—the explosions—was gone, except for very low levels of fallout from the bombs themselves, radiation levels in the area returned to normal.

32. Ozone Depletion

During spring in the Southern Hemisphere, meteorological conditions concentrate wind, humidity, and temperature over Antarctica and cause the ozone hole. The depleted ozone layer spreads out, and during several weeks, weather forecasts in New Zealand and Australia include specific details of the concentration of ozone because of the increased danger from UVB exposure. Residents of Punta Arenas, a city of 120,000 in southern Chile which is the closest major population center to the ozone hole, also receive detailed ozone reports. Few studies have been done on UVB levels in Punta Arenas, or on the effects that ozone depletion may have on human health there. One, published in 1995, found no effects, though local dermatologists report more cases of sunburn in years when ozone levels are lowest.

33. Pesticides

In 2001, 90 countries signed a treaty designed to ban the use of "persistent organic pollutants," also known as POPs, most of which are pesticides. The POPs treaty outlines control measures for the production, import, export, disposal, and use of: aldrin, chlordane, DDT, dieldrin, dioxins, endrin, furans, heptachlor, hexachlorobenzene, mirex, PCBs, and toxaphene. Most of the chemicals on this list were subject to an immediate ban. DDT use was granted an exemption in countries where it is used to control the transmission of malaria via mosquitoes pending development of effective alternatives. PCBs were also given an exemption. Manufacture of PCBs has ended, but countries can continue to use equipment containing PCBs until 2025.

Debate over the treaty centered on whether adoption in this case of the Precautionary Principle—the idea that in the face of scientific uncertainty public health policy should err on the side of caution—would set a precedent for other global health issues. Most European nations supported this position. Several nations, including the United States, said that such a blanket approach ignores the large body of scientific evidence that shows the risks from many of these chemicals is actually very low at the levels to which most people are exposed. In the end, a compromise advocated both precaution and scientific evidence. In the case of these chemicals, the decision to ban them or phase them out was largely based not on remaining speculation about their health risks but on their environmental persistence, which is the same reason many of them have already been banned in the United States.

37. Water Pollution

At the time John Snow solved the cholera riddle of the Broadway Street pump in Soho, London, most scientists still believed that the disease was spread by "miasma," mysterious vapors in the air. This belief persisted despite Snow's essay "The Mode of Communication of Cholera," published five years earlier, describing that it must come from some other source, probably food or water, since he had attended to many cholera patients and hadn't gotten sick.

The critical clues in Snow's detective work came from three sources: two women who lived miles away and got cholera who said they traveled to the affected neighborhood and drank from the well each day, inmates in a nearby workhouse that had its own well (they didn't get sick), and workers in a neighborhood brewery who didn't get sick because they drank only beer and never touched the well water.

38. Antibiotic Resistance

Part of the problem of antibiotic resistance is that bacteria reproduce so prolifically. But another major issue is that bacteria have developed several ways of undergoing genetic mutation. These multiple paths of mutation also allow bacteria to evolve new drug-resistance traits very rapidly.

Bacteria can swap large chunks of their DNA on carriers called plasmids, little round loops of DNA carrying several complete genes. For example, an E. coli can send a plasmid with part of its DNA into a salmonella, or vice versa.

Bacteria can also swap smaller chunks of DNA, called transposons, which carry all or parts of a few genes.

Bacteria also have the ability to simply absorb a chunk of leftover DNA from another bacterium that died. They literally can grab pieces of used DNA that float by, and incorporate those pieces into their own DNA.

And a bacterium can undergo mutation to its own DNA just one chemical link in one gene at a time. Bacteria don't undergo that kind of mutation very often, just once in every 1 million to 100 million times a cell divides. But because bacteria reproduce so prolifically, these mutations are far more common in bacteria than in almost any organism.

Not only do these new traits arise quickly, but they can counter the effects of antibiotic drugs in several ways. Sometimes the new resistance trait keeps the antibiotic from binding to the target bacterium, so the drug can't do any damage. Sometimes the bacteria develop the ability to degrade the antibiotic *after* it has bonded. So the drug can bond to the germ and do *some* damage, but less than it does to bacteria that have no resistance at all. Even this modest resistance gives that strain a competitive advantage over its weaker relatives. Sometimes the bacteria develop the ability to pump out a drug that gets inside the bacterium cell before the drug can do any damage. And sometimes bacteria grow thicker skins, in the sense that they develop cell walls with reduced permeability. So the drug can latch onto the outside but can't get inside the cell to attack it.

40. Cancer

A fuller explanation of the biology of cancer follows. Recall from the chapter that cancer arises when normal controls limiting cellular reproduction don't work.

Healthy cells make copies of themselves only when told to do so by molecules called growth factors, which latch onto receptors on the outside of the cells' membranes. The "grow" signal is transmitted down into the cells and through a series of proteins that adjust the signal before it gets to the DNA in the nuclei. Genes with names like *ras* and *myc* turn up the signal, calling for more growth. Think of them as accelerators. If mutated, they are called oncogenes and produce faulty control proteins that will spur too much growth, which can lead to the formation of tumors.

But some other growth-regulating genes, like the p53 gene (which has lots of different jobs in the body), do just the opposite. To keep things in balance, these so-called suppressor genes turn the grow signal down. They restrict growth. Think of them as brakes. They keep unrestricted growth from happening by making a cell senescent so it doesn't reproduce anymore, or by killing the cell altogether. So mutating these growth-limiting genes is like taking your foot off the brake while your other foot is pushing on the accelerator. These are two key conditions for creating the uncontrolled cell growth that is cancer.

But for most cancers, those changes alone aren't enough. Several more mutations contribute to the creation of most forms of the disease. For instance, when something in a cell's internal chemistry goes wrong, including mutations to the DNA, the cell kills itself, a process called apoptosis (pronounced a-pa-TOE-sis), or cellular suicide. This phenomenon is bad for the individual cell, but good for the overall organism, which obviously doesn't want mutant cells floating around and collecting further mutations, which can lead to cancer. But in cancer cells, the genes that control apoptosis (which include p53) are also mutated and don't work right. The defective cell and its offspring survive, living on to pick up the rest of the mutations necessary for cancer.

One of those additional mutations renders the cell "immortal," allowing it and its offspring to reproduce forever. Normal cells have a built-in life expectancy, controlled by little sections of DNA on the end of each double helix molecule called telomeres. Think of them as the plastic tips on the ends of shoelaces that keep them from unraveling. Each time a normal cell makes a copy of itself, the telomeres get a little shorter. After 50 or 60 growth cycles, the telomeres are gone, the DNA unravels, and the cell can't make copies of itself anymore. Cancer cells develop the ability to produce the protein telomerase, which keeps the telomeres from shrinking. That cell can make copies of itself, and its mutated DNA, forever.

But to become cancerous, a cell needs still one more trick to survive. When those first mutations to these critical genes accumulate in a single cell, a process that can take decades, a tumor starts to grow. But when the tumor gets big enough it will starve to death unless it can make blood vessels grow to feed it. Cancer cells develop the abnormal ability to secrete growth factors that make blood vessels grow throughout the tumor.

Still more mutations are necessary before the tumor can become malignant and spread, a process known as metastasis. Each organ or type of tissue in the body is contained in a special membrane, a kind of sac. Cancer cells develop the ability to dissolve their way through those membranes, and through the walls of blood or lymph vessels, so they can travel on those fluid highways to anyplace in the body.

Cancer cells have to develop another unique ability too. Most normal cells can't survive unless they stick together. Literally. They have chemicals on their membranes that let them adhere to one another. If they detach from other cells, it triggers apoptosis. In addition to all their other tricks, cancer cells develop the abnormal ability to break that adhesion *without* causing apoptosis, so clumps of just a few cells can separate from the tumor and survive on their own.

It is these two final biological abilities that make cancer so deadly. If a tumor stayed in one place, it would be much easier to remove surgically. But because cancer cells have the ability to break off from one another and to burrow in another part of the body and spread elsewhere, they can escape easy treatment and interfere with various bodily functions, enough to be fatal. This ability to spread, in fact, is embodied in the word *cancer.* It's based on the Greek word *karkinos,* for crab. Hippocrates first used the word to describe advanced tumors and the way they spread out like the legs of a crab.

41. Heart Disease

The survival rate for people suffering heart attacks 25 years ago was approximately 50 percent. Now it's about 59 percent.

42. Human Immunodeficiency Virus

Clinicians disagree on whether HAART—"the AIDS cocktail"—is more effective if given to people with HIV before they develop AIDS.

43. Mammography

The percentage of women age 40 or above who are having mammograms has gone up from 29 percent in 1987 to 67 percent in 1998.

46. Sexually Transmitted Disease

The advent of Pap tests has reduced cervical cancer deaths in the United States by at least 74 percent since 1955, and the rate continues to decline 1.6 percent per year. Other tests for HPV that can identify the DNA of the high-risk strains are just becoming available and may be even more accurate and effective than Pap technology.

48. X Rays

Roentgen made his discovery in November 1895. By the next January, word had spread around the world. "NEW LIGHT SEES THROUGH FLESH TO BONES!" said one headline in a U.S. newspaper. "HIDDEN SOLIDS REVEALED!!" said another. Thomas Edison tried unsuccessfully to make a commercial "X-ray lamp" as a consumer product. But the apparatus to generate X rays wasn't hard to build, and thousands flocked to studios set up to take "bone portraits." Detectives touted the use of Roentgen devices in following unfaithful spouses, and lead underwear was manufactured to foil attempts at peeking with "X-ray glasses." Poems about this fascinating new technology appeared in popular journals, including this one:

> The Roentgen Rays, The Roentgen Rays
> What is this craze?

The town's ablaze
With the new phase
Of X-ray's waves

I'm full of daze,
Shock and amaze;
For nowadays
I hear they'll gaze
Thro' cloak and gown—and even stays
Those naughty, naughty Roentgen rays

—Wilhema, in *Photograp*

INDEX